THE STORY OF PROOF

LOGIC AND THE HISTORY OF MATHEMATICS

John Stillwell

PRINCETON UNIVERSITY PRESS

PRINCETON AND OXFORD

Requests for permission to reproduce material from this work should be sent to permissions@press.princeton.edu

Published by Princeton University Press
41 William Street, Princeton, New Jersey 08540
99 Banbury Road, Oxford OX2 6JX

press.princeton.edu

All Rights Reserved
ISBN 9780691234366
ISBN (e-book) 9780691234373

Library of Congress Control Number: 2022932472

British Library Cataloging-in-Publication Data is available

Editorial: Diana Gillooly, Kristen Hop, Kiran Pandey
Production: Jacqueline Poirier
Publicity: Matthew Taylor, Carmen Jimenez
Copyeditor: Patricia J. Watson

Jacket images: (*Top*): Leonardo da Vinci (1452–1519), Codex Atlanticus, sheet 518 recto, © Veneranda Biblioteca Ambrosiana / Metis e Mida Informatica / Mondadori Portfolio / Bridgeman Images. (*Left*): Five tetrahedra in a dodecahedron, courtesy of the author. (*Right*): Order-3 heptakis heptagonal tiling, courtesy of Claudio Rocchini.

This book has been composed in MinionPro

Printed on acid-free paper. ∞

Printed in the United States of America

1   3   5   7   9   10   8   6   4   2

To Elaine, my *sine qua non*

■ ■ ■ ■ ■

Contents

· · · · ·

Preface

■ ■ ■ ■ ■

Proof is the glory of mathematics—and its most characteristic feature—
yet proof itself is not considered an interesting topic by many mathematicians. In the United States proof is not deemed an essential part of
mathematics education until upper university level, where "introduction
to proof" courses are offered. Yet by withholding the concept of proof, we
prevent students from seeing how mathematics actually *works*. I thought
about naming this book "How Mathematics Works," before deciding on
a more modest but accurate title. It is about proof—not just about what
proof is but about where it came from, and perhaps where it is going.

We know that mathematics has a logical structure, but we also know
that this structure is ever-changing, reflecting its evolution in the col-
lective human mind. There is generally more than one way to prove a
given theorem or to develop a given theory. Often the way first discovered
is not the simplest or most natural, but vestiges of the old ways survive
because of historical inertia or because they appeal to human senses or
psychology. For example, geometry continues to appeal to human visual
intuition even though it can be done by the symbolic methods of algebra or analysis. Because of this, the mathematical experience is greatly
enriched by awareness of historical and logical issues, and we owe it
to our students to present mathematics as a rich experience. Even pro-
fessional mathematicians will be enlightened, I believe, by seeing the
evolution of proof in mathematics, because advances in mathematics are
often advances in the concept of proof.

A major theme of the book is the relation between logic and computation, where "computation" is understood broadly to include classical
algebra. In ancient Greece logic was strong (though deployed mainly in
geometry) and computation was weak. In ancient China and India, computation ruled, as it did in Europe when algebra arrived from India via
the Muslim world. Then in the seventeenth century Europe made the

further step to *infinitesimal* algebra, namely calculus, which dominated mathematics (and physics) for the next two centuries. Leibniz, in unpublished work, dreamed of reducing logic itself to algebraic calculation. The dream of Leibniz began to take shape when Boole in 1847 created what we now call *Boolean algebra*, thereby reducing a significant fragment of logic to genuine calculation.

But the full logic of mathematics, and the full concept of computation, was not well understood until the twentieth century. In 1879 Frege described a logic adequate for mathematics, but the very idea that logic and computation were mathematical *concepts*, rather than just mathematical methods, did not arise until the 1920s. When this happened, through the work of Post, Gödel, Turing, and others, logic and computation became actual branches of mathematics—indeed, essentially the *same* branch.

Unfortunately, logic and computation developed largely in isolation from the rest of mathematics, so they are less well known than they should be in the mathematical community.[1] This book tries to remedy this situation, by presenting logic as it developed in mainstream mathematics. The history of mathematics can be viewed as a history of proof, because mathematics presents the most extreme challenges to proof: the Pythagorean discovery of irrational numbers, the sixteenth-century encounter with imaginary numbers, the seventeenth-century controversies over infinitesimals, and the nineteenth-century struggles with infinity, to mention just a few.

A second and related theme of the book is the development of concepts, since proofs can often be articulated only when suitable abstract concepts and notation are available to express them. This is seen most clearly in the development of algebra, where many abstractions originated and later spread to other parts of mathematics. But concept development is also a key to geometry and analysis, where concepts that now seem obvious, such as "area" and "limit," emerged only after long struggles with provisional concepts that failed to capture exactly what was intended.

In fact, the network of mathematical concepts is about as complex as the network of theorems, and I have tried to highlight both theorems and concepts by writing them in bold where they make key appearances in

---

1. Weil (1950) described logic as the "hygiene of the mathematician, it is not his source of food," as though logicians were sanitation workers.

the story. In the early chapters, new concepts are simple and infrequent enough to be defined informally, but later chapters make more formal definitions, particularly when several new concepts arise together and depend on each other.

I hope this book clarifies the role of logic, computation, and abstraction in mathematics for a general mathematical audience and hence gives a better understanding of the nature of proof. It is not an introduction to proof as much as a panoramic view of proof in basic mathematics. Many of the perennial concerns of all mathematicians—such as the relations among geometry, algebra, and analysis and their seemingly different styles of proof—are seen afresh from the viewpoint of logic and history. We see the intuitive origins of concepts, the search for axioms that capture intuition, new intuitions that emerge from axioms, and the connections among geometry, algebra, and analysis that axioms bring to light. It is fairly well known, for example, that Hilbert in the 1890s filled the gaps in Euclid's axioms for geometry. It is less well known that in doing so Hilbert found new connections among geometry, algebra, and even analysis. These connections are explained in chapters 3 and 11.

The arrangement of the book is partly chronological, partly by topic. Fields of mathematics are introduced in chronological order: geometry and number theory, algebra, algebraic geometry, calculus, and so on. But sometimes we follow a particular topic over a long period, so as not to break a train of thought, before turning to the next topic in chronological order. For example, in chapter 4 the story of algebra is told from ancient times until the nineteenth century, because it is mostly self-contained. The influence of algebra on other fields of mathematics, such as geometry, calculus, and number theory, is then told in chapters 5, 6, and 7.

Arranging material by topic also serves to arrange methods of proof, because of the different methods of proof in different fields mentioned above. Today, these methods are so different that practitioners in one field often fail to understand those in another. Among other things, I hope this book will contribute to mutual understanding by explaining the methods of proof characteristic of different fields. It should be accessible to senior undergraduates and also of interest to their teachers—possibly serving as a bridge between my two previous books, *Elements of Mathematics* and *Reverse Mathematics* (Stillwell 2016, 2018).

As always, I thank my wife, Elaine, for her eagle eye in proofreading the manuscript. I also thank Mark Hunacek and the anonymous reviewers for helpful suggestions and corrections.

John Stillwell
South Melbourne, 2021

CHAPTER 1

Before Euclid

The signature theorem of mathematics is surely the **Pythagorean theorem**, which was discovered independently in several cultures long before Euclid made it the first major theorem in his *Elements* (book 1, proposition 47). All the early roads in mathematics led to the Pythagorean theorem, no doubt because it reflects both sides of basic mathematics: number and space, or arithmetic and geometry, or the discrete and the continuous.

The arithmetic side of the Pythagorean theorem was observed in remarkable depth as early as 1800 BCE, when Babylonian mathematicians found many triples $\langle a, b, c \rangle$ of natural numbers such that $a^2 + b^2 = c^2$. Whether they viewed each triple a, b, c as sides of a right-angled triangle has been questioned; however, the connection was not missed in ancient India and China, where there were also geometric demonstrations of particular cases of the theorem.

Nevertheless, the Pythagoreans are rightly associated with the theorem because of their discovery that $\sqrt{2}$, the hypotenuse of the triangle with unit sides, is **irrational**. This discovery was a turning point in Greek mathematics, even a "crisis of foundations," because it forced a reckoning with *infinity* and, with it, the need for *proof*. In India and China, where irrationality was overlooked, there was no "crisis," hence no perceived need to develop mathematics in a deductive manner from self-evident axioms.

The nature of irrational numbers, as we will see, is a deep problem that has stimulated mathematicians for millennia. Even in antiquity, with Eudoxus's theory of proportions, the Greeks took the first step from the discrete toward the continuous.

1.1 THE PYTHAGOREAN THEOREM

For many people, the Pythagorean theorem is where geometry begins, and it is where proof begins too. Figure 1.1 shows the pure geometric form of the theorem: for a right-angled triangle (white), the square on the hypotenuse (gray) is equal to the sum of the squares on the other two sides (black).

Figure 1.1 : The Pythagorean theorem

What "equality" and "sum" mean in this context can be explained immediately with the help of figure 1.2. Each half of the picture shows a large square with four copies of the triangle inside it. On the left, the large square minus the four triangles is identical with the square on the hypotenuse. On the right, the large square minus four triangles is identical with the squares on the other two sides. Therefore, the square on the hypotenuse *equals* the sum of the squares on the other two sides.

Thus we are implicitly assuming some "common notions," as Euclid called them:

1. Identical figures are equal.
2. Things equal to the same thing are equal to each other.
3. If equals are added to equals the sums are equal.
4. If equals are subtracted from equals the differences are equal.

These assumptions sound a little like algebra, and they are obviously true for numbers, but here they are being applied to geometric objects.

Figure 1.2 : Seeing the Pythagorean theorem

In that sense we have a purely geometric proof of a geometric theorem. The reasons why the Pythagoreans wanted to keep geometry pure will emerge in section 1.3 below.

Although figure 1.2 is as convincing as a picture can be, some might quibble that we have not really explained why the gray and black regions are squares. The Greeks who came after Pythagoras did indeed quibble about details like this, due to concerns about the nature of geometric objects that will also emerge in section 1.3. The result was Euclid's *Elements*, produced around 300 BCE, a system of proof that placed geometry on a firm (but wordy) logical foundation. Chapter 2 expands figure 1.2 into a proof in the style of Euclid. We will see that the saying "a picture is worth a thousand words" is pretty close to the mark.

Origins of the Pythagorean Theorem

As noted above, the Pythagorean theorem was discovered independently in several ancient cultures, probably earlier than Pythagoras himself. Special cases of it occur in ancient India and China, and perhaps earliest of all in Babylonia (part of modern Iraq). Thus the theorem is a fine example of the universality of mathematics. As we will see in later chapters, it recurs in different guises throughout the history of geometry, and also in number theory.

It is not known how it was first proved. The proof above is one suggestion, given by Heath (1925, 1:354) in his edition of the *Elements*. The Chinese and Indian mathematicians were more interested in triangles whose sides had particular numerical values, such as 3, 4, 5 or 5, 12, 13.

As we will see in the next section, the Babylonians developed the theory of numerical right-angled triangles to an extraordinarily high level.

1.2 PYTHAGOREAN TRIPLES

If the sides of a right-angled triangle are a, b, c, with c the hypotenuse, then the Pythagorean theorem is expressed by the equation

$$a^2 + b^2 = c^2,$$

in the algebraic notation of today. Indeed, we call a^2 "a squared" in memory of the fact that a^2 represents a square of side a. We also understand that a^2 is found by multiplying a by itself, and the Pythagoreans would have agreed with us when a is a whole number. What made the Pythagorean theorem interesting to them are the whole-number triples $\langle a, b, c \rangle$ satisfying the equation above. Today, such triples are known as **Pythagorean triples**. The simplest example is of course $\langle 3, 4, 5 \rangle$, because

$$3^2 + 4^2 = 9 + 16 = 25 = 5^2,$$

but there are infinitely many Pythagorean triples. In fact, the right-angled triangles whose sides are Pythagorean triples come in infinitely many shapes because the slopes b/a of their hypotenuses can take infinitely many values.

The most impressive evidence for this fact appears on a Babylonian clay tablet from around 1800 BCE. The tablet, known as Plimpton 322 (its catalog number in a collection at Columbia University), contains columns of numbers that Neugebauer and Sachs (1945) interpreted as values of b and c in a table of Pythagorean triples. Part of the tablet is broken off, so what remains are pairs $\langle b, c \rangle$ rather than triples. Some have questioned whether the Babylonian compiler of the tablet really had right-angled triangles in mind. In my opinion, yes, because all the values $c^2 - b^2$ are perfect squares *and* the pairs $\langle b, c \rangle$ are listed in order of the values b/a—the slopes of the corresponding hypotenuses. Figure 1.3 is a completed table that includes the values of a and b/a and also a fraction x that I explain below.

The column of a values reveals something else interesting. These values are all divisible only by powers of 2, 3, and 5, which makes them particularly "round" numbers in the Babylonian system, which was based on the number 60 (some of their system survives today, with 60 minutes in a hour and 60 seconds in a minute).

We do not know how the Babylonians discovered these triples. However, the amazingly complex values of b and c can be generated from the

a	b	c	b/a	x
120	119	169	0.9917	12/5
3456	3367	4825	0.9742	64/27
4800	4601	6649	0.9585	75/32
13500	12709	18541	0.9414	125/54
72	65	97	0.9028	9/4
360	319	481	0.8861	20/9
2700	2291	3541	0.8485	54/25
960	799	1249	0.8323	32/15
600	481	769	0.8017	25/12
6480	4961	8161	0.7656	81/40
60	45	75	0.7500	2
2400	1679	2929	0.6996	48/25
240	161	289	0.6708	15/8
2700	1771	3229	0.6559	50/27
90	56	106	0.6222	9/5

Figure 1.3 : Pythagorean triples in Plimpton 322

fractions x, which are fairly simple combinations of powers of 2, 3, and 5. In terms of x, the whole numbers a, b, and c are denominator and numerators of the fractions

$$\frac{b}{a} = \frac{1}{2}\left(x - \frac{1}{x}\right) \quad \text{and} \quad \frac{c}{a} = \frac{1}{2}\left(x + \frac{1}{x}\right).$$

For example, with $x = 12/5$ we get

$$\frac{1}{2}\left(x - \frac{1}{x}\right) = \frac{1}{2}\left(\frac{12}{5} - \frac{5}{12}\right) = \frac{119}{120} \quad \text{and} \quad \frac{1}{2}\left(x + \frac{1}{x}\right) = \frac{1}{2}\left(\frac{12}{5} + \frac{5}{12}\right) = \frac{169}{120}.$$

The huge triple $\langle 13500, 12709, 18541 \rangle$ is similarly generated from the fraction $125/54 = 5^3/2 \cdot 3^3$, which has roughly the same complexity as $13500 = 2^2 \cdot 3^3 \cdot 5^3$. Thus, it is plausible that the Babylonians could have generated complex Pythagorean triples by relatively simple arithmetic. At the same time, the link with geometry is hard to deny when the triples are seen to be arranged in order of the slopes b/a—an order that could not be guessed from the arrangement of a, b, c, or x values! And when one sees that these slopes cover a range of angles, roughly equally spaced, between 30° and 45° (figure 1.4), it looks as though the Babylonians were collecting triangles of different shapes.

It is also conspicuous which shape is *missing* from this collection of triangles: the one with equal sides a and b, shown in red in figure 1.4.

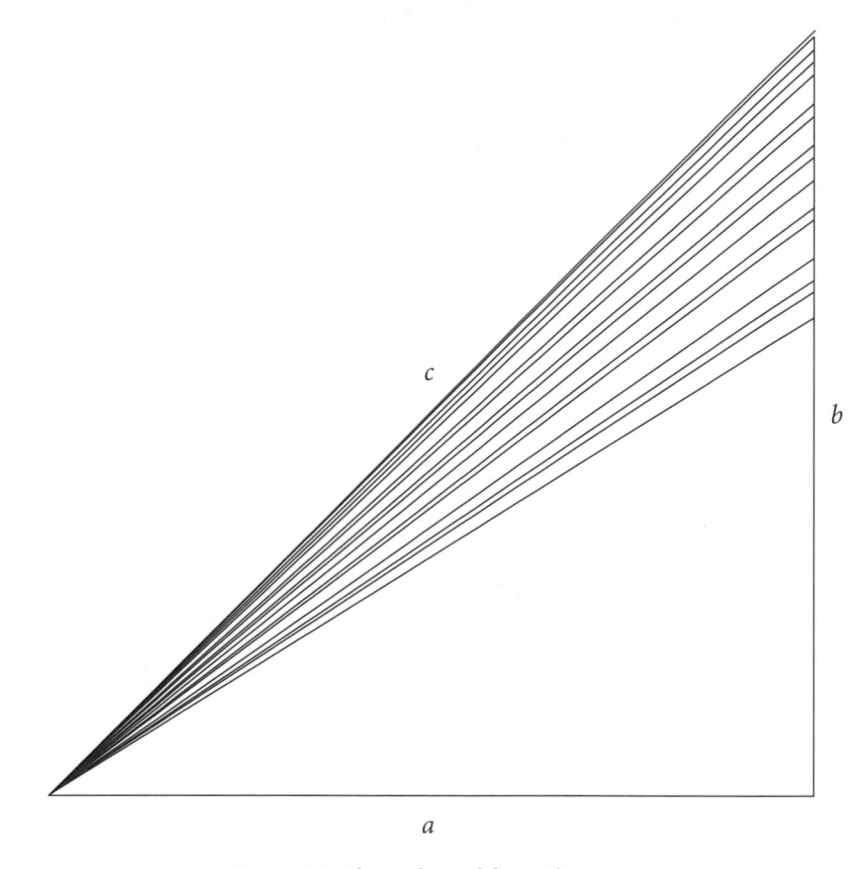

c

b

a

Figure 1.4 : Slopes derived from Plimpton 322

As we now know, because the Pythagoreans discovered it, this shape is missing because the hypotenuse of this triangle is *irrational*.

1.3 IRRATIONALITY

Irrationality follows naturally from the Pythagorean theorem, but apparently it was found by the Pythagoreans alone. Like other discoverers of the theorem, the Pythagoreans knew special cases with whole-number values of a, b, c. But, apparently they were the only ones to ask, Why do we find no such triples with $a = b$? The question points to its own answer: *it is contradictory to suppose there are whole numbers a and c such that* $c^2 = 2a^2$.

The argument of the Pythagoreans is not known, but the result must have been common knowledge by the time of Aristotle (384–322 BCE), as he apparently assumes his readers will understand the following brief hint:

> The diagonal of the square is incommensurable with the side, because odd numbers are equal to evens if it is supposed commensurable.

(Aristotle, *Prior Analytics*, bk. 1, chap. 23)

Here "commensurable" means being a whole number multiple of a common unit of measure, so we are supposing that $c^2 = 2a^2$, where the side of the square is a units and its diagonal is c units. We reach the contradiction "odd = even" as follows.

First, by choosing the unit of measure as large as possible, we can assume that the whole numbers c and a have no common divisor (except 1). In particular, at most one of them can be even.

Now $c^2 = 2a^2$ implies that the number c^2 is even. Since the square of an odd number is odd, c must also be even, say $c = 2d$. Substituting $2d$ for c gives

$$(2d)^2 = 2a^2 \quad \text{so} \quad 2d^2 = a^2.$$

But then a similar argument shows a is even, which is a contradiction.

So it is wrong to suppose there are whole numbers a and c with $c^2 = 2a^2$.

The usual way to express this fact today is that *there are no natural numbers c and a such that $\sqrt{2} = c/a$* or, more simply, that $\sqrt{2}$ *is irrational*.

1.4 FROM IRRATIONALS TO INFINITY

The argument for irrationality of $\sqrt{2}$ is very short and transparent in modern algebraic symbolism. Judging by the excerpt from Aristotle, it was also comprehensible enough when equations were written out in words, as the ancient Greeks did.

But there was also a geometric approach to incommensurable quantities that the Greeks called *anthyphaeresis*. It gives a different and deeper insight into the nature of $\sqrt{2}$ and, indeed, a different proof that it is irrational. Anthyphaeresis is a process that can be applied to two quantities, such as lengths or natural numbers, by repeatedly subtracting the smaller from the larger. Since it was later used to great effect by Euclid, it is today called the **Euclidean algorithm**.

More formally, given two quantities a_1 and b_1 with $a_1 > b_1$, one forms the new pair of quantities b_1 and $a_1 - b_1$ and calls the greater of them a_2

and the lesser b_2. Then one does the same with the pair a_2, b_2, and so on. For example, if $a_1 = 5$, $b_1 = 3$ we get

$$\langle a_1, b_1 \rangle = \langle 5, 3 \rangle$$
$$\langle a_2, b_2 \rangle = \langle 3, 2 \rangle$$
$$\langle a_3, b_3 \rangle = \langle 2, 1 \rangle$$
$$\langle a_4, b_4 \rangle = \langle 1, 1 \rangle,$$

at which point the algorithm terminates because $a_4 = b_4$. The Euclidean algorithm always terminates when a_1 and b_1 are natural numbers, because subtraction produces smaller natural numbers and natural numbers cannot decrease forever. Conversely, *a ratio for which the Euclidean algorithm runs forever is irrational.*

In section 2.6 we will see the consequences of the Euclidean algorithm for natural numbers, but for the Greeks before Euclid the process of anthyphaeresis was most revealing for pairs of incommensurable quantities, such as $a_1 = \sqrt{2}$ and $b_1 = 1$. In this case the numbers a_n, b_n can and do decrease forever. In fact, we have

$$\langle a_1, b_1 \rangle = \langle \sqrt{2}, 1 \rangle$$
$$\langle a_2, b_2 \rangle = \langle 1, \sqrt{2} - 1 \rangle$$
$$\langle a_3, b_3 \rangle = \langle 2 - \sqrt{2}, \sqrt{2} - 1 \rangle = \langle (\sqrt{2} - 1)\sqrt{2}, (\sqrt{2} - 1)1 \rangle,$$

so $\langle a_3, b_3 \rangle$ is the same as $\langle a_1, b_1 \rangle$, just scaled down by the factor $\sqrt{2} - 1$. Two more steps will give $\langle a_5, b_5 \rangle$, again the same as $\langle a_1, b_1 \rangle$ but scaled down by the factor $(\sqrt{2} - 1)^2$, and so on. Thus the numbers $\langle a_n, b_n \rangle$ decrease forever, but they return to the same ratio every other step.

Since this cannot happen for any pair $\langle a, b \rangle$ of natural numbers, it follows that $\sqrt{2}$ and 1 are not in a natural number ratio; that is, $\sqrt{2}$ is irrational. Moreover, we have discovered that the pair $\langle \sqrt{2}, 1 \rangle$ behaves *periodically* under anthyphaeresis, producing pairs in the same ratio every other step. It turns out, though this was not understood until algebra was better developed, that periodicity is a special phenomenon occurring with square roots of natural numbers.

Visual Form of the Euclidean Algorithm

If a and b are lengths, we can represent the pair $\{a, b\}$ by the rectangle with adjacent sides a and b. If, say, $a > b$, then the pair $\{b, a - b\}$ is represented by the rectangle obtained by cutting a square of side b from the original rectangle, shown in light gray in figure 1.5. The algorithm then repeats the process of cutting off a square in the light gray rectangle, and so on.

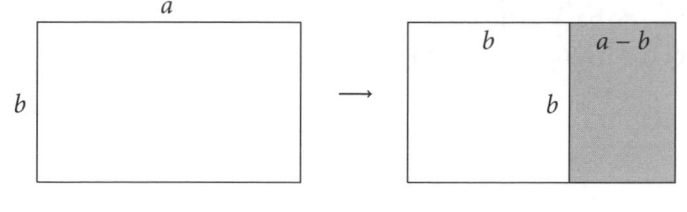

Figure 1.5 : First step of the Euclidean algorithm

When $a = \sqrt{2}$ and $b = 1$, two steps of the algorithm give the light gray rectangle shown in figure 1.6, which is the *same shape* as the original rectangle. This is because its sides are again in the ratio $\sqrt{2} : 1$, as we saw in the calculation above. Since the new rectangle is the same shape as the old, it is clear that the process of cutting off a square will continue forever.

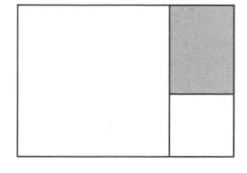

Figure 1.6 : After two steps of the algorithm on $\sqrt{2}$ and 1

The Greeks were fascinated by geometric constructions in which the original figure reappears at a reduced size. The simplest example is the so-called *golden rectangle* (see figure 1.7), in which removal of a square leaves a rectangle the same shape as the original. It follows that the Euclidean algorithm runs forever on the sides a and b of the golden rectangle, and hence these sides are in irrational ratio. This particular ratio is called the **golden ratio**.

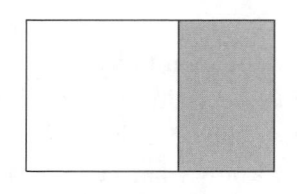

Figure 1.7 : The golden rectangle

The golden ratio is also the ratio of the diagonal to the side of the regular pentagon, where the recurrence of the original figure at reduced size can be seen in figure 1.8.

It is believed that the study of the golden ratio and the regular pentagon may go back to the Pythagoreans, in which case they were probably aware of the irrationality of the golden ratio as well as that of $\sqrt{2}$.

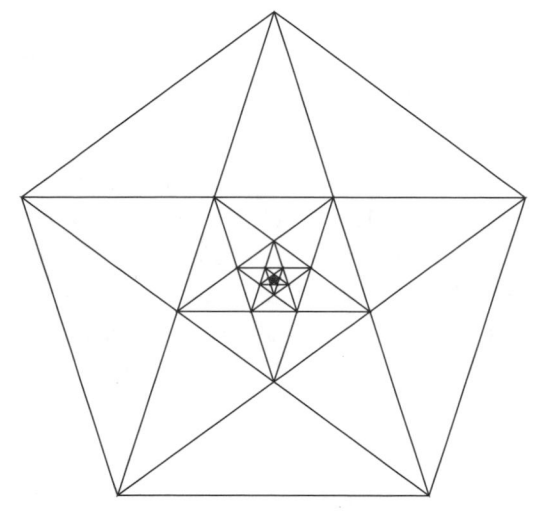

Figure 1.8 : Infinite series of pentagrams

1.5 FEAR OF INFINITY

As we have just seen, irrationality brings infinite processes to the attention of mathematicians, albeit processes of a simple and repetitive kind. At an even more primitive level, the natural numbers 0, 1, 2, 3, . . . themselves represent the kind of infinity where a simple process—in this case, adding 1—is repeated without end. An infinity that involves endless repetition was called by the Greeks a *potential infinity*. They contrasted it with *actual infinity*—a somehow completed infinite totality—which was considered unacceptable or downright contradictory.

The legendary opponent of infinity was Zeno of Elea, who lived around 450 BCE. Zeno posed certain "paradoxes of the infinite," which we know only from Aristotle, who described the paradoxes only to debunk them, so we do not really know what Zeno meant by them or how they were originally stated. It will become clear, however, that Zeno accepted potential infinity while rejecting actual infinity.

A typical Zeno paradox is his first, the *paradox of the dichotomy*, in which he argues that motion is impossible because

> before any distance can be traversed, half the distance must be traversed [and so on], that these half distances are infinite in number, and that it is impossible to traverse distances infinite in number. (Aristotle, *Physics*, bk. 8, chap. 8, 263a)

Apparently, Zeno is arguing that the infinite sequence of events

reaching 1/2 way

reaching 1/4 way

reaching 1/8 way

...

cannot be completed. Aristotle answers, a few lines below this statement, that

> the element of infinity is present in the time no less than in the distance.

In other words, if one can conceive an infinite sequence of places

$$1/2 \text{ way}, 1/4 \text{ way}, 1/8 \text{ way}, \ldots.$$

then one can conceive an infinite sequence of times at which

$$1/2 \text{ way is reached}, 1/4 \text{ way is reached}, 1/8 \text{ way is reached}, \ldots.$$

Thus if Zeno is willing to admit the potential infinity of places, he has to admit the potential infinity of times. It is not a question of *completing* an infinity but only of correlating one potential infinity with another. We claim only that each of the places can be reached at a certain time; we do not have to consider the totality of places or the totality of times.

At any rate, after Zeno, Greek mathematicians handled questions about infinity by this style of argument—dealing with members of a potential infinity one by one rather than in their totality. The "actual infinity scare" was nevertheless productive, because it led to a very subtle understanding of the relation between the continuous and the discrete.

1.6 EUDOXUS

Eudoxus of Cnidus, who lived from approximately 390 BCE to 330 BCE, was a student of Plato and is believed to have taught Aristotle. His most important accomplishments are the **theory of proportions** and the **method of exhaustion**. Together, they form the summit of the Greek treatment of infinity, and they come down to us mainly through the exposition in book 5 of Euclid's *Elements*. In particular, the theory of proportions was the best treatment of rational and irrational quantities available until the nineteenth century. Indeed, it is probably the best treatment possible as long as one rejects actual infinity, which most mathematicians did until the 1870s.

The theory of proportions deals with "magnitudes" (typically lengths) and their relation to "numbers," which are natural numbers. It thereby builds a bridge between the two worlds separated by the Pythagoreans: the world of magnitudes, which vary *continuously*, and the world of counting, where numbers jump *discretely* from each number to its successor.

The theory is complicated somewhat because the Greeks thought in terms of ratios of magnitudes and ratios of numbers, without having the algebraic machinery of fractions that makes ratios easy to handle. We can understand the ratio of natural numbers m and n as the fraction m/n, so we will write the ratio of lengths a and b as the fraction a/b.[1] The key idea of Eudoxus is that ratios of lengths, a/b and c/d, are equal if and only if, for *each* natural number ratio m/n,

$$\frac{m}{n} < \frac{a}{b} \quad \text{if and only if} \quad \frac{m}{n} < \frac{c}{d}.$$

Equivalently (and this is how Eudoxus put it), for each natural number pair m and n,

$$mb < na \quad \text{if and only if} \quad md < nc.$$

Thus the infinity of natural number pairs m, n is behind the definition of equality of length ratios, but only potentially so, because equality

---

1. It may seem unwieldy to work with ratios of lengths rather than just lengths, but in fact length is a *relative* concept and only the ratio of lengths is absolute. When we say length $a = 3$, for example, we really mean that 3 is the ratio of a to the unit length. In chapter 9 we will see that the relative concept of length is a specific characteristic of Euclidean geometry.

depends on a single (though arbitrary) pair m, n. In defining unequal length ratios, infinity can be avoided completely, because one *particular* pair can witness inequality. Namely, if $a/b < c/d$ then there is a particular m/n such that

$$\frac{a}{b} < \frac{m}{n} < \frac{c}{d},$$

and likewise, if $c/d < a/b$ then there is a particular m/n between c/d and a/b. Today we would say that ratios of lengths are *separable* by ratios of natural numbers.

The Archimedean Axiom

The assumption that natural number ratios separate ratios of lengths is equivalent to a property later called the *Archimedean property*: if $a/b > 0$ then $a/b > m/n > 0$ for some natural numbers m and n. It follows, obviously, that in fact $a/b > 1/n$, so $na > b$. This gives the usual statement of the **Archimedean axiom**: *if a and b are any nonzero lengths, then there is a natural number n such that na > b.*

Another statement of the Archimedean axiom is: *there is no ratio a/b so small that $0 < a/b < 1/n$ for each natural number n*, or more concisely, *there are no infinitesimals*. This property was assumed by Euclid and Archimedes (hence the name), but some later mathematicians, such as Leibniz, thought that infinitesimals exist. We will see in chapter 4 that the existence of infinitesimals was a big issue in the development of calculus.

Mathematical practice today has translated Eudoxus's theory into our concept of the **real number system** $\mathbb{R}$. The ratios of lengths are the nonnegative real numbers, and among them lie the nonnegative **rational numbers**, which are the ratios m/n of natural numbers. Any two distinct real numbers are separated by a rational number, so there are no infinitesimals in $\mathbb{R}$. Conversely, each real number is determined by the rational numbers less than it and the rational numbers greater than it. Exactly how this came about, and what the real numbers *are*, is explained in chapter 11. It turns out that separation by rational numbers is the key to answering this question.

The Method of Exhaustion

We discuss the method of exhaustion only briefly here, because it is a generalization of the theory of proportions. Also, the best examples of the method occur in the work of Euclid and Archimedes, discussed in chapter 2. The basic idea is to approximate an "unknown quantity," such as the area or volume of a curved region, by "known quantities" such as areas of triangles or volumes of prisms. This generalizes the idea of approximating a ratio of lengths by ratios of natural numbers. Generally, there is a potential infinity of approximating objects, but as long as they come "arbitrarily close" to the unknown quantity it is possible to draw conclusions without appealing to actual infinity.

An example is approximation of the circle by polygons, shown in figure 1.9, which allows us to draw the conclusion that the area of the circle is proportional to the square of its radius.

Figure 1.9 shows polygons approximating the circle from inside and outside. Only the first two approximations are shown, but one can imagine a continuation of the sequence by repeatedly doubling the number of sides. It is clear that the area of the gap between inner and outer polygons becomes arbitrarily small in the process, and hence both inner and outer polygons come arbitrarily close to the circle in area.

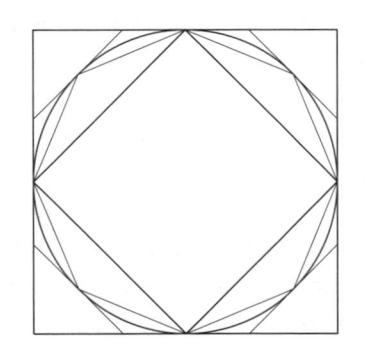

Figure 1.9 : Approximating the circle by polygons

Also, the area of each polygon P_n is a sum of triangles, whose area $P_n(R)$ for radius R is known and proportional to R^2. Now comes a typical example of reasoning "by exhaustion": suppose that the area $C(R)$ of the circle of radius R is *not* proportional to R^2. Thus, if we compare circles of radius R and R' we have either

$$C(R)/C(R') < R^2/R'^2$$

or

$$C(R)/C(R') > R^2/R'^2.$$

If $C(R)/C(R') < R^2/R'^2$, then by choosing n so that $P_n(R)$ is sufficiently close to $C(R)$ and $P_n(R')$ is sufficiently close to $C(R')$, we will get

$$P_n(R)/P_n(R') < R^2/R'^2,$$

which is a contradiction. If $C(R)/C(R') < R^2/R'^2$ we get a similar contradiction. Therefore *the only possibility is that* $C(R)/C(R') = R^2/R'^2$.

We have established what we want by *exhausting* all other possibilities. This is what "exhaustion" means in the method of exhaustion. Notice also that we used only the potential infinity of polygons by going only far enough to contradict a given inequality. This is typical of the method.

1.7 REMARKS

We have seen in the development of Greek mathematics many topics considered tricky in undergraduate mathematics today, such as proof by contradiction, the use of infinity, and the idea of choosing a "sufficiently close" approximation. This just goes to show, in my opinion, that ancient mathematics is good training in the art of proof.

At the same time, we have seen that ancient arguments can often be streamlined by the use of algebraic symbolism, and the art of algebra was missing in ancient times.

The other thing missing, in what we know of this early stage, was the systematic deduction of theorems from axioms. The art of **axiomatics** also began in ancient times, as we will see in the next chapter.

CHAPTER 2

Euclid

In the story of proof, Euclid comes near the beginning because his *Elements* was composed around 300 BCE and few earlier examples of proof survive. Unfortunately, this means plunging the reader into deep water immediately, because Euclid did so much that the *Elements* became the model of proof until quite recent times. There was no major advance in the technique of proof until algebraic symbolism was added in the sixteenth century, and no advance in logic itself until the nineteenth century.

Also, the *Elements* is conceptually subtle in separating the continuous (geometry) from the discrete (number theory), following the Pythagorean separation of quantity and "number." The difficult book 5 begins to build a bridge between the two, with the theory of proportions, and by admitting (limited) use of infinity. Infinity is also used in an elegant determination of the volume of the tetrahedron.

Because of the many difficulties in the *Elements*, readers may prefer to skim the next two chapters and come back to the details later. Still, *some* acquaintance with Euclid is needed to understand the later development of mathematics. The *Elements* influenced not only mathematics but also philosophy (Spinoza's *Ethics*) and law (Abraham Lincoln was an admirer). However, philosophy and law could not attain a standard of proof *higher* than that of the *Elements*, whereas mathematics could.

Eventually, in the nineteenth century, mathematicians became aware of gaps in Euclid's reasoning and of alternatives to his axioms, which led to more rigorous foundations of mathematics by the end of the nineteenth century. At this point another "crisis of foundations" emerged and transformed the concept of proof in many ways, some of which are still being worked out.

2.1 DEFINITION, THEOREM, AND PROOF

Euclid's *Elements* is the oldest mathematics book that looks "modern," in the sense of containing definitions, theorems, and proofs, arranged in logical order. On closer inspection one sees some flaws—Euclid tries to define terms that should remain undefined, and he tries to prove some statements that should be axioms—but nevertheless, the *Elements* is a masterpiece that set the standard of mathematical proof for over 2,000 years. Perhaps the most important lesson taught by the *Elements* is that mathematics can be built, cumulatively, by deduction from self-evident *axioms*.

The *Elements* is founded on simple objects such as points, lines, and circles, the associated quantities of length and angle, and certain axioms about them (which are traditionally called *postulates*). These axioms are, in the classic translation of Heath (1925):

P1. To draw a straight line from any point to any point.
P2. To produce a finite straight line continuously in a straight line.
P3. To describe a circle with any center and distance.
P4. That all right angles are equal to one another.
P5. That, if a straight line falling on two straight lines make the interior angles on the same side less than two right angles, the two straight lines, if produced indefinitely, meet on that side on which are the angles less than the two right angles.

One sees immediately a peculiarity of Euclid's language, which favors **construction** over mere **existence**. Postulate 1 does not say there *exists* a straight line (segment) between any two points but, rather, that the line segment *can be drawn*. And Postulate 2 does not say a straight line is infinite but (more conservatively) that a line segment can be *produced continuously*, that is, extended indefinitely. The question of construction is an important secondary theme of the *Elements*, and many of his theorems state that a certain figure can be constructed by the instruments that draw straight lines and circles ("straightedge" and "compass"). Unfortunately, Euclid's very first construction, on which several others depend, does *not* follow from his axioms. We will therefore assume existence in cases where Euclid makes a construction and postpone the question of extra axioms until later.

Perhaps the oddest postulate, to modern ears, is "all right angles are equal." To make sense of this, one has to realize that for Euclid an angle is merely a pair of half lines with a common endpoint—it does not come with a measure in degrees or radians. One can only say whether angles are equal or not, and a right angle *ABC* is one for which the two angles *ABC* and *ABD* in figure 2.1 are equal. Postulating that all right angles are equal then gives a standard unit of angle measure, *the* right angle. Indeed, one finds throughout the *Elements* that angles (or sums of angles) are always given as multiples of the right angle.

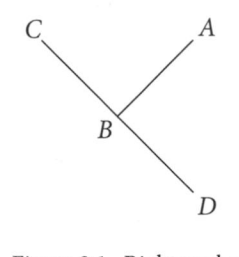

Figure 2.1 : Right angles

Postulate 5, known as the **parallel axiom**, actually states the condition for lines *not* to be parallel. If line $\mathcal{N}$ falls on lines $\mathcal{L}$ and $\mathcal{M}$ and makes angles α and β as shown in figure 2.2, and if $\alpha + \beta$ is less than two right angles, then postulate 5 says that $\mathcal{L}$ and $\mathcal{M}$ will meet somewhere on the right.

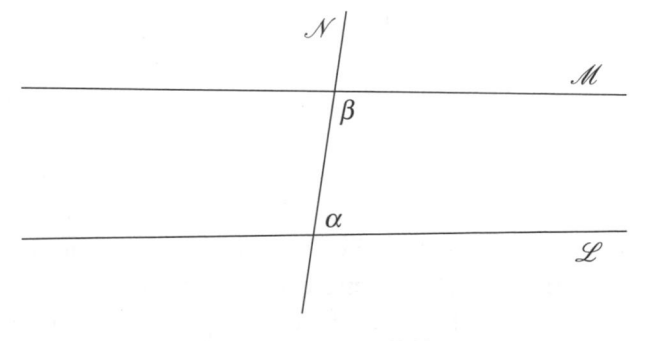

Figure 2.2 : Nonparallel lines

Therefore, if $\mathscr{L}$ and $\mathscr{M}$ do *not* meet—that is, are parallel—then $\alpha + \beta$ equals two right angles. This is one of many equivalent statements of the parallel axiom, convenient for proving that the angle sum of any triangle is two right angles. There are also equivalents of the axiom that do not mention the concept of angle, for example, the so-called **Playfair's axiom** saying that for any line $\mathscr{L}$ and a point P outside it there is exactly one line $\mathscr{M}$ through P that does meet $\mathscr{L}$ (see section 2.3).

As we will see later, many mathematicians were dissatisfied with the parallel axiom and hoped that it could be proved from postulates 1–4. However, this turned out to be impossible, for very interesting reasons we will see in chapter 9.

The principles of deduction were not explicitly stated in the *Elements*, except for the following, which Euclid called *common notions*. They can be viewed as properties of equality (and inequality), addition, and subtraction. (The first four were used in section 1.1 for a visual proof of the Pythagorean theorem.)

Common notion 1. Things that are equal to the same thing are equal to each other.

Common notion 2. If equals are added to equals, the wholes are equal.

Common notion 3. If equals are subtracted from equals, the remainders are equal.

Common notion 4. Things that coincide with one another are equal.

Common notion 5. The whole is greater than the part.

When the common notions are written in modern symbolism, they look rather like principles of algebra:

1. If $A = B$ and $C = B$ then $A = C$.
2. If $A = B$ then $A + C = B + C$.
3. If $A = B$ then $A - C = B - C$.
4. $A = A$.
5. If $A \subset B$ then $A < B$.

However, despite this promising start, algebra failed to materialize in the *Elements*. We take up the question of algebra again later. Now let us look at Euclid's theorems, or "propositions" as they are traditionally called. It is enough to look at book 1, which already contains some remarkable deductions.

2.2 THE ISOSCELES TRIANGLE THEOREM AND SAS

As a simple example of a deduction in Euclid's system, let us show that his proposition 5 follows from his proposition 4. Concisely stated, these propositions in book 1 of the *Elements* are:

Proposition 4. If two triangles agree in two sides and the included angle, then they agree in all corresponding sides and angles.
Proposition 5. If a triangle has two equal sides, then its two angles, other than the angle included by the equal sides, are also equal.

In modern geometry, proposition 4 is often abbreviated SAS, for "side angle side," and is considered an axiom. The triangle described in proposition 5 is called *isosceles*, from the Greek for "equal sides." Euclid's proof that SAS implies the isosceles triangle theorem was a traditional stumbling block for students of the *Elements*, known as the "ass's bridge" because asses could not get past it (or possibly because of Euclid's diagram, which consists of five lines resembling a bridge).

A much shorter proof was given by the later Greek geometer Pappus, so let's look at the Pappus proof instead. The reader should be warned, however, that the Pappus proof is almost too clever, because it takes the two triangles in SAS to be the *same* triangle. This is OK because no one said that the two triangles have to be different!

Suppose, then, that *ABC* is a triangle with *AB* = *AC*, as in figure 2.3. Notice that we may also take this to be a picture of triangle *ACB*.

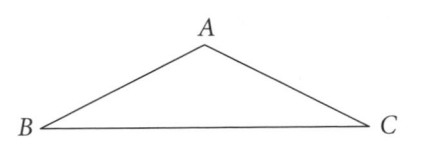

Figure 2.3 : An isosceles triangle

Now, the triangles *ABC* and *ACB* agree in two sides and the included angle, because *AB* = *AC*, *AC* = *AB*, and the angle at *A* is the included angle in both. Therefore, by proposition 4, the triangles agree in all corresponding angles. In particular, the angle at *B* (in triangle *ABC*) equals the angle at *C* (in triangle *ACB*, which of course is the same triangle). That's it!

SAS Implies ASA

SAS states a condition for triangles to be *congruent*—meaning they agree in all side lengths and all corresponding angles. Another such condition is ASA (for "angle side angle"): *if triangles agree in two angles and the side common to these angles, then they are congruent.* ASA is part of Euclid's proposition 26. It follows from SAS by a logical device already used in section 1.3: prove a statement false by showing that it leads to contradiction.

Suppose that $A_1B_1C_1$ and $A_2B_2C_2$ are two triangles, with equal angles α and β as shown in figure 2.4, and $A_1B_1 = A_2B_2$. Thus $A_1B_1C_1$ and $A_2B_2C_2$ agree in two angles and their common side, so ASA holds. Now suppose, for the sake of contradiction, that $A_1B_1C_1$ and $A_2B_2C_2$ are *not* congruent.

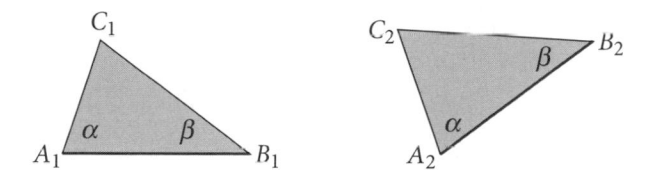

Figure 2.4 : Triangles satisfying ASA

Then not all corresponding sides are equal, else SAS holds and the triangles are congruent, contrary to our supposition. So, some corresponding sides are *unequal*, and (renaming if necessary) we can assume $A_1C_1 < A_2C_2$.

But then we can choose a point C on A_2C_2, between A_2 and C_2, so that $A_2C = A_1C_1$. Hence drawing the line B_2C creates an angle β' that is only part of β (figure 2.5), so $\beta > \beta'$ because "the whole is greater than the part."

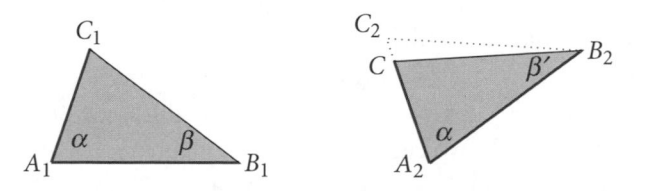

Figure 2.5 : Hypothetical triangles satisfying SAS

Yet the triangles $A_1B_1C_1$ and A_2B_2C satisfy SAS, since $A_2C = A_1C_1$, and hence they are congruent. In particular, β' in A_2B_2C equals the corresponding angle β in $A_1B_1C_1$, which is again a contradiction.

Therefore, it is false to suppose $A_1B_1C_1$ and $A_2B_2C_2$ are not congruent.

2.3 VARIANTS OF THE PARALLEL AXIOM

The proof of ASA above has the bonus feature that it holds even if the point C_2 does not exist! That is, we need only assume that the second "triangle" consists of the segment A_2B_2 and lines out of A_2 and B_2 at angles α and β, respectively. On the line through A_2 we can still choose the point C so that $A_2C = A_1C_1$ and arrive a contradiction as above.

This strong version of ASA enables us to prove the following variant of the parallel axiom P5: *If a straight line $\mathscr{N}$ falling on two straight lines $\mathscr{L}$ and $\mathscr{M}$ makes angles α and β, respectively, on the same side, with $\alpha + \beta =$ two right angles, then $\mathscr{L}$ and $\mathscr{M}$ are parallel.*

Suppose we have a line $\mathscr{N}$ that crosses two lines $\mathscr{L}$ and $\mathscr{M}$, making angles α and β as shown in figure 2.6, so $\alpha + \beta =$ two right angles.

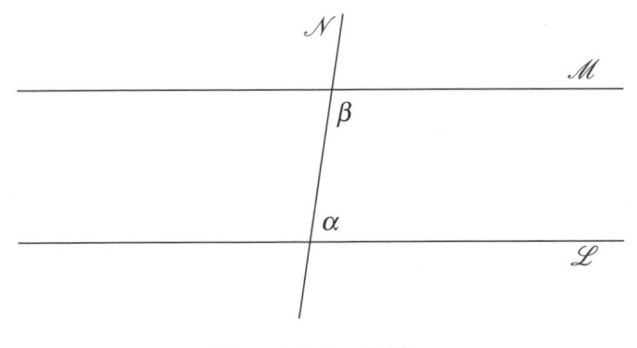

Figure 2.6 : Parallel lines

We can then find all the angles in figure 2.6 with the help of Euclid's book 1, proposition 13, which states: *If α and β together make a straight angle, then $\alpha + \beta =$ two right angles.* This proposition is proved by considering two angles α and β that make a straight angle at point P and comparing them with two right angles ρ that meet at P (figure 2.7). Since all right angles are equal, we can call them all ρ and, by subtraction, get the three angles shown on the right of figure 2.7. Since the right angle on

Figure 2.7 : Lines making a straight angle

the right consists of $\alpha - \rho$ and β, we get $\rho = \alpha - \rho + \beta$ and therefore

$$\alpha + \beta = 2\rho.$$

It follows from proposition 13, by "subtracting equals from equals," that the angles α and β formed by $\mathscr{L}$, $\mathscr{M}$, and $\mathscr{N}$ in figure 2.6 recur as shown in figure 2.8.

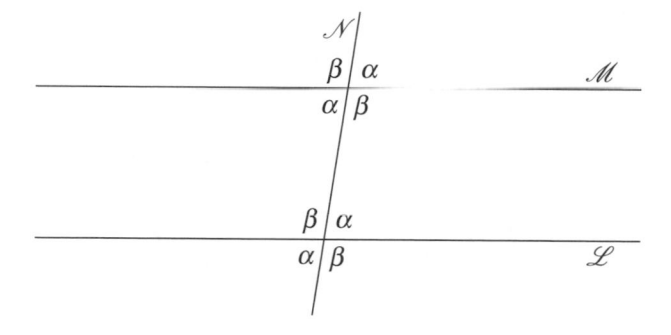

Figure 2.8 : Angles related to parallels

However, we do not yet know that $\mathscr{L}$ and $\mathscr{M}$ are parallel! We prove that they are with the help of the strong ASA. If in fact the lines $\mathscr{L}$ and $\mathscr{M}$ in figure 2.8 meet (say, on the right), then they form a triangle in combination with the segment of $\mathscr{N}$ between them. The same segment and angles occur on the left, forming a triangle congruent to the one on the right, by ASA. But then the lines $\mathscr{L}$ and $\mathscr{M}$ meet at two points—one on the right and one on the left. This is contrary to axiom P1, which, although Euclid does not say so explicitly, gives a *unique* line between any two points.

This contradiction shows that $\mathscr{L}$ and $\mathscr{M}$ do not meet; that is, they are parallel.

It follows that *there is a unique parallel $\mathscr{M}$ to a given line $\mathscr{L}$ through a given point P outside $\mathscr{L}$*. This is because for any such P we can choose a line $\mathcal{N}$ through P that crosses $\mathcal{L}$ at angle α, say. We then choose the line $\mathcal{M}$ through P that crosses $\mathcal{N}$ at angle β, where $\alpha + \beta =$ two right angles.

The italicized sentence above is another equivalent of the parallel axiom, often the most convenient one, because it avoids the concept of angle and it says that parallels exist. It is called **Playfair's axiom**, after the Scottish mathematician John Playfair (1748–1819). It appeared in his book Playfair (1795).

Parallelograms and Triangles

With the existence of parallels we get the existence of *parallelograms*, the four-sided figures whose opposite sides are parallel. Figure 2.8 gives us some equal angles. Figure 2.9 shows some of the equal angles, all of which can be deduced from figure 2.8 by choosing which lines to interpret as $\mathcal{L}$, $\mathcal{M}$, and $\mathcal{N}$.

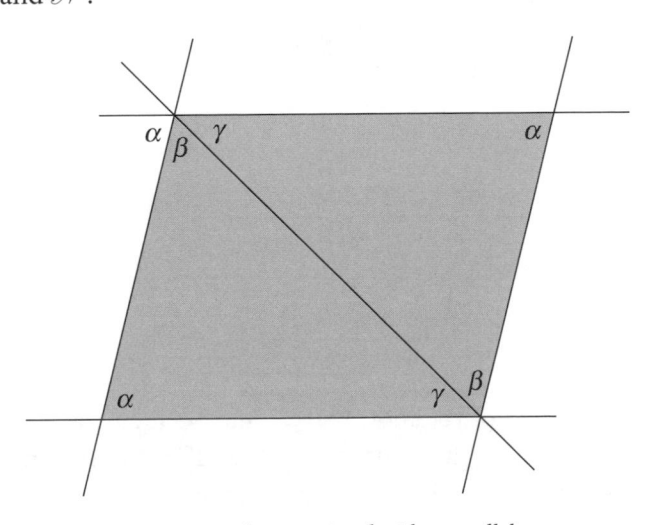

Figure 2.9 : Angles associated with a parallelogram

Notice that the gray parallelogram consists of two triangles with the diagonal as their common side and corresponding equal angles β and γ. By ASA, these triangles are congruent; hence *opposite sides of a parallelogram are equal*. Notice also in figure 2.9 that the angle sum $\alpha + \beta + \gamma$ of each triangle equals the straight angle at the top left, and therefore *the angle sum of any triangle is two right angles*. The latter statement is actually Euclid's proposition 32 of book 1. It too is an equivalent of the parallel axiom.

To sum up, we have found the following consequences of Euclid's parallel axiom (which in fact also imply it, and hence are equivalent to it):

Playfair's axiom. For any line $\mathscr{L}$ and a point P outside it, there is a unique parallel to $\mathscr{L}$ through P.

Angle sum of a triangle. The angles of any triangle have sum equal to two right angles.

2.4 THE PYTHAGOREAN THEOREM

> He was 40 years old before he looked on Geometry; which happened accidentally. Being in a Gentleman's Library, Euclid's Elements lay open, and 'twas the 47 El. libri I. He read the Proposition. *By* G—sayd he (he would now and then sweare an emphaticall Oath by way of emphasis) *this is impossible*! So he reads the Demonstration of it, which referred him back to such a Proposition; which proposition he read. That referred him back to another, which he also read …that at last he was demonstratively convinced of that trueth. This made him in love with Geometry.

This quotation is about the philosopher Thomas Hobbes (1588–1679), from *Brief Lives* by Aubrey (1898: 332). It neatly and concisely captures the effectiveness of the deductive method: by seeing how a proposition depends on previous propositions, and ultimately on self-evident propositions called *axioms*, anyone can be convinced of its truth. As long as each proposition is a logical consequence of the previous propositions, and hence of the axioms, it does not matter how long and complex the chain of consequences may be. The proposition that so impressed Hobbes is the Pythagorean theorem: proposition 47 in book 1 of the *Elements*.

Aubrey's account, incidentally, describes how a proof should first be read, which is *backward*: first find what propositions the theorem depends on, and then observe how it follows from these propositions. In the process, one learns what concepts are involved and how they are connected. This is not to say that it is easy to construct a proof in the first place. In fact, the proof that Hobbes loved is incredibly complicated when analyzed in detail. It involves dozens of connections. However, if one knows enough connections, one can string them together to make proofs. In the last two sections we have already seen most of the connections needed to prove the Pythagorean theorem.

Figure 2.10 shows again a figure from section 1.1, this time with some sides and angles labeled to guide the steps of a proof. The square on the left, with each side $a + b$, has inside it four copies of the right-angled triangle with perpendicular sides a, b and angles α and β opposite to them. The light gray region therefore has each side equal to the hypotenuse c of the triangle. Also, the angle γ at each corner of the light gray region makes a straight angle with the angles α and β. Therefore,

$$\alpha + \beta + \gamma = \text{two right angles.}$$

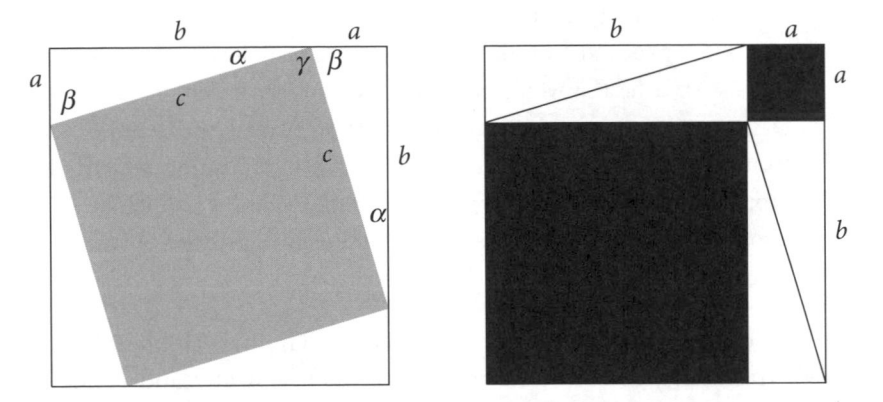

Figure 2.10 : Seeing how to prove the Pythagorean theorem

On the other hand, the angle sum of each triangle is also two right angles, so γ must be a right angle, and therefore the light gray region is the square on the hypotenuse. Thus the square on the hypotenuse equals the big square minus four times the triangle.

Turning now to the square on the right, which also has side $a + b$, we find similarly that the black regions are squares on the sides a and b. The sum of the black squares is again equal to the big square minus four times the triangle; hence it equals the square on the hypotenuse.

2.5 GLIMPSES OF ALGEBRA

As mentioned in section 2.1, Euclid's "common notions" look like algebra in the way they deal with equality, addition, and subtraction. Indeed, the proof above made extensive use of "adding equals to equals," "subtracting equals from equals," and the principle that "things equal to the same thing are equal to each other." However, this is still not algebra as we know it because a full notion of **multiplication** is missing. Admittedly, we did say "four times the triangle," but this really meant

$$\text{triangle} + \text{triangle} + \text{triangle} + \text{triangle}.$$

We did not multiply one length (or area) by another.

This is because Euclid followed the Pythagoreans in denying geometric quantities such as length and area some of the attributes of numbers. Lengths can be added and subtracted, and we can decide whether two lengths are equal or not. But defining what **area** is and what it has to do with the product of lengths is already a complicated problem. Yet it is a problem we need to solve in order to understand what the "sum of two squares" means in the Pythagorean theorem. In the proof of the Pythagorean theorem we were able to solve it by showing the sum of two squares was equal to a single square by addition and subtraction of clearly equal areas.

Euclid solved the problem of defining area in general with great ingenuity, but unfortunately in a way that stymied the development of algebra in geometry until the seventeenth century.

Simply put, Euclid's solution was to define the product of line segments a and b to be the *rectangle* with adjacent sides a and b. This definition is compatible with multiplication when a and b are whole numbers—because the rectangle then consists of ab unit squares—but it is also meaningful when a or b is an irrational length, which the Greeks did not consider to be a number. The immediate difficulty with this definition is to decide **equality**; for example, is a rectangle with sides $\sqrt{2}$ and $\sqrt{3}$ equal to a rectangle with sides $\sqrt{6}$ and 1?

Before answering this question, let us consider some simple examples of polygons that might be considered "equal," by "addition" and "subtraction," according to Euclid's common notions 1, 2, and 3. First, a rectangle of width a and height b is equal to any parallelogram with the same base and height, as figure 2.11 suggests. This is because the rectangle results from the parallelogram by subtraction, then addition, of equal triangles.

The triangles are equal by the result from section 2.4 that opposite sides and opposite angles of a parallelogram are equal. Equality of parallelogram sides implies, by subtraction and addition again, that the

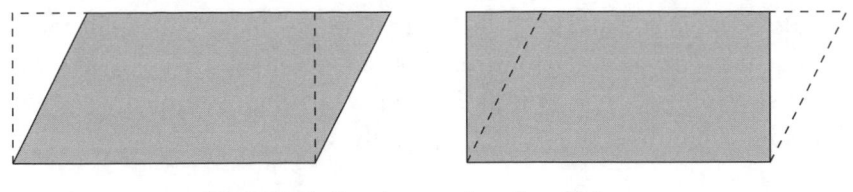

Figure 2.11 : Equal rectangle and parallelogram

triangles have equal width, and then equality of angles implies they are congruent, by SAS.

Next, one notices that any triangle, added to a copy of itself, makes a parallelogram (figure 2.12).

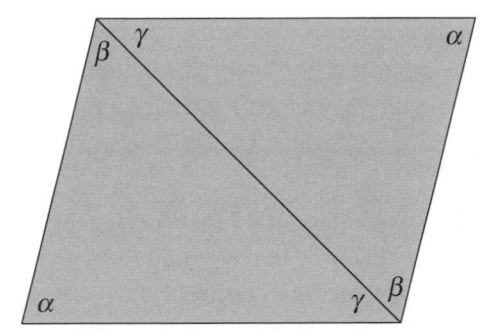

Figure 2.12 : Triangle and parallelogram

It follows that the area of a triangle is half that of a parallelogram with the same base and height, a result Euclid uses in his proof of the Pythagorean theorem. (This makes for a somewhat longer path to the theorem than the one described in the previous section.)

In general, Euclid considers regions "equal" if one can be converted to the other by addition and subtraction of finitely many equal figures. Remarkably, this definition coincides with the modern concept of "equal area" for polygons. However, the "product" ab has very limited algebraic properties. One has the *commutative law*

$$ab = ba,$$

because the rectangle with adjacent sides a and b is the *same* as the rectangle with adjacent sides b and a. And one has the *distributive law*

$$a(b+c) = ab + ac,$$

which is Euclid's proposition 1 of book 2. However, there is very little else. A product *ab* of two lengths is not a length, so if *c* is a length then *ab* + *c* does not make sense. Also, while *abc* is considered meaningful (it is a box with adjacent edges *a*, *b*, and *c*), *abcd* is not, because the Greeks did not believe there could be mutually perpendicular lengths *a*, *b*, *c*, and *d*.

Another limitation is that finitely many additions and subtractions do *not* generally work for curved regions; for example, one would not expect to find a square equal in area to a circle by this method. More disappointing, the method does not generally work for polyhedra either. In particular, Dehn (1900) proved that a regular tetrahedron and a cube of the same volume are not "equal" by addition and subtraction of finitely many equal polyhedra.

Because of this, one is led to consider cutting polyhedra into *infinitely many* pieces. Euclid himself found the volume of a tetrahedron by cutting it into infinitely many prisms, as we will see in section 2.7.

2.6 NUMBER THEORY AND INDUCTION

In books 7–9 of the *Elements* Euclid develops something entirely different from the geometry in the first six books: it is what we would now call elementary number theory. His development looks superficially similar to the geometry—with geometric terminology such as "measures" instead of "divides," and careful step-by-step proofs—but no axioms are stated. This, perhaps, is because there were no doubts about the foundations of number theory as there were about the foundations of geometry. Nevertheless, we will see that Euclid was fleetingly aware of **induction**, which is recognized today as a fundamental principle—indeed, an *axiom*—of number theory.

Also similar to the first six books, Euclid's number theory propositions are a mixture of theorems and constructions. Proposition 1 of book 7, indeed, applies what we now call the **Euclidean algorithm** (section 1.4) to test whether two given numbers are relatively prime, that is, whether their greatest common divisor is 1. His proposition 2 shows, more generally, that the algorithm gives the **greatest common divisor** of any two numbers. In proposition 1 Euclid states the algorithm in its simplest form: "Two unequal numbers being set out, and the less being continually subtracted in turn from the greater. . . ." This is the form that also applies to geometric quantities such as lengths and that will run forever if the lengths are in irrational ratio, as we saw in section 1.3.

In propositions 1 and 2 of book 7 Euclid assumes that the algorithm will terminate when the two quantities are natural numbers. Assuming also that any common divisor is preserved by subtraction, the greatest common divisor will survive (and be obvious) when the algorithm terminates with equal numbers. Both of these assumptions rest on induction.

1. Termination occurs because the algorithm produces smaller numbers, and *natural numbers cannot decrease forever*. This is the form of induction often called **infinite descent**.
2. The persistence of the greatest common divisor can be proved by the "base step, induction step" form of induction. Suppose the initial numbers a and b, with $a > b$, have a common divisor d, so

$$a = da' \quad \text{and} \quad b = db'.$$

Then the next pair $b = db'$ and $a - b = d(a' - b')$ also have the common divisor d. This is the base step. The induction step is similar. If the pair at step n, a_n and b_n, have common divisor d, then so have the pair at step $n + 1$, a_{n+1} and b_{n+1}, by the same argument as at step 1.

As we will see below, Euclid sometimes recognized when he was using infinite descent and pointed it out. But the "base step, induction step" idea does not occur in the *Elements*, if only because Euclid does not have notation (such as subscripts) to talk about sequences of arbitrary length. Instead of writing, say, $a_1, a_2, \ldots, a_n$ he would write a short sequence such as A, B, C and leave the reader to adapt the argument for the short sequence to one for a sequence of arbitrary length. This happens in his famous proof that there are infinitely many primes.

Primes

Euclid proved two famous theorems about prime numbers:

- *There are infinitely many primes.* Euclid actually proves a stronger result, which avoids mentioning infinity: given any finite collection of primes, we can (in principle) find another (book 9, proposition 20).

- *If a prime p divides a product ab of natural numbers a and b, then p divides a or p divides b* (book 7, proposition 30). We will call this the **prime divisor property**. It easily implies what we now call the **fundamental theorem of arithmetic**: each natural number > 1 has a prime factorization that is unique up to the order of factors.

To prove that there are infinitely many primes, Euclid needs the preliminary result that every natural number $k > 1$ has a prime divisor. This is proposition 31 in book 7 and is proved as follows. If k is not already prime, then k factorizes as ab, where $1 < a, b < k$. If one of a or b is prime, we are done. If not, continue splitting the factors into smaller factors. Since natural numbers cannot decrease forever (Euclid says it is "impossible in numbers"), eventually we find a prime factor of k.

Now the proof that there are infinitely many primes goes as follows. Suppose we are given some primes $p_1, p_2, \ldots, p_n$ (Euclid calls them just A, B, C). Consider the number

$$k = (p_1 p_2 \cdots p_n) + 1.$$

Then k is *not* divisible by any of $p_1, p_2, \ldots, p_n$. If p_i divides k, then p_i also divides $k - (p_1 p_2 \cdots p_n) = 1$, which is absurd. On the other hand, *some* prime p divides k, as we have just seen; hence p is a prime different from the given primes $p_1, p_2, \ldots, p_n$.

Comments. Unlike the variation of Euclid's proof often given today, his is *not* by contradiction. He does not suppose that there are finitely many primes and then look for a contradiction. Instead, he proves directly (and as finitely as possible) that there are infinitely many primes, by showing how to *increase* any given finite collection of primes. Also, strictly speaking, Euclid does not say take the product of the given primes and add 1. He actually says take the least common multiple of the given primes and add 1, but the least common multiple *is* the product in the case of distinct primes.

The second property, about a prime dividing a product, comes at the end of a rather lengthy sequence of consequences of the Euclidean algorithm. Today, using better notation and allowing negative integers, we can break the argument down to a shorter sequence of steps. We abbreviate "greatest common divisor" by gcd.

1. The Euclidean algorithm on a pair a, b preserves any common divisor d of a and b at each step. Therefore, when the algorithm terminates, necessarily with a pair of equal numbers a_k, b_k, we must have $\gcd(a, b) = \gcd(a_k, b_k) = a_k = b_k$.
2. Each of the numbers $a, b, a_1, b_1, \ldots, a_k, b_k$ occurring in the running of the Euclidean algorithm is of the form $ma + nb$, where m and n are *integers* (possibly negative). This follows by induction: it is obviously true at the beginning, and if true at step n, for a_n and b_n, then it is true at step $n + 1$ because one of a_{n+1}, b_{n+1} equals a_n or b_n, and the other equals their difference.

   It follows in particular that $\gcd(a, b) = ma + nb$ for some integers m and n.
3. Now suppose that a prime p divides ab but that p does *not* divide a. Thus we want to show that p divides b.

   Notice, since p does not divide a, that $1 = \gcd(a, p) = ma + np$ for some integers m and n. Multiplying both sides of this equation by b, we get

   $$b = mab + npb.$$

   We see that p divides the first term on the right side because p divides ab, and obviously p divides np. Therefore p divides the sum of these terms, which is b.

As mentioned above, **unique prime factorization** now follows. However, Euclid did not immediately draw this conclusion, though he proved a result close to it as proposition 14 of book 9. In fact, unique prime factorization was not considered interesting until the nineteenth century, when mathematicians sought to generalize it to other kinds of numbers, so we will postpone further discussion of it until later.

2.7 GEOMETRIC SERIES

The geometric series, which we write as $a + ar + ar^2 + ar^3 + \cdots$, occurs very early in Greek mathematics. The special case $\frac{1}{2} + \frac{1}{4} + \frac{1}{8} + \cdots$ is implicit in Zeno's paradox of the dichotomy, discussed in section 1.5, and both finite and infinite geometric series occur in Euclid's *Elements*. Euclid uses them in two noteworthy theorems: his theorem on even perfect numbers in book 9, proposition 36, and his theorem on the volume of the tetrahedron in book 12, proposition 4. Since perfect numbers involve only finite geometric series, we discuss them first.

Perfect Numbers

Euclid, among his definitions at the beginning of book 7, defines a **perfect number** to be a number "equal to the sum of its own parts." By "parts" of a natural number n he means the natural number divisors of n, other than n itself. For example, the "parts" of 6 are 1, 2, and 3. Since $1 + 2 + 3 = 6$, the number 6 is perfect. The next perfect number is

$$28 = 1 + 2 + 4 + 7 + 14,$$

and the next two are 496 and 8128. These were the only perfect numbers known in ancient times, but Euclid was able to prove a theorem that covers all the perfect numbers known today and, very likely, all perfect numbers that exist. It is the final proposition in book 9 of the *Elements*, proposition 36. We will state and prove it in modern notation, though the general idea is similar to Euclid's.

Euclid's perfect number theorem. *If p is a prime number of the form $2^n - 1$, then $2^{n-1}p$ is perfect.*

The number $2^{n-1}p$ has the obvious divisors

$$1, 2, 2^2, \ldots, 2^{n-1} \quad \text{and} \quad p, 2p, 2^2p, \ldots, p2^{n-2},$$

other than itself. And the prime divisor property in the previous section implies these are the *only* such divisors. So

$$\text{sum of divisors} = 1 + 2 + 2^2 + \cdots + 2^{n-1} + p\left(1 + 2 + 2^2 + \cdots + 2^{n-2}\right),$$

which involves two geometric series of the form

$$S_k = 1 + 2 + 2^2 + \cdots + 2^k.$$

We notice that

$$2S_k = 2 + 2^2 + 2^3 + \cdots + 2^k + 2^{k+1} = S_k - 1 + 2^{k+1},$$

and therefore, subtracting S_k from both sides,

$$S_k = 2^{k+1} - 1.$$

Substituting this (for $k = n - 1$ and $n - 2$) in the formula for the sum of divisors, we get

$$\text{sum of divisors} = 2^n - 1 + p\left(2^{n-1} - 1\right)$$
$$= 2^n - 1 + 2^{n-1}p - p$$
$$= 2^{n-1}p \qquad \qquad \text{because } p = 2^n - 1,$$

which proves the theorem.

No further progress on the nature of perfect numbers was made until the eighteenth century, when Euler proved that all *even* perfect numbers are of Euclid's form. We do not yet know whether there are any odd perfect numbers. We also do not know much about the primes of the form $2^n - 1$, which are called **Mersenne primes**. All the large primes found by computers in recent decades are in fact Mersenne primes, but we do not know whether there are infinitely many primes of this form.

Volume of a Tetrahedron

In book 9, proposition 35, Euclid actually found a rule for summing any finite geometric series, not just the one required for his perfect number theorem. In modern notation the rule is

$$a + ar + ar^2 + \cdots + ar^k = \frac{a\left(1 - r^{k+1}\right)}{1 - r} \qquad \text{for } r \neq 1,$$

and it is proved by an argument similar to the one above. This formula is a stepping stone to summing the *infinite* geometric series $a + ar + ar^2 + \cdots$ when $0 < r < 1$.

When $|r| < 1$ the term r^{k+1} in the formula above becomes as small as we please by taking k sufficiently large. Therefore, the sum $a + ar + ar^2 + \cdots + ar^k$ can be made to exceed any number *less* than $\frac{a}{1-r}$. It also, obviously, cannot exceed $\frac{a}{1-r}$. So, by the method of exhaustion, the value of the infinite sum is

$$a + ar + ar^2 + \cdots = \frac{a}{1 - r},$$

because we have exhausted all other possibilities.

Both Euclid and Archimedes solved geometric problems by reducing them to the summation of infinite geometric series.[1] One solved by

---

1. There is no good reason for calling some series "arithmetic" and others "geometric," but it may help to remember that the infinite series used by Euclid and Archimedes were actually related to geometry.

Euclid was finding the volume of the tetrahedron, in book 12, proposition 4. To do so, he divided the tetrahedron into infinitely many *prisms*, the solid analogues of triangles, whose volume can be found by adding and subtracting equal figures as we did for areas in section 2.5. (That section mentioned that finitely many additions and subtractions do not work for the tetrahedron.)

The first step of the subdivision, in his proposition 3, finds two prisms inside the tetrahedron, as shown in figure 2.13. When these two prisms are removed, what remains is two tetrahedra similar to the original, but half its height, width, and depth. The small tetrahedra are dissected in the same way (figure 2.14), and so on. The infinitely many prisms "exhaust" the tetrahedron in the sense that any point inside the tetrahedron is inside one of the prisms. Also, the first pair of prisms have volume equal to 1/4 base area × height (of the tetrahedron), the next two pairs have total volume 1/4 that of the first pair, and so on.

In proposition 4, Euclid gives precisely this construction of infinitely many prisms inside the tetrahedron. He does not directly correlate them

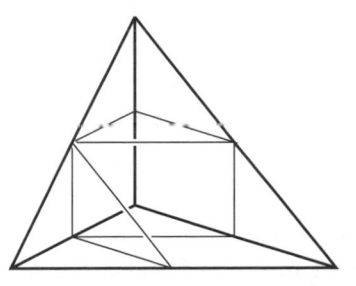

Figure 2.13 : Euclid's dissection of the tetrahedron

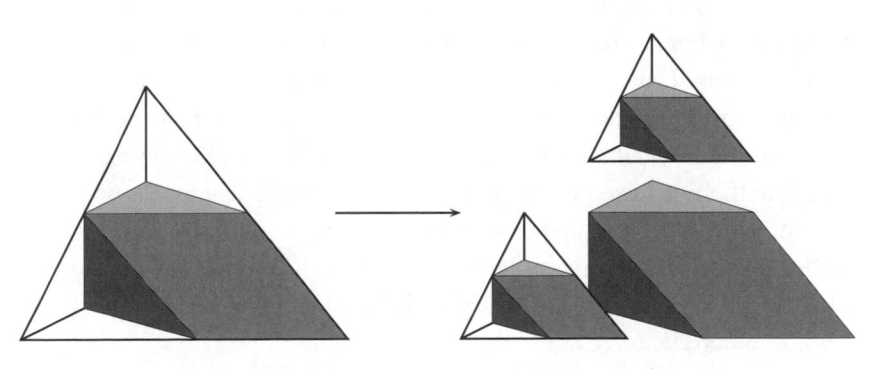

Figure 2.14 : Repeated dissection of the tetrahedron

with an infinite geometric series, but it is easy to do so with modern notation. This brings us very quickly to a formula for the volume of the tetrahedron that Euclid found by more geometric reasoning.

Altogether, the total volume of the prisms, and hence of the tetrahedron, equals the infinite sum

$$\left(1 + \frac{1}{4} + \frac{1}{4^2} + \frac{1}{4^3} + \cdots\right) \times \text{base area} \times \text{height},$$

which equals

$$\frac{1}{3} \times \text{base area} \times \text{height}$$

by the formula for the sum of an infinite geometric series.

2.8 REMARKS

Most of Euclid's definitions are definitions in the modern sense, that is, abbreviations. For example he defines a "triangle" in terms of "lines," and any occurrence of the term "triangle" could in principle be replaced by the definition. All definitions, if they are not circular, must ultimately be based on certain undefined terms, just as theorems must be based on unproved propositions (axioms). Euclid violates this principle in his first few definitions, which attempt to "define" what are really undefined terms, such as "point" and "line." These "definitions" perhaps help to assure readers that their mental images of these terms are also what Euclid has in mind, but they are not used in the rest of the *Elements*.

Another weakness of the *Elements* is that there are gaps in Euclid's reasoning, where a statement does not follow from preceding statements or his axioms. Typically, such gaps occur where the statement is visually obvious, so the reader accepts it without thinking. Most of the first six books of the *Elements* are in fact so visual that Byrne (1847) was able to translate them into sequences of pictures with just a few words of explanation. Figure 2.15, for instance, is Byrne's version of book 1, proposition 1, the construction of an equilateral triangle. (The picture is from a beautiful modern recreation of Byrne's book, by Nicholas Rougeux, at https://www.c82.net/euclid/.)

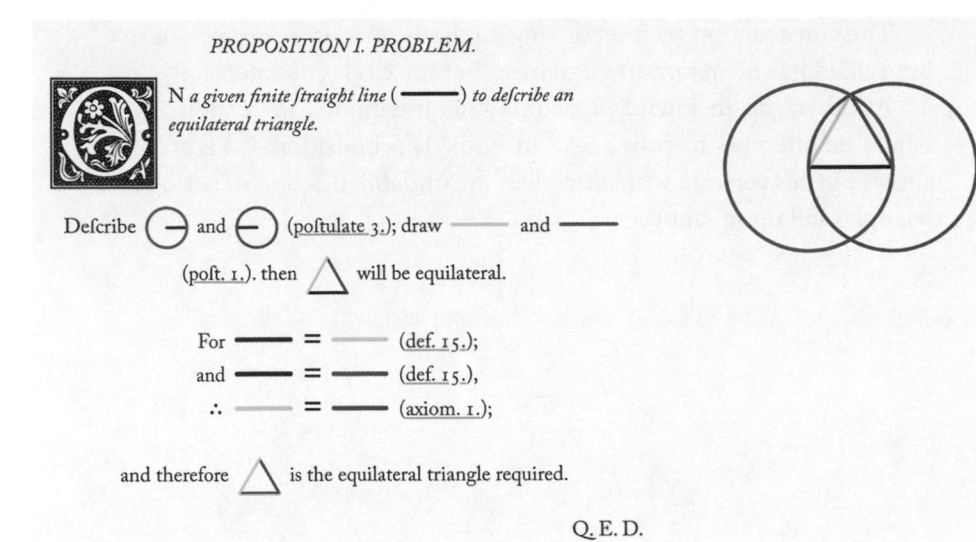

PROPOSITION I. PROBLEM.

O N *a given finite straight line* (———) *to describe an equilateral triangle.*

Describe (— and (— (postulate 3.); draw ——— and ———

(post. 1.). then △ will be equilateral.

For ——— = ——— (def. 15.);
and ——— = ——— (def. 15.),
∴ ——— = ——— (axiom. 1.);

and therefore △ is the equilateral triangle required.

Q. E. D.

Figure 2.15 : Byrne's version of book 1, proposition 1

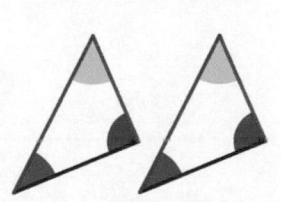

PROPOSITION IV. THEOREM.

I F *two triangles have two sides of the one respectively equal to two sides of the other,* (——— *to* ——— *and* ——— *to*

———) *and the angles* (▲ *and* ▲) *contained by those equal sides also equal; then their bases or their sides* (——— *and*

———) *are also equal: and the remaining and their remaining angles opposite to equal sides are respectively equal* (▲ = ▲ *and* ◀ =

◀): *and the triangles are equal in every respect.*

Let the two triangles be conceived, to be so placed, that the vertex of one of the equal angles, ▲ or ▲ ; shall fall upon that of the other, and ——— to coincide with ——— , then will ——— coincide with ——— if applied: consequently ——— will coincide with ——— , or two straight lines will enclose a space, which is impossible (ax. 10),

therefore ——— = ——— , ▲ = ▲ and ◀ = ◀ ,

and as the triangles △ and △ coincide, when applied, they are equal in every respect.

Q. E. D.

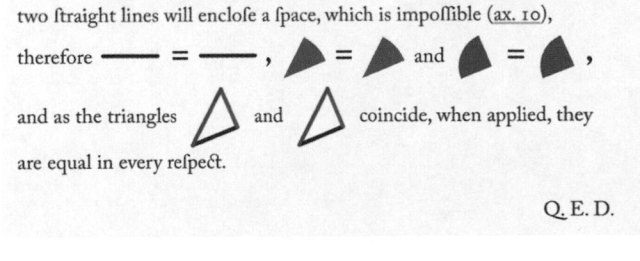

Figure 2.16 : Byrne's version of book 1, proposition 4

The construction by intersecting circles is of course very persuasive, but Euclid has no axiom to guarantee that the circles *do* intersect!

Another gap in Euclid's reasoning, as mentioned in section 2.5, is where he attempts to prove SAS in book 1, proposition 4. Figure 2.16 shows Byrne's version, which involves the undefined action of letting one triangle "fall upon" another.

CHAPTER 3

After Euclid

As we saw in chapter 1, Euclid's geometry can be seen as a response to the "crisis in foundations" provoked by the discovery of irrational numbers. 2,000 years later, Euclid's geometry provoked a crisis of its own, when *non-Euclidean* geometry was discovered. We explore non-Euclidean geometry in detail in chapter 9. This chapter outlines how Euclid's geometry was rebuilt in the nineteenth century.

Rebuilding would probably have happened even without awareness of non-Euclidean geometry, because there are some serious gaps in Euclid's system. This chapter discusses these gaps and how they were filled by Hilbert (1899). In filling the gaps in Euclid, Hilbert clarified the nature of *axioms* in geometry. On the one hand, the role of the **parallel axiom** became clear: simply by changing this axiom, while leaving the remaining axioms the same, one switches from Euclidean to non-Euclidean geometry.

On the other hand, Hilbert's new axioms reveal new and richer structure within geometry (both Euclidean and non-Euclidean), which includes the real numbers and algebra. The different facets of this structure correspond to different groups of axioms.

Axioms of a special type, called **incidence axioms**, give another important kind of geometry, called **projective geometry**. Projective geometry originated in the study of projective drawing in the fifteenth century, but its axioms throw new light not only on geometry but also on algebra.

3.1 INCIDENCE

The first group of Hilbert's axioms is about the relationship of points and lines. When a point P belongs to line $\mathcal{L}$, we say that P and $\mathcal{L}$ are *incident*, and their relationship is called **incidence**.

The first **incidence axiom**, listed below, is Euclid's axiom that two points determine a line, and the last is the parallel axiom (in Playfair's form; see section 2.1). The other two state that there are "enough" points and lines to be interesting. Euclid assumed that points and lines already exist, so he did not have any other incidence axioms. His axioms attempt only to describe the nature of points and lines. For Hilbert, "points" and "lines" can be *any objects* related to each other in the way stated by his axioms, so one cannot assume, for example, that infinitely many points and lines exist.

I1. For any two points A, B there is a unique line containing A, B.
I2. Every line contains at least two points.
I3. There exist three points not all in the same line.
I4. For each line $\mathcal{L}$ and point P not on $\mathcal{L}$ there is a unique line containing P but with no point in common with $\mathcal{L}$.

In Hilbert's system, the incidence axioms for points and lines interact with other axioms to give an *algebraic* structure to the points on a line. Namely, it is possible to define "sum," "product," "negative," and "inverse" of points, with special points 0 and 1, and to prove the following properties, which are the same as the properties of sum and product for numbers:

$$A + B = B + A \qquad A \cdot B = B \cdot A$$
$$A + (B + C) = (A + B) + C \qquad A \cdot (B \cdot C) = (A \cdot B) \cdot C$$
$$A + 0 = A \qquad A \cdot 1 = A$$
$$A + (-A) = 0 \qquad A \cdot A^{-1} = 1 \text{ when } A \neq 0$$
$$A \cdot (B + C) = A \cdot B + A \cdot C.$$

These properties, viewed as axioms, define a **field**, an algebraic structure we discuss further in section 4.6. Among their consequences is $(-1)(-1) = 1$. If one adds some more complicated incidence axioms (see section 3.8), it is possible to derive the field structure from incidence axioms alone.

3.2 ORDER

The next group of axioms is about *betweenness* or **order**—a concept overlooked by Euclid, probably because it is too "obvious." The first to draw attention to betweenness was the German mathematician Moritz Pasch, in the 1880s. We write $A * B * C$ to denote that B is between A and C.

B1. If $A * B * C$ then A, B, C are three points on a line, and also $C * B * A$.

B2. For any two points A and B there is a point C with $A * B * C$.

B3. Of three points on a line, exactly one is between the other two.

B4. Suppose that A, B, C are three points not in a line and that $\mathscr{L}$ is a line not passing through any of A, B, C. If $\mathscr{L}$ contains a point D between A and B, then $\mathscr{L}$ contains either a point between A and C or a point between B and C, but not both (**Pasch's axiom**).

The first three of these axioms imply that points on a line are "ordered" as one expects, namely, like numbers. If we take two distinct points A and B on a line $\mathscr{L}$ and declare arbitrarily that $A < B$, then the points $C \neq A, B$ are divided into three distinct classes:

$\{C : C * A * B\}$, called the points less than A

$\{C : A * C * B\}$, called the points between A and B

$\{C : A * B * C\}$, called the points greater than B.

Any $C \neq A, B$ must fall into one of these classes by axiom B3. Then for any point $D \neq C$ we can define $C < D$ or $D < C$ depending on the classes to which C and D belong. For example, $C < D$ if C is among the points less than A and if D is between A and B; also $C < D$ if C, D are both between A and B but $C * D * B$.

Finally, by a similar consideration of possible cases, we can prove that this $<$ relation is a **linear order** like the ordering of numbers:

1. For any two distinct points C, D either $C < D$ or $D < C$ (but not both).
2. For no point C is $C < C$.
3. If $C < D$ and $D < E$ then $C < E$.

By combining the order axioms with the incidence axioms we can prove that the field of points on a line is an **ordered field**. That is, as well as the field properties enumerated in the previous section, there is a linear order < such that

- if $A < B$ then $A + C < B + C$, for any C; and
- if $0 < A$ and $0 < B$ then $0 < A \cdot B$.

Thus the order axioms bring us closer to creating the real numbers within geometry, but we have not yet pinned them down completely. More axioms are needed before we can be sure this field is $\mathbb{R}$ and not, say, the rational numbers, which also form an ordered field.

The last axiom of order, B4 or Pasch's axiom, can be depicted as shown in figure 3.1. Another way to put it is that a line passing through the interior of one side of a triangle cannot avoid hitting one of the other sides. This axiom is assumed unconsciously in many of Euclid's proofs that rely on diagrams of triangles. Pasch's axiom is also needed to prove some seemingly obvious properties of the plane, such as each line $\mathscr{L}$ *separates* the plane. That is, the points of the plane not on $\mathscr{L}$ are divided into two classes, $\mathscr{A}$ and $\mathscr{B}$, and any line from a point in $\mathscr{A}$ to a point in $\mathscr{B}$ meets $\mathscr{L}$. A proof of this (which is quite lengthy) may be found in Hartshorne (2000: 75–76).

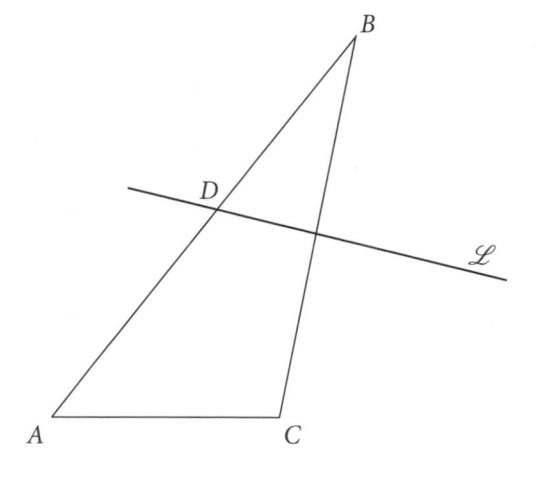

Figure 3.1 : Pasch's axiom

Properties of Ordered Fields

Together with the field properties stated in section 3.1, the defining properties

Property 1. If $A < B$ then $A + C < B + C$,
Property 2. If $A, B > 0$ then $A \cdot B > 0$

are a very concise description of ordered fields. However, these two properties have a series of consequences that are more enlightening, since they show how closely an ordered field resembles the ordinary number system.

Property 3. If $A > 0$ then $-A < 0$ and, conversely,
Property 4. $0 < 1$.
Property 5. If $A < 0 < B$ then $A \cdot B < 0$.
Property 6. If $A < B$ and $C > 0$ then $AC < BC$.
Property 7. If $C > 0$ then $C^{-1} > 0$.
Property 8. If $A < B$ and $C > 0$ then $A/C < B/C$.

For example, $A > 0$ implies $-A < 0$ because if $-A > 0$ we get $0 > A$ by adding A to both sides of the inequality (property 1). This contradicts $A > 0$, so the assumption that $-A > 0$ is wrong. Property 4 follows from this because if $1 < 0$ we get $-1 > 0$, hence $0 < (-1)(-1) = 1$ by property 2. This contradicts $1 < 0$, so the latter assumption is also wrong.

The other properties on the list can be proved by similar reasoning.

We now use these properties to show that an ordered field $\mathbb{F}$ contains copies not just of 0 and 1 but of all natural numbers, integers, and rational numbers. We get the remaining natural numbers in $\mathbb{F}$ as

$$2 = 1 + 1, \quad 3 = 1 + 1 + 1, \quad 4 = 1 + 1 + 1 + 1, \quad \ldots,$$

and we find, by repeatedly adding 1 to both sides of the inequality $0 < 1$, that

$$0 < 1 < 2 < 3 < 4 < \cdots.$$

The negative integers, $-N$ for each natural number N, are similarly shown to be in correct order. Finally, the rational numbers in $\mathbb{F}$ are obtained as quotients M/N of integers with $N \neq 0$.

As with ordinary fractions, we can prove (since $\mathbb{F}$ is a field) that, for integers $A, B, C, D > 0$,

$$A/B < C/D \text{ if and only if } AD < BC.$$

Thus the ordering of (positive) rationals in $\mathbb{F}$ is determined by the ordering of positive integers in $\mathbb{F}$, so it agrees with that of the ordinary positive rationals.

Finally, it can be checked that correct ordering extends to the negative rationals of $\mathbb{F}$. It is also true that sums and products correspond to those of the ordinary rational numbers, since $\mathbb{F}$ is a field. Thus *the rational numbers of $\mathbb{F}$ are a faithful copy of the ordinary rational numbers.*

3.3 CONGRUENCE

The next group of axioms is about **equality of line segments** and **equality of angles**, both denoted by $\cong$. Angles are defined as pairs of *rays* with a common origin. They and line segments are defined as follows:

- Each pair A, B of distinct points define the **line segment** AB, whose points are A, B, and the points between them.
- Each pair of distinct points C, D define a **ray** $\overrightarrow{CD}$ with **origin** C, whose points are C, D, the points between them, and the points E such that $C * D * E$.

Then $AB \cong CD$ means that AB and CD have equal length, and $\angle ABC \cong \angle DEF$ means that $\angle ABC$ and $\angle DEF$ are equal angles. The notion of equality for lengths and angles is called congruence, and it is governed by the following axioms C1–C6. Notice that C2 and C5 contain versions of Euclid's common notion 1: "Things equal to the same thing are equal to each other."

C1. For any line segment AB, and any ray $\mathscr{R}$ originating at a point C, there is a unique point D on $\mathscr{R}$ with $AB \cong CD$.

C2. If $AB \cong CD$ and $AB \cong EF$ then $CD \cong EF$. For any AB, $AB \cong AB$.

C3. Suppose $A * B * C$ and $D * E * F$. If $AB \cong DE$ and $BC \cong EF$ then $AC \cong DF$. (Addition of lengths is well defined.)

C4. For any angle $\angle BAC$ and any ray $\overrightarrow{DF}$ there is a unique ray $\overrightarrow{DE}$ with $\angle BAC \cong \angle EDF$.

C5. For any angles α, β, γ, if $\alpha \cong \beta$ and $\alpha \cong \gamma$ then $\beta \cong \gamma$. Also, $\alpha \cong \alpha$.

C6. Suppose ABC and DEF are triangles with $AB \cong DE$, $AC \cong DF$, and $\angle BAC \cong \angle EDF$. Then the two triangles are congruent, namely, $BC \cong EF$, $\angle ABC \cong \angle DEF$, and $\angle ACB \cong \angle DFE$.

These axioms include the well-known congruence axiom side-angle-side (SAS) as axiom C6. As mentioned in section 2.5, Euclid did not state SAS as an axiom, but he tried to prove it in proposition 4 of book 1 using an undefined operation transporting one triangle to "fall upon" another. Hilbert's axioms C1 and C4 also take the place of operations used but not defined by Euclid in his ruler and compass constructions: transporting a length and transporting an angle.

The congruence axioms capture the concepts of distance (or length) and angle by giving conditions for line segments to have the same length and for pairs of rays to have the same angle. They also allow us to say when AB has *greater* length than CD, namely, when $AB \cong CE$, where E is a point such that $C * D * E$ (the whole is greater than the part).

Now that we have the concept of distance, we can state an axiom about the incidence of circles. This is the notorious unstated assumption in Euclid's proof of proposition 1, book 1, mentioned in section 2.8.

CI. Two circles meet if one of them contains points both inside and outside the other.

Axiom CI is not a pure incidence axiom because it involves the concepts **inside** and **outside** for a circle. These are defined in terms of the concept of distance: the points inside a circle $\mathscr{C}$ are those whose distance from its center is less than the radius; those outside are those whose distance from the center is greater than the radius.

3.4 COMPLETENESS

Hilbert's last two axioms are not strictly needed for Euclid's geometry, which requires only the points constructible with straightedge and compass. The axioms describe the large- and small-scale structure of the line, to ensure that the line behaves as a *number line*. Although Euclid did not state any axioms of this nature, to some extent they are foreshadowed by his theory of proportions in book 5 of the *Elements*.

The first is the so-called **Archimedean axiom**, which says that no length can be "infinitely large" relative to another.

Ar. For any line segments AB and CD there is a natural number n such that n copies of AB are together greater than CD.

Another way to say this is that there are *no infinitesimals*. As mentioned in section 1.6, an infinitesimal is a nonzero length, n copies of which do not exceed the unit length for any natural number n. We will see in chapter 6 that at times certain mathematicians have entertained the existence of infinitesimals, although they do not exist in $\mathbb{R}$.

Finally, there is the so-called **Dedekind axiom**, which says that the line is **complete** and implies that its points correspond to real numbers. Hilbert wanted an axiom like this to force the line in Euclidean geometry to be the same as the line $\mathbb{R}$ of real numbers.

> De. Suppose the points of a line $\mathscr{L}$ are divided into two nonempty subsets $\mathscr{A}$ and $\mathscr{B}$ in such a way that no point of $\mathscr{A}$ is between two points of $\mathscr{B}$ and no point of $\mathscr{B}$ is between two points of $\mathscr{A}$. Then there is a unique point P, either in $\mathscr{A}$ or $\mathscr{B}$, which lies between any other two points, of which one is in $\mathscr{A}$ and the other is in $\mathscr{B}$.

This axiom is modeled on Dedekind's definition of the real numbers as so-called **Dedekind cuts** in the rational numbers, which in turn was inspired by the Eudoxus "theory of proportions" in the *Elements*, book 5 (see section 1.6). We say more about Dedekind cuts in chapter 11.

It follows from Hilbert's axioms Ar and De that the ordered field of points on a line is complete and Archimedean. This is enough to characterize the real numbers, because any field with these properties is **isomorphic** to $\mathbb{R}$. That is, if $\mathbb{F}$ is a complete, Archimedean, ordered field, we can find a one-to-one correspondence between members of $\mathbb{F}$ and members of $\mathbb{R}$, with the properties that sums in $\mathbb{F}$ correspond to sums in $\mathbb{R}$, and products in $\mathbb{F}$ correspond to products in $\mathbb{R}$.

Characterization of $\mathbb{R}$. *Any complete, Archimedean, ordered field $\mathbb{F}$ is isomorphic to* $\mathbb{R}$.

Outline of proof. We find a correspondence between $\mathbb{F}$ and $\mathbb{R}$ by building, in turn, replicas of the natural numbers, rational numbers, and real numbers in $\mathbb{F}$ and then showing that the replicas exhaust all the members of $\mathbb{F}$.

Section 3.2 explained how to build replicas of the rational numbers in $\mathbb{F}$ and why their ordering is the same as that of the ordinary rationals. Now we show that each member X of $\mathbb{F}$ is determined by the rational numbers of $\mathbb{F}$ less than X and the rational numbers of $\mathbb{F}$ greater than X.

If, on the contrary, there are *two* members X, X' of $\mathbb{F}$ between the same sets of rational numbers, then the difference $X' - X > 0$ but $X' - X$ is less than any positive rational. This contradicts the Archimedean axiom, so in fact X *is* determined by the pair of sets of rationals, respectively, less than X and greater than X. But the same sets determine a unique real number x, as noted in section 1.6, so we have a one-to-one correspondence $X \leftrightarrow x$ between members of $\mathbb{F}$ and members of $\mathbb{R}$.

Sums and products also correspond, since this is already arranged for the rational members of $\mathbb{F}$ and $\mathbb{R}$ in section 3.2, and the correspondence carries over to sets of rationals. (For more details, see the treatment of Dedekind cuts in chapter 11.) □

Somewhat surprising, Hilbert seems not to have noticed that the completeness axiom De *implies* the Archimedean axiom Ar. The proof is very simple. Suppose that $\mathbb{F}$ is a complete ordered field and that $\mathbb{F}$ has infinitesimal elements. The axiom De gives a least upper bound b to these infinitesimals, as we will see in chapter 11, and it is easy to see that b cannot exist. If b is infinitesimal, then so is $2b$, contrary to the upper bound property; if b is not infinitesimal, neither is $b/2$, contrary to the least upper bound property. Thus there are no infinitesimals. It follows that the characterization theorem can be strengthened to state that *any complete ordered field is isomorphic to $\mathbb{R}$*.

Because of this characterization theorem, we can define the **real number system** $\mathbb{R}$, with its sum and product operations, by starting with the algebraic concept of a field and imposing the order and completeness properties. Curiously, this takes almost as many axioms as Hilbert's approach to $\mathbb{R}$ through geometry. The algebraic approach takes 15 axioms:

$$9 \text{ (for the field)}$$
$$+3 \text{ (for the linear order)}$$
$$+2 \text{ (for the ordered field)}$$
$$+1 \text{ (for completeness)},$$

in place of the 17 Hilbert axioms above (or 16, because Ar is redundant). Nevertheless, the algebraic approach is generally preferred today, since the field and linear order axioms are well known throughout mathematics.

3.5 THE EUCLIDEAN PLANE

An issue that will arise often in this book is the question of **consistency**, which asks whether the axiom system is *free from contradiction*. Usually we prove consistency of a system Σ by constructing a **model** of it: a structure $\mathcal{M}$ in which the axioms of Σ are true when interpreted as statements about $\mathcal{M}$. A model guarantees consistency, since contradictory statements cannot be true of an actually existing structure.

Hilbert proved consistency of his axiom system for geometry by building a model of it from the real number system $\mathbb{R}$. We will describe this model only briefly here, since most readers will know it as the *coordinate geometry* from high school, but we say more about it in chapter 5.

The model is the set $\mathbb{R}^2$ of **ordered pairs**[1] $\langle a, b \rangle$ of real numbers, written

$$\mathbb{R}^2 = \{\langle a, b \rangle : a, b \in \mathbb{R}\}.$$

This set is the *plane* of the model, and ordered pairs $P = \langle a, b \rangle$ are its *points*. We visualize P relative to lines called the x- and y-axes on the diagram shown in figure 3.2. The numbers a and b, the horizontal and vertical distances of P from the **origin** O, are called the **coordinates** of P (figure 3.2).

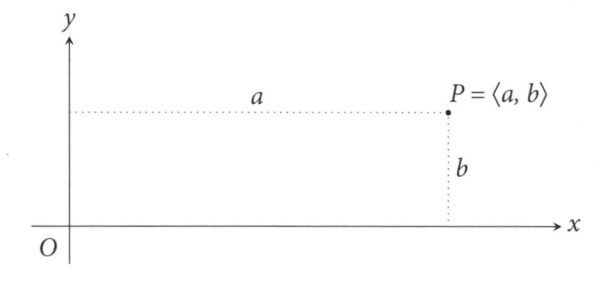

Figure 3.2 : Coordinates of a point

The definition of **distance** is motivated by the Pythagorean theorem. We define the distance between $P_1 = \langle a_1, b_1 \rangle$ and $P_2 = \langle a_2, b_2 \rangle$ by

$$\text{dist}(P_1, P_2) = \sqrt{(a_2 - a_1)^2 + (b_2 - b_1)^2},$$

which is the length of the hypotenuse of the right-angled triangle $P_1 Q P_2$ shown in figure 3.3. Finally, the definition of **angle measure** is motivated by trigonometry. In particular, the angle θ shown in figure 3.3 is given by

$$\tan \theta = \frac{b_2 - b_1}{a_2 - a_1}.$$

---

1. We will denote the ordered pair of a and b in this book by $\langle a, b \rangle$ rather than the rival notation (a, b) because we use the latter for the open interval $(a, b) = \{x \in \mathbb{R} : a < x < b\}$.

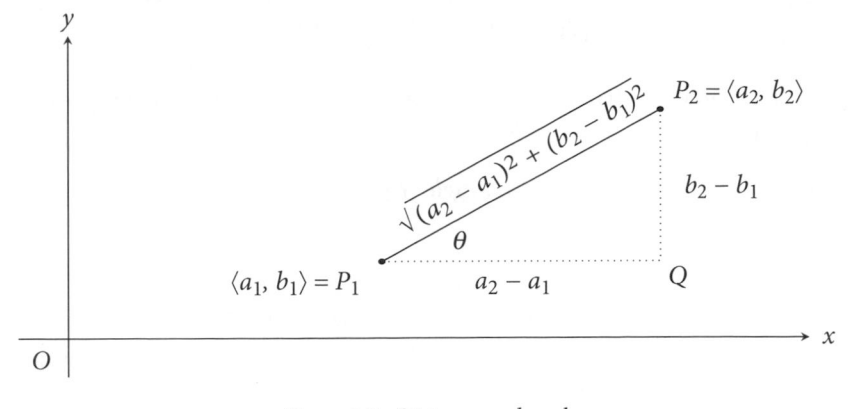

Figure 3.3 : Distance and angle

It is not actually necessary to use the trigonometric function tan—its definition requires calculus, as we will see in chapter 6. To satisfy the congruence axioms we need only define **equality** of angles, and this boils down to equality of ratios such as $(b_2 - b_1)/(a_2 - a_1)$. (By the same token, we can define equality of distances without defining the square root function, via equality of sums of squares.)

Lines and **circles**, as readers surely know, are defined by equations in x and y: lines by equations of the form $ax + by + c = 0$ (hence the name "linear" for such equations) and circles by equations of the form

$$(x - a)^2 + (y - b)^2 = r^2.$$

In the latter equation, $\langle a, b \rangle$ is the center of the circle and r is its radius. Indeed, the equation simply says that the distance from the center $\langle a, b \rangle$ to a point $\langle x, y \rangle$ on the circle is equal to r. With these definitions one can prove, for example, Hilbert's axiom CI, by finding the intersection of the circles as a common solution of their equations.

In fact, all of Hilbert's axioms can be confirmed by algebraic calculations, so $\mathbb{R}^2$ is a model of them. More interesting, $\mathbb{R}^2$ is essentially the *only* model of the Euclidean plane, because $\mathbb{R}$ is essentially the only model of the line, since $\mathbb{R}$ is essentially the only complete ordered field. (As we say today: any complete ordered field is **isomorphic** to $\mathbb{R}$.)

From the typical mathematician's point of view, which takes $\mathbb{R}$ as given, this is very satisfying. However, it is possible to question *the consistency of $\mathbb{R}$ itself*, as Hilbert realized. In later chapters we will dig more deeply into this question.

3.6 THE TRIANGLE INEQUALITY

One property of triangles is so simple that one might suppose it axiomatic:

Triangle inequality. *In any triangle, the sum of any two sides is greater than the third.*

However, for both Euclid and Hilbert, this proposition is not an axiom but a theorem. (In the *Elements* it is proposition 20 of book 1.) With coordinates, we can prove the triangle inequality by some simple but not particularly elegant algebra.

By suitable choice of axes and unit length, we can assume that the three vertices of the triangle are $(0,0)$, $(1,0)$, and (a,b). Then we want to prove

$$1 + \sqrt{(a-1)^2 + b^2} > \sqrt{a^2 + b^2}$$

or, squaring both sides,

$$1 + (a-1)^2 + b^2 + 2\sqrt{(a-1)^2 + b^2} > a^2 + b^2.$$

Expanding and rearranging, the required inequality becomes

$$2\sqrt{(a-1)^2 + b^2} > 2a - 2.$$

Canceling the 2 and squaring again, the required inequality is now

$$(a-1)^2 + b^2 > (a-1)^2, \quad \text{or just} \quad b^2 > 0.$$

And this is true since $b \neq 0$ (because the three points are not in a line).

The Shortest Path Property of Reflection

Around 100 CE the Greek mathematician Heron found an elegant consequence of the triangle inequality that contains the germ of a crucial idea in physics: *when light passes from A to B via reflection in a line $\mathscr{L}$ it follows the shortest possible path.* This idea was later to grow into an immensely influential idea in mechanics called the *principle of least action*.

Figure 3.4 shows the situation: a ray of light leaves point A, is reflected by line $\mathscr{L}$ at P, and arrives at point B. We know, by the nature of reflection, that AP makes the same angle with $\mathscr{L}$ as does PB, but why is APB the shortest path from A to B via $\mathscr{L}$?

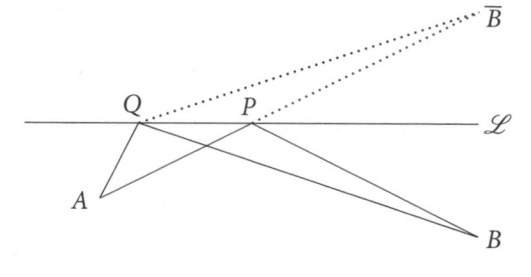

Figure 3.4 : Shortest path property of reflection

To see why, consider the mirror image $\overline{B}$ of B in the line $\mathscr{L}$. Because of the equal angles, A, P, and $\overline{B}$ lie along the line segment $A\overline{B}$ whose length equals the length of the path of the light ray APB. Any *other* path AQB from A to B via $\mathscr{L}$ is longer because it has the same length as $AQ\overline{B}$, which is longer than $A\overline{B}$ by the triangle inequality.

3.7 PROJECTIVE GEOMETRY

Projective geometry is a branch of geometry that arose from discoveries of Italian Renaissance artists in perspective drawing. A typical problem, often solved incorrectly, was to draw a tiled floor. Figure 3.5 is an example, taken from the 1496 book *The Art of Dying Well*, by Savonarola. The

Figure 3.5 : Incorrect perspective

floor looks wavy because the artist has drawn a set of equally spaced sloping lines and a set of horizontal lines to make rows of floor tiles. Each
tile is a parallelogram divided by its diagonal into two triangles. But,
since the horizontals are poorly spaced, the diagonals are not straight.
The artist was evidently not up-to-date on drawing technique, because
a method for correctly drawing tiled floors had already been published
in 1436, in the book *On Painting* by Leon Battista Alberti. Figure 3.6,
from Alberti's book, shows how the diagonals control the spacing of the
horizontal lines.

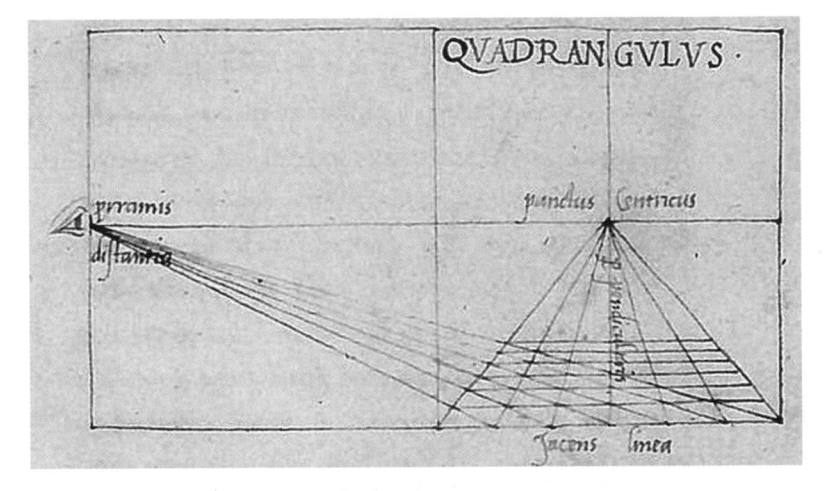

Figure 3.6 : Correct perspective

To see the method step-by-step, assume that one row of tiles touches
the bottom edge of the picture, marking it with equally spaced points.
Alberti chooses any horizontal line as the horizon and draws lines from
the marked points on the bottom edge to a single point on the horizon.
These depict the parallel lines of tiles perpendicular to the bottom edge
(figure 3.7), because *parallels appear to meet on the horizon.* Another
horizontal line, near the bottom edge, completes the first row of tiles.

To find the horizontal lines for the second, third, fourth, . . . rows of
tiles, draw the *diagonal of any tile in the bottom row* (shown in red in
figure 3.8). It crosses successive parallels at the corners of tiles in the second, third, fourth, . . . rows, so these rows can be depicted by drawing
horizontal lines through the successive crossings, as shown in figure 3.8.

The construction works because certain things remain the same in
any view of the plane:

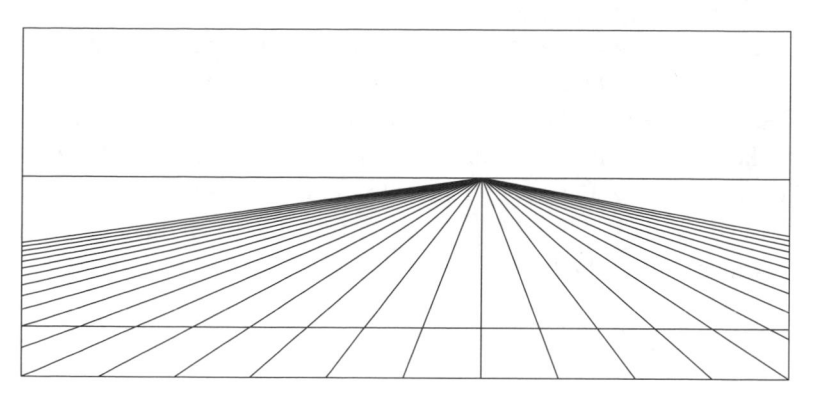

Figure 3.7 : Beginning Alberti's construction

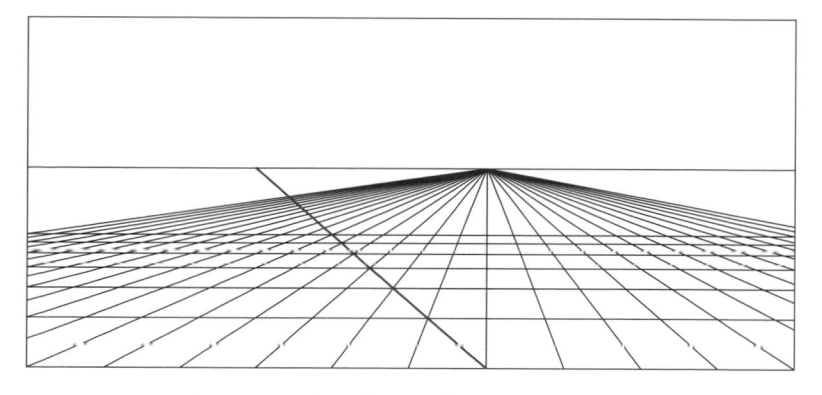

Figure 3.8 : Completing Alberti's construction

- Straight lines remain straight.
- Intersections remain intersections.
- Parallel lines remain parallel or meet on the horizon.

Perspective Drawing without Measurement

The properties of a perspective view do *not* include length and angle, since equal lengths or angles need not remain equal in the perspective view. Nevertheless, length measurement is involved in Alberti's construction: to mark the equally spaced points on the baseline and to make lines parallel to it. Artists of course do not mind this, but a mathematician might wonder if a perspective view can be made *without* measurement.

To do so we give up the idea of using actual parallels and embrace the idea that "parallels" are lines that meet on an arbitrarily chosen line called the **horizon**. We can then begin with a single "tile" whose sides are "parallel" (so they meet on our chosen horizon) and construct the other "tiles" one by one. Figure 3.9 shows the first few steps. We use the fact that the diagonals of the tiling are also parallel, so each diagonal ends at the same point on the horizon.

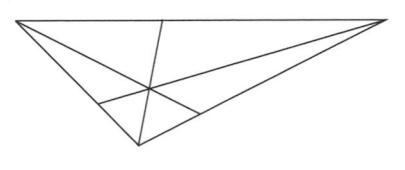

Draw diagonal of first tile
extended to the horizon

Draw diagonal of second tile
to the endpoint of the first diagonal

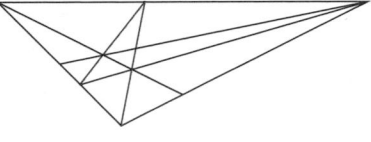

Draw side of second tile
through the new intersection

Draw side of two more tiles
through the new intersection

Figure 3.9 : Constructing the tiled floor

It should be clear, I hope, that each new line creates new intersections, through which we can draw new lines, creating new tiles, ad infinitum.

The construction shows that complex and interesting geometric figures can be constructed using only a device for drawing straight lines. (The device is often called a "ruler" but is better called a **straightedge**, since it need not have any marks on it.) It also leads us to believe that there is a complex and interesting geometry involving only points and lines in the plane. Because of its historical association with perspective drawing, which involves the projection of one plane onto another (see figure 3.10, from Bosse 1653), this geometry is called **projective geometry**.

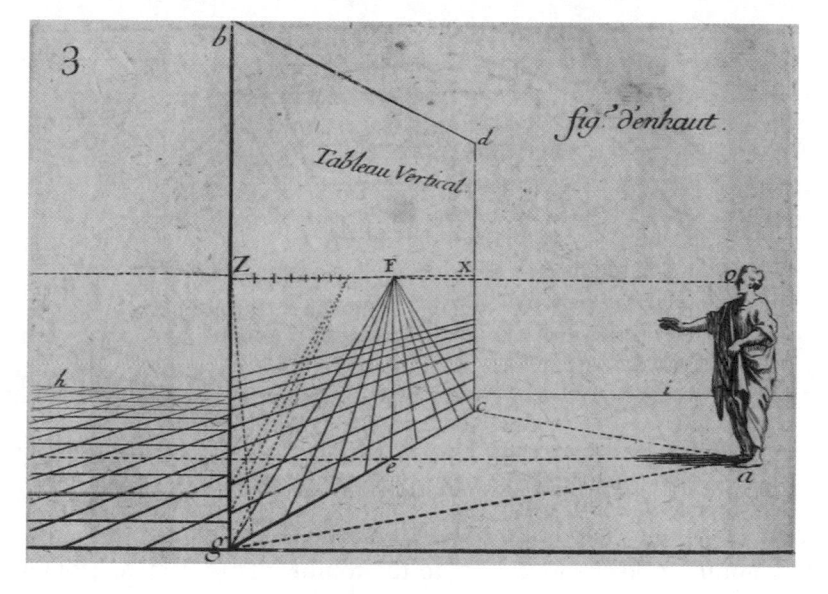

Figure 3.10 : Projection of one plane onto another

3.8 THE PAPPUS AND DESARGUES THEOREMS

The fundamental idea of projective geometry is that *any two lines meet*. To make this possible we add a "horizon" line to the Euclidean plane to form the so-called **real projective plane**. The horizon line is no different from any other line, so in fact we can choose any line in the extended plane to be the "horizon" and then call lines "parallel" if they meet on the horizon or "at infinity." The formal name for the horizon is the *line at infinity*.

Since the objects of projective geometry are "points" and "lines," its theorems are statements that mention only points and lines. An interesting example of such a theorem has been known since ancient times:

Theorem of Pappus. *If A, B, C, D, E, F are points that lie alternately on two lines, then the intersections of AB and DE, BC and EF, and CD and FA lie on a line.*

The theorem is illustrated in figure 3.11.

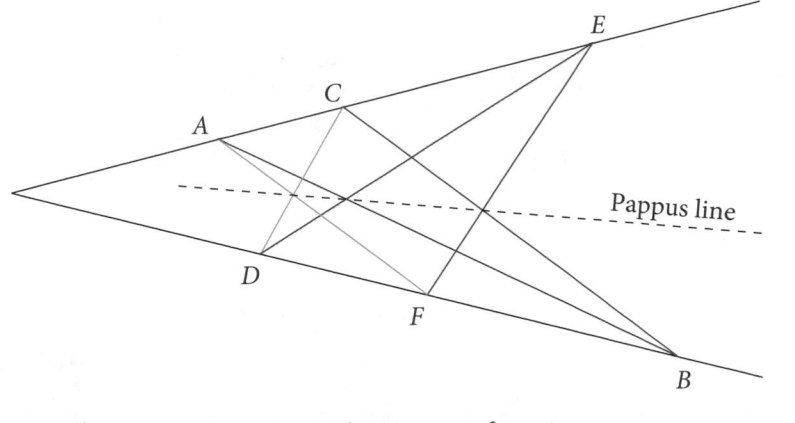

Figure 3.11 : The Pappus configuration

The classical proof of this theorem involves the concept of length, so one might wonder whether there is a "projective" proof: using only axioms for points and lines. Before we answer that question, let us consider another old theorem, due to the founder of projective geometry, Girard Desargues, around 1640:

Theorem of Desargues. *If triangles ABC and A′B′C′ are in perspective from a point P (that is, the pairs A, A′ and B, B′ and C, C′ are on lines through P) then the intersections of the corresponding sides AB and A′B′, BC and B′C′, and CA and C′A′, lie on a line.*

This theorem is illustrated in figure 3.12, with triangles *ABC* and *A′B′C′* filled in and colored red and blue. *X*, *Y*, and *Z* are the intersections of their corresponding sides. Coloring the triangles like this hopefully suggests that they lie in space, in which case *the planes containing the two triangles meet along a line.* The line of intersection of the two planes is of course the line containing *X*, *Y*, and *Z*. This is a "projective" proof of the Desargues theorem, because it involves only the intersections of planes and lines.

The catch is that the proof involves intersections of planes as well as lines, so it is about projective *space*, not just the projective *plane*. It needs an axiom about intersection of planes, whereas plane projective geometry should have axioms about points and lines only. Surprisingly, the proof of the Desargues theorem in the plane, like that of the Pappus theorem, involves the concept of length. There are projective planes in which

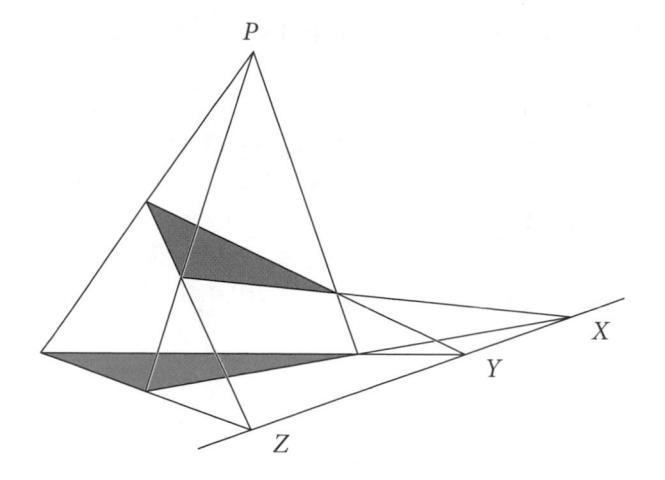

Figure 3.12 : Desargues's theorem

the Desargues theorem does not hold. But what, precisely, *is* a projective plane?

Axioms and Models for Projective Planes

Looking back at our motivating example for projective geometry—drawing a tiled floor in perspective—we see that we have made the following assumptions about points and lines:

1. Any two points lie in a unique line.
2. Any two lines meet in a unique point.
3. There are four points, no three of which lie in a line.

The first was already in Euclid, the second is to ensure that even "parallel" lines meet, and the third is to ensure *planarity*, specifically, the existence of a "tile" with four distinct sides. These three statements are called the **axioms for a projective plane**.

They seem to reflect the world of perspective drawing, so we expect that they are consistent, but to be sure we construct a **model** of them. This model is in fact the real projective plane mentioned earlier but defined in a mathematically precise manner. We now call it $\mathbb{RP}^2$ for short.

The "points" of $\mathbb{RP}^2$ are lines through the origin in ordinary three-dimensional space $\mathbb{R}^3$, and the "lines" of $\mathbb{RP}^2$ are planes through the origin. That's all! In particular, each **projective line**, modeled by a plane through the origin, has a **point at infinity**, which is the horizontal line in this plane.

This model captures what is seen by an "all-seeing eye" at the origin, as suggested by figure 3.13, but it is easy to verify the three axioms directly. The first two simply restate the facts that two lines through O determine a plane through O and that two planes through O meet in a line through O. Four "points" satisfying the third axiom are, for example, the lines through O and the points $\langle 1, 0, 0 \rangle$, $\langle 0, 1, 0 \rangle$, $\langle 0, 0, 1 \rangle$, and $\langle 1, 1, 1 \rangle$.

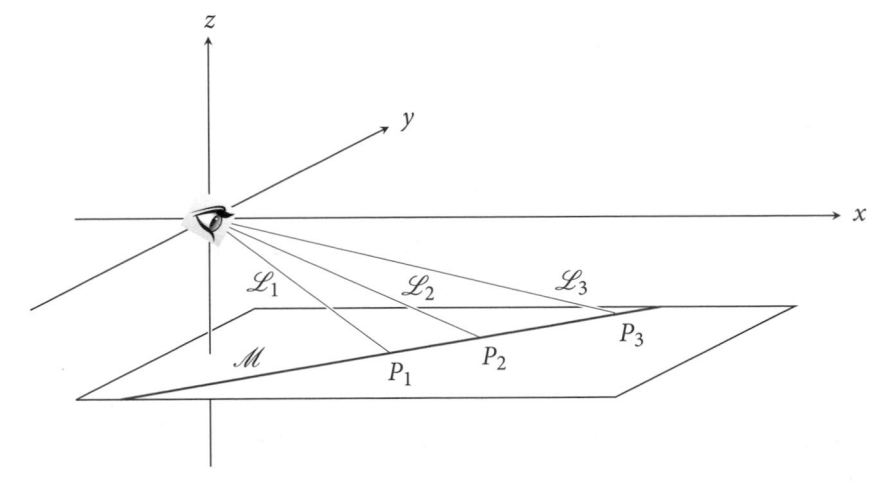

Figure 3.13 : Viewing a plane projectively

In figure 3.13 points P_1, P_2, P_3 of the ordinary plane $z = -1$ correspond to lines $\mathscr{L}_1, \mathscr{L}_2, \mathscr{L}_3$ through O; a line $\mathscr{M}$ of the plane $z = -1$ corresponds to a plane through O (containing $\mathscr{L}_1, \mathscr{L}_2, \mathscr{L}_3$), but the eye also sees the "point at infinity" of $\mathscr{M}$ as the horizontal line in the latter plane.

Additional Axioms

Axioms 1, 2, and 3 above are taken as the defining axioms of a projective plane, presumably because no other properties of points and lines seem sufficiently simple and obvious. However, the axioms do not have many interesting consequences. Not only do they fail to prove the Pappus and Desargues theorems, they do not even imply that there are infinitely many points!

The axioms cannot prove that there are more than seven points because they have a model with exactly seven points. This is called the

Fano plane and is shown in figure 3.14. Its "points" are the seven dots, and its "lines" are the seven curves connecting triples of points, including the circle. It is easy to check that the first two axioms hold in the Fano plane, and the third one is verified by the three corner points together with the point in the center.

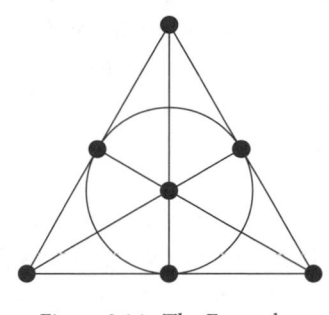

Figure 3.14 : The Fano plane

This raises the question, What are the best axioms to add to the projective plane axioms? Hilbert (1899) suggested adding the theorems of Pappus and Desargues as axioms because this allows us to define sum and product operations for points on a line that satisfy the field properties, much as we can do on the basis of incidence and congruence axioms in Euclidean geometry. The beauty of the projective axioms is that they are entirely axioms of incidence, so they provide a much simpler basis for the field concept.

Indeed, Hessenberg (1905) proved that the Pappus theorem implies that of Desargues, so in fact we can derive the nine field properties from just four statements about points and lines: the three projective plane axioms and the Pappus theorem. This discovery has admittedly not sent many algebraists scrambling for geometry books, but it does show that a common mathematical structure can arise in extraordinarily different ways, potentially offering new insights. We say more about the algebraic consequences, and models, of projective plane axioms in section 4.9.

3.9 REMARKS

In this chapter we have seen various ways in which geometry became more algebraic over recent centuries. For this to happen, algebra had to develop, first as an efficient means of computation and proof and later as a source of mathematical structures to emulate. We have not yet gone into many of the algebraic details because we do not yet wish to assume much familiarity with algebra. The details will become clearer in the next chapter, when we describe the evolution of algebra and its methods of proof.

However, it is worth mentioning a curious correspondence between the axioms used unconsciously by Euclid and the axioms used unconsciously by the early algebraists. Euclid took for granted the incidence axioms, which later turned out to control the field properties of sum and product. These properties were used by algebraists for centuries before being explicitly recognized in the nineteenth century. As we will see in section 4.6, the field properties came to light only after the discovery of algebraic structures that are *not* fields, despite having the operations $+$, $-$, $\cdot$, and $\div$. Likewise, the missing axioms in Euclid came to light only after the discovery of geometries that are not Euclidean. We will see how this came about in chapter 9.

CHAPTER 4

Algebra

Algebra today is a self-contained discipline that involves, among other things, computation with abstract objects. We can recognize algebra in the work of Indian and Islamic mathematicians from the sixth century onward, yet Euclid's geometry was algebra's foundation and method of proof until the sixteenth century. A typical example is the solution of the quadratic equation. But the sixteenth-century solution of the *cubic* equation created a problem that geometry could not explain.

Bombelli solved this problem by *symbolic calculation* with $\sqrt{-1}$, and symbolic calculation (with help from Viète and Descartes) became a new method of proof. In symbolic calculation, as in Euclid's geometry, there were hidden assumptions not explicitly stated until the nineteenth century. However, the hidden assumptions in algebra (such as $a + b = b + a$) were more easily enunciated than the hidden assumptions in Euclid's geometry.

In the nineteenth century algebra moved to higher levels of abstraction when the process of symbolic calculation itself came under mathematical scrutiny. Galois (1831) was the first to see that solvability of polynomial equations was a problem about *symmetry*, and he developed a theory of symmetry—**group theory**—to solve it. Group theory, like traditional algebra itself, involves computation with symbols, but of a more difficult kind. In particular, "multiplication" no longer needs be commutative.

Other problems from traditional algebra involve **algebraic numbers**: the numbers that satisfy polynomial equations with rational coefficients. These problems led to the theory of **rings** and **fields**. Finally, algebraists found new depths in the oldest equations of all—linear equations—creating the **linear algebra** that today permeates almost all fields of mathematics.

4.1 QUADRATIC EQUATIONS

In chapter 3 we observed how algebra became increasingly noticeable in geometry in the nineteenth century, a time when algebra was expanding its role throughout mathematics. With hindsight, we can see that algebra was part of geometry from ancient times and, indeed, that some geometric problems solved by the ancients would be classified as "algebra" today. A case in point is the solution of quadratic equations.

Perhaps the oldest examples are from ancient Mesopotamia, around 3,500 to 4,000 years ago. One of them, given in Katz and Parshall (2014: 24), essentially solves the equation $x^2 + 2x = 120$ by geometrically "completing the square." The left-hand side is given as a square plus a rectangle with one side equal to that of the square and another side equal to 2 (figure 4.1).

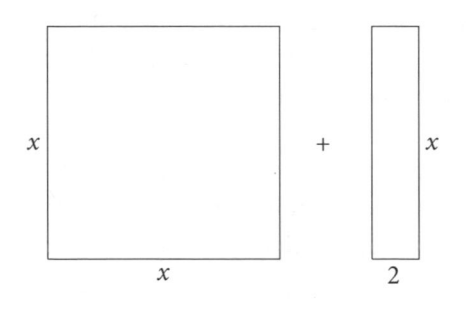

Figure 4.1 : Square plus rectangle

Splitting the rectangle down the middle and attaching the two halves to the square gives figure 4.2, which we are told has area 120. The problem is to find the side x of the original square.

Filling the gap at the top right of figure 4.2 with a square of side 1 gives the complete square, necessarily of area 121, shown in figure 4.3.

Conveniently, $121 = 11 \times 11$, so the side $x + 1$ of the completed square is 11, and therefore $x = 10$. (We have named the unknown side x for simplicity in this account, but of course it could be called anything. The essence of the argument is the manipulation of squares and rectangles.)

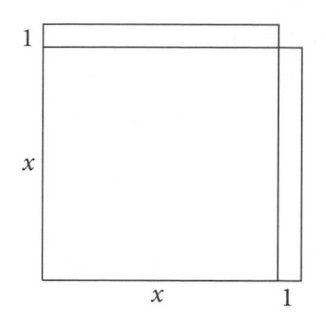

Figure 4.2 : Half rectangles attached to the square

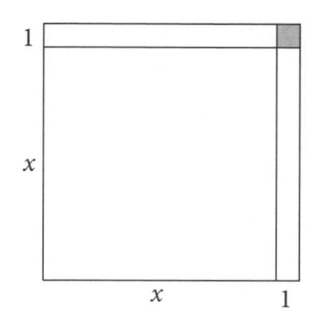

Figure 4.3 : Completed square

Problems like this about areas of squares and rectangles recurred down the ages, in Greek as well as Indian and other Asian cultures. A famous example, because it marked a turn toward algebra as we know it, is the treatment of quadratic equations by the Persian mathematician Muḥammad ibn Mūsā al-Khwārizmī. In a book written around 825 CE, about what he called *al-jabr*, al-Khwārizmī solves quadratic equations by rearranging, adding, and subtracting areas. The idea is the same as in the Mesopotamian example above but with added emphasis on *al-jabr* and *al-mūqabala*, which may be translated as "restoring broken parts" and "balancing," respectively.

Because of his emphasis on manipulation—even though he was manipulating areas and not symbols—al-Khwārizmī came to be seen as the father of "algebra," a word derived from his "al-jabr." Indeed, for a long time the word "algebra" had a broader meaning in Europe, which included the resetting of broken bones. (The name al-Khwārizmī, incidentally, gave us the words "algorism" and later "algorithm.")

Examples of completing the square typically were contrived to give a perfect square, such as 121, but the general algorithm for solving quadratic equations was known before al-Khwārizmī's time. The equivalent of our formula

$$x = \frac{\sqrt{b^2 - 4ac} - b}{2a},$$

for one solution of $ax^2 + bx + c = 0$, was given by the Indian mathematician Brahmagupta in 628 CE. Like everyone else in his time, Brahmagupta did not write anything as concise as a formula but, rather, wrote an algorithm expressed in words. The very idea of quadratic equations had to wait until the sixteenth century, when it became possible to write them concisely enough to be taken in at a glance.

4.2 CUBIC EQUATIONS

Two breakthroughs led to algebra as a form of symbolic calculation with equations, and all its consequences in modern mathematics. They were, first, the method of symbolic calculation in arithmetic brought to Europe by the book *Liber abaci (Book of Calculation)* by Fibonacci (1202); and second, the solution of the cubic equation by Italian mathematicians in the early sixteenth century.

Written calculation in arithmetic brought the Indian numerals

$$0, 1, 2, 3, 4, 5, 6, 7, 8, 9$$

to Europe, but it was not really a *mathematical* innovation, because calculation with written numerals is formally identical with calculation on the abacus, which Europeans had used for centuries. It is not more efficient either, because abacus calculation is, if anything, more rapid than written calculation. Nor was the *Liber abaci* innovative in methods of proof, since Fibonacci followed the style of Euclid's *Elements* throughout.

The idea of doing algebra by written calculation with symbols did not occur to him, as can be seen from the *Liber quadratorum (Book of Squares)* of Fibonacci (1225). In this book Fibonacci proves an important result about sums of squares, which we would write as

$$\left(a^2 + b^2\right)\left(c^2 + d^2\right) = \left(ac - bd\right)^2 + \left(ad + bc\right)^2$$

by expanding both sides and comparing the results. But the expansion is done *in words* and takes five pages (see Fibonacci 1225: 23–28)! Perhaps because there was no abacus counterpart of this calculation, Fibonacci saw no need for speed or brevity, which symbolic calculation can provide in algebra better than it can in arithmetic. Something else had to happen to bring symbolism into algebra. The galvanizing event was the solution of the cubic equation.

Solution of the Cubic Equation

Before we see how the cubic equation was solved, let us record the reaction of one of those on the scene:

> In our own days Scipione del Ferro of Bologna has solved the case of the cube and first power equal to a constant, a very elegant and admirable accomplishment. Since this art surpasses all human subtlety and the perspicuity of mortal talent and is a truly celestial gift and a very clear test of the capacity of men's minds, whoever applies himself to it will believe that there is nothing that he cannot understand.
>
> (Cardano 1545: 8)

Cardano's excitement is understandable but, alas, his 1545 book *Ars magna (Great Art)* is painful reading today because it does not use symbolic calculation. It does not even use negative numbers, so what we would write as $x^3 - px - q = 0$ was viewed as $x^3 = px + q$ and written "cube equal to first power and the number." Thus, for Cardano there were as many different types of cubic equation as there were ways of assigning $+$ and $-$ signs to the coefficients. Each has its own chapter in his book. The proofs are lengthy, in ancient Greek style, with references to Euclid and even Plato.

At the risk of making the solution look easier than it was at the time, it is presented here in modern notation, and we consider only the case $x^3 = px + q$, to which any cubic is easily reducible by a simple change of variable. This case is notationally simplest, and it exposes the ingenuity of the solution most clearly.

If we let $x = u + v$ then the equation $x^3 = px + q$ becomes

$$(u+v)^3 = p(u+v) + q. \tag{*}$$

Now

$$(u+v)^3 = u^3 + 3u^2v + 3uv^2 + v^3 = 3uv(u+v) + u^3 + v^3,$$

so (*) is satisfied if

$$p = 3uv \quad \text{and} \quad q = u^3 + v^3.$$

These two equations imply

$$v = p/3u \quad \text{and hence} \quad q = u^3 + (p/3u)^3 = u^3 + p^3/(3u)^3. \qquad (**)$$

The second of these equations is *quadratic in* u^3, namely,

$$qu^3 = \left(u^3\right)^2 + \left(\frac{p}{3}\right)^3 \quad \text{or} \quad \left(u^3\right)^2 - qu^3 - \left(\frac{p}{3}\right)^3 = 0.$$

By the quadratic formula it has two solutions:

$$u^3 = \frac{q}{2} \pm \sqrt{\left(\frac{q}{2}\right)^2 - \left(\frac{p}{3}\right)^3}.$$

Going back to the equations (**), we see that they lead to the *same quadratic equation* for v^3. So, of the two solutions just found, one is u^3 and the other is v^3:

$$u^3, v^3 = \frac{q}{2} + \sqrt{\left(\frac{q}{2}\right)^2 - \left(\frac{p}{3}\right)^3}, \frac{q}{2} - \sqrt{\left(\frac{q}{2}\right)^2 - \left(\frac{p}{3}\right)^3}.$$

Taking cube roots we get

$$u, v = \sqrt[3]{\frac{q}{2} + \sqrt{\left(\frac{q}{2}\right)^2 - \left(\frac{p}{3}\right)^3}}, \sqrt[3]{\frac{q}{2} - \sqrt{\left(\frac{q}{2}\right)^2 - \left(\frac{p}{3}\right)^3}},$$

so finally,

$$x = u + v = \sqrt[3]{\frac{q}{2} + \sqrt{\left(\frac{q}{2}\right)^2 - \left(\frac{p}{3}\right)^3}} + \sqrt[3]{\frac{q}{2} - \sqrt{\left(\frac{q}{2}\right)^2 - \left(\frac{p}{3}\right)^3}}.$$

This way of writing the solution of $x^3 = px + q$ is known as the *Cardano formula*, though it should really be credited to Scipione del Ferro, as Cardano himself admitted.[1]

The derivation shows how much the *Ars magna* can be streamlined in modern notation. It also neatly takes care of the two possible values,

---

1. The solution was discovered independently by Niccolò Fontana, known as Tartaglia. He disclosed his solution to Cardano, who used it in the *Ars magna* after learning of del Ferro's prior discovery.

+ and −, of the square root. However, it glosses over the fact that a cube root has *three* possible values. Taking account of the three values of the cube root requires complex numbers, which Cardano dismissed as "mental tortures." The first to take them seriously was Cardano's compatriot, Rafael Bombelli.

Bombelli and the Equation $x^3 = 15x + 4$

Bombelli (1572) noticed a problem with the Cardano formula

$$x = \sqrt[3]{\frac{q}{2} + \sqrt{\left(\frac{q}{2}\right)^2 - \left(\frac{p}{3}\right)^3}} + \sqrt[3]{\frac{q}{2} - \sqrt{\left(\frac{q}{2}\right)^2 - \left(\frac{p}{3}\right)^3}}.$$

Not only can the term under the square root sign, $\left(\frac{q}{2}\right)^2 - \left(\frac{p}{3}\right)^3$, be negative—this can happen for equations with a straightforward solution. An example is $x^3 = 15x + 4$. It has the obvious solution $x = 4$, yet the Cardano formula gives $x = u + v$ where

$$u = \sqrt[3]{2 + \sqrt{-121}}, \quad v = \sqrt[3]{2 - \sqrt{-121}}.$$

Bombelli guessed that in fact

$$u = 2 + \sqrt{-1}, \quad v = 2 - \sqrt{-1},$$

so that $x = u + v = 4$ as required.

He concealed the calculation that led him to this conclusion, but it is easily checked that

$$u^3 = 2 + \sqrt{-121}, \quad v^3 = 2 - \sqrt{-121}$$

if we assume that the symbol $\sqrt{-1}$ obeys the ordinary rules of calculation. If so, then we can use $(a + b)^3 = a^3 + 3a^2 b + 3ab^2 + b^3$ to conclude that

$$u^3 = (2 + \sqrt{-1})^3 = 2^3 + 3 \cdot 2^2 \sqrt{-1} + 3 \cdot 2(\sqrt{-1})^2 + (\sqrt{-1})^3$$
$$= 8 + 12\sqrt{-1} - 6 - \sqrt{-1}$$
$$= 2 + 11\sqrt{-1}$$
$$= 2 + \sqrt{-121},$$

and similarly $v^3 = 2 - \sqrt{-121}$. So indeed,

$$x = u + v = \sqrt[3]{(2 + i)^3} + \sqrt[3]{(2 - i)^2} = 2 + i + 2 - i = 4.$$

Bombelli broke new ground by calculating with the undefined symbol $i = \sqrt{-1}$, which at the time had no geometric interpretation. Algebra was about to be freed from its dependence on geometry and to become an independent system of calculation, like arithmetic.

4.3 ALGEBRA AS "UNIVERSAL ARITHMETICK"

Between Bombelli (1572) and Descartes (1637), algebraic notation developed rapidly, with symbols for unknowns; for sums, differences, products, quotients, square roots, and equals; and the superscript notation for powers of a variable. Aside from Bombelli and Descartes, important contributions were made by Stevin (1585b), Viète (1591), and Harriot (1631). Descartes (1637) has almost the notation used today, except that x^2 is written xx (but x^3 and higher powers are as written today),[2] and there is a different equality sign.

With the new notation, mathematicians could handle equations with a fluency previously unimaginable. For example, in 1596 Adrien van Roomen could pose an equation of 45th degree to Viète—and Viète could solve it. By the early eighteenth century algebraic computation was so powerful as to be accepted as a *method of proof*, whose results are *theorems*. Here is Newton in his *Universal Arithmetick* (second English edition of 1728):

> COMPUTATION is either perform'd by Numbers, as in Vulgar Arithmetick, or by Species, as usual among Algebraists. They are both built on the same Foundations, and aim at the same End, viz. Arithmetick Definitely and Particularly, Algebra Indefinitely and Universally; so that almost all expressions that are to be found out by this Computation, and particularly Conclusions, may be called *Theorems*. But Algebra is particularly excellent in this, that whereas in Arithmetick Questions are only resolv'd by proceeding from given Quantities to Quantities sought, Algebra proceeds in a retrograde Order, from the Quantities sought, as if they were given, to the Quantities given, as if they were sought, to the end that we may some way or other come to a Conclusion or Equation, from which one may bring out the Quantity sought. And after this Way the most difficult Problems are resolv'd, the Resolutions whereof would be

---

2. Bressoud (2019: 72) points out that xx is actually simpler than x^2 from a typesetter's point of view—both require two symbols, but xx is slightly easier to set.

sought in vain from only common Arithmetick. Yet Arithmetick in all its operations is so subservient to Algebra, as that they seem both but to make one perfect Science of Computing; and therefore I will explain them both together.

(Newton 1728: 1)

But if equations are theorems, what are the axioms? This question is answered in section 4.6, but first we should recognize the limitations of algebra that is not far removed from arithmetic. By using symbols for variable numbers, algebra achieved greater generality than of ordinary arithmetic. But symbolic arithmetic is still arithmetic, since letters obey the same rules as ordinary numbers. The symbolism was more powerful and efficient, yes, but still not powerful enough to get far beyond the cubic equation.

The solution of the cubic was quickly followed by a solution of the quartic (fourth-degree) equation, by Cardano's student Lodovico Ferrari, and also published in Cardano (1545). But all attempts to solve the *quintic* (fifth-degree) equation failed, and by 1800 an explanation of this failure was sought in a *mathematical analysis of algebra itself*: an analysis of the relationships among equations, roots, and processes of solution.

4.4 POLYNOMIALS AND SYMMETRIC FUNCTIONS

Around 1600, Viète explored a new approach to the theory of equations by looking at the relationship between roots and coefficients. If we take the cubic equation

$$x^3 + e_1 x^2 + e_2 x + e_3 = 0$$

and suppose it has roots x_1, x_2, x_3, then

$$x^3 + e_1 x^2 + e_2 x + e_3 = (x - x_1)(x - x_2)(x - x_3) \qquad (*)$$
$$= x^3 - (x_1 + x_2 + x_3)x^2 + (x_1 x_2 + x_2 x_3 + x_3 x_1)x - x_1 x_2 x_3,$$

so the coefficients are expressed in terms of the roots by

$$e_1 = -(x_1 + x_2 + x_3),$$
$$e_2 = x_1 x_2 + x_2 x_3 + x_3 x_1,$$
$$e_3 = -x_1 x_2 x_3.$$

The functions on the right side of these equations are called **symmetric functions** of the variables x_1, x_2, x_3 because they remain unchanged when x_1, x_2, x_3 are permuted in any way. This could be anticipated from their defining equation (*), since the product $(x - x_1)(x - x_2)(x - x_3)$ is unchanged when its factors are permuted. The particular functions e_1, e_2, e_3 are called the **elementary symmetric (polynomial) functions** of x_1, x_2, x_3.

The Factor Theorem

Viète's idea assumes that roots correspond to factors, which is true but was first recognized as a theorem by Descartes (1637). Before looking at symmetric functions in general we therefore state and prove the theorem.

Factor theorem. *If $p(x)$ is a polynomial and $p(a) = 0$, then $x - a$ is a factor of $p(x)$, that is, $p(x) = (x - a)q(x)$ for some polynomial $q(x)$.*

Proof. Let $p(x) = a_n x^n + a_{n-1} x^{n-1} + \cdots + a_1 x + a_0$. Then for any number a

$$p(x) - p(a) = a_n(x^n - a^n) + a_{n-1}(x^{n-1} - a^{n-1}) + \cdots + a_1(x - a).$$

Each term on the right-hand side has a factor $x - a$ because

$$x^m - a^m = (x - a)\left(x^{m-1} + ax^{m-2} + \cdots + a^{m-2}x + a^{m-1}\right),$$

so $x - a$ divides $p(x) - p(a)$.

In particular, if $p(a) = 0$ then $x - a$ divides $p(x)$. □

From this theorem it follows that:

1. For each root $x = a$ of the equation $p(x) = 0$, $p(x)$ has a factor $x - a$.
2. A polynomial equation $p(x) = 0$ of degree n has at most n roots.
3. *If* every polynomial equation with real coefficients has a complex root (the so-called **fundamental theorem of algebra**) and $p(x)$ has degree n, then

$$p(x) = (x - x_1)(x - x_2)\cdots(x - x_n) \times \text{constant},$$

   where $x_1, x_2, \ldots, x_n$ are the roots of $p(x) = 0$ (not necessarily distinct).
4. The above proof of the factor theorem holds for polynomials with coefficients that are *themselves* polynomials (in some other variables) since we assume only that the coefficients obey the usual rules for addition, subtraction, and multiplication. We will use this

fact below to conclude that a polynomial $p(x)$ has a factor x when $p(0) = 0$, even when the coefficients of $p(x)$ involve other variables.

Now we are ready to return to the relations between roots and coefficients for polynomials of arbitrary degree. Without loss of generality we can take the case where the polynomial equation is

$$x^n + e_1 x^{n-1} + \cdots + e_{n-1} x + e_n = 0$$

with roots $x_1, x_2, \ldots, x_n$. Factorizing the polynomial by the factor theorem, we similarly find

$$e_1 = -(x_1 + x_2 + \cdots + x_n),$$
$$\vdots$$
$$e_n = (-1)^n x_1 x_2 \cdots x_n,$$

and $e_1, e_2, \ldots, e_n$ are called the **elementary symmetric (polynomial) functions**[3] of $x_1, x_2, \ldots, x_n$. The polynomial e_k is $(-1)^k$ times the sum of all products of k distinct variables x_i, and hence it has degree k. (This is the reason for the ordering of subscripts on the e_i.)

The elementary symmetric polynomials are not the only symmetric polynomials; for example, $x_1^2 + x_2^2 + x_3^2$ is symmetric in x_1, x_2, x_3. Girard (1629), and later Newton, noticed that power sums $x_1^p + x_2^p + \cdots + x_n^p$ of the roots could be expressed as polynomials in the elementary symmetric functions, and hence as polynomials in the *coefficients* of the original equation. In this case, we notice that

$$e_1^2 = (x_1 + x_2 + x_3)^2 = x_1^2 + x_2^2 + x_3^2 + 2x_1 x_2 + 2x_2 x_3 + 2x_3 x_1,$$
$$= x_1^2 + x_2^2 + x_3^2 + 2e_2,$$

so

$$x_1^2 + x_2^2 + x_3^2 = e_1^2 - 2e_2.$$

The Fundamental Theorem of Symmetric Polynomials

The power sum results are instances of a general theorem about symmetric polynomials, which seems much more subtle.[4] Possibly Newton was

---

3. One sometimes considers symmetric *rational* functions, but here we are concerned only with the symmetric polynomials.

4. I believe I can confirm that the power sum results are not hard, having discovered them by myself when I was in high school.

aware of this theorem, but I have been unable to locate a published proof before the nineteenth century.

Fundamental theorem of symmetric polynomials. *Any symmetric polynomial in $x_1, x_2, \ldots, x_n$ is a polynomial in the elementary symmetric polynomials $e_1, e_2, \ldots, e_n$.*

Proof. The proof is by a double induction, on the number n of variables and the degree d of the polynomial. When both equal 1 the only elementary symmetric function is $e_1 = -x_1$, and every polynomial of degree 1 in x_1 is a polynomial in e_1.

Now suppose that the theorem is true for polynomials of $n-1$ variables, and also for those of n variables and degree up to $d-1$. Given a symmetric polynomial $f(x_1, \ldots, x_{n-1}, x_n)$ of degree d, we first relate $f(x_1, \ldots, x_{n-1}, 0)$ to the polynomials

$$e_i^0(x_1, \ldots, x_{n-1}) = e_i(x_1, \ldots, x_{n-1}, 0).$$

These polynomials are, up to $\pm$ sign, the elementary symmetric polynomials in $x_1, \ldots, x_{n-1}$. So, by our induction hypothesis on the number of variables,

$$f(x_1, \ldots, x_{n-1}, 0) = g(e_1^0, \ldots, e_{n-1}^0) \quad \text{for some polynomial } g,$$

since $f(x_1, \ldots, x_{n-1}, 0)$ is symmetric in $x_1, \ldots, x_{n-1}$.

Now consider the polynomial

$$h(x_1, \ldots, x_{n-1}, x_n) = f(x_1, \ldots, x_{n-1}, x_n) - g(e_1, \ldots, e_{n-1}), \qquad (*)$$

which is symmetric since f and the e_i are. When we substitute 0 for x_n the right side of $(*)$ becomes 0 by the definition of g. So, by the factor theorem of the previous section, x_n divides $h(x_1, \ldots, x_{n-1}, x_n)$. Since h is symmetric, it is divisible by each x_i, and hence by $x_1 \cdots x_n = \pm e_n$. Since e_i and e_i^0 both have degree i, $g(e_1, \ldots, e_{n-1})$ has the same degree as $g(e_1^0, \ldots, e_{n-1}^0)$, so the degree of h is no greater than the degree of f.

We therefore have

$$h(x_1, \ldots, x_{n-1}, x_n) = x_1 \cdots x_n k(x_1, \ldots, x_{n-1}, x_n) = \pm e_n k(x_1, \ldots, x_{n-1}, x_n),$$

for some polynomial k of degree less than that of h, and hence of f. Then, by our induction hypothesis on degree, k is a polynomial in the e_i, hence so is h, and so is f by $(*)$. This completes the induction. $\qquad \square$

Given the subtle induction argument involved in the fundamental theorem of symmetric polynomials, it seems unlikely that Newton had a general proof. But the theorem seems to have been known by the eighteenth century, if not rigorously proved, and it makes an interesting appearance in a proof of the **fundamental theorem of algebra** that we will see in section 8.5.

4.5 MODERN ALGEBRA: GROUPS

To see where groups came from, and why their discovery was so remarkable, we should begin with the theory of equations as it stood around 1800.

Classical Theory of Equations

In the previous section we observed that the *factor theorem* of Descartes (1637) implies a factorization of each polynomial $p(x)$ into n factors of the form $x - x_i$, where the x_i are the roots of the equation $p(x) = 0$, assuming roots exist. This amounts to assuming the fundamental theorem of algebra, which we take up in chapter 8. For now we assume it in order to concentrate on the implications of the factorization of $p(x)$. By dividing the equation $p(x) = 0$ by the leading coefficient of $p(x)$ we can assume

$$p(x) = x^n + a_{n-1}x^{n-1} + \cdots + a_1 x + a_0 = (x - x_1)(x - x_2)\cdots(x - x_n).$$

This gives the relations between the roots and coefficients we saw in the previous section. For example, for the quadratic equation we have

$$x^2 + bx + c = (x - x_1)(x - x_2) = x^2 - (x_1 + x_2)x + x_1 x_2,$$

so

$$b = -(x_1 + x_2),$$
$$c = x_1 x_2.$$

Now, if the solutions of the quadratic, cubic, and quartic equations are any guide, we expect to solve $p(x) = 0$ by the operations $+, -, \times, \div, \sqrt{\ }, \sqrt[3]{\ }$ (and possibly other nth roots) on the coefficients of $p(x)$. These operations, on the *symmetric* functions that are the coefficients, produce the totally *asymmetric* functions that are the roots. How can this be?

We can see how it happens in the case of the quadratic equation

$$x^2 - (x_1 + x_2)x + x_1 x_2 = 0.$$

According to the quadratic formula, the roots are

$$x = \frac{x_1 + x_2 \pm \sqrt{(x_1 + x_2)^2 - 4x_1 x_2}}{2}$$

$$= \frac{x_1 + x_2 \pm \sqrt{x_1^2 - 2x_1 x_2 + x_2^2}}{2}$$

$$= \frac{x_1 + x_2 \pm \sqrt{(x_1 - x_2)^2}}{2}$$

$$= \frac{x_1 + x_2 \pm (x_1 - x_2)}{2}$$

$$= x_1, x_2.$$

Here $\sqrt{}$ is the symmetry-breaking operation, and any other case is similar, though more complicated. The **radicals** $\sqrt{}, \sqrt[3]{}, \ldots$ can break symmetry because of their multiple values, in this case the $\pm$ values of the square root.

The above analysis exposes the problem of **solution by radicals** but does not go far toward solving it. What is lacking is a "theory of symmetry" and an understanding of the role of radicals in symmetry breaking. Such a theory is a huge leap in abstraction, one of the greatest in the history of mathematics, and it was made by the twenty-year-old Evariste Galois in 1831.

Galois Theory

As we saw in the previous subsection, the theory of equations raised two questions that were new to mathematics: what is symmetry, and how is it broken? The short answer is that symmetry is captured by the *group* concept and (in the case of equations) that symmetry breaking corresponds to formation of a *quotient group*. The group concept is extremely general, yet miraculously appropriate for many mathematical purposes, as we will see elsewhere in this book.

However, for the symmetry involved in the theory of equations, we can specialize the answer somewhat: symmetry is captured by *finite permutation groups*, and symmetry breaking corresponds to a quotient that is a *cyclic* group. To simplify even further, we will confine the discussion to the permutations involved in the quintic equation, which are permutations of its five roots x_1, x_2, x_3, x_4, x_5. But first, let us define the general group concept as simply as possible.

Definitions. A **permutation** is a bijective function on some domain, and the **product** of permutations f and g is their composite $f \circ g$ defined by $f \circ g(x) = f(g(x))$. A **permutation group** is a set of bijective functions on some domain, that includes the identity function, the inverse of each member, and the composite of any two members. In particular, the $n!$ permutations of an n-element set form the **symmetric group**, denoted by S_n.

The permutations of any domain obviously include the *identity* function $\mathbf{1}(x) = x$, and since they are bijective functions, each permutation f has an *inverse* f^{-1} such that $f \circ f^{-1} = f^{-1} \circ f = \mathbf{1}$. Finally, the product of permutations is *associative* (as for all functions):

$$f \circ (g \circ h) = (f \circ g) \circ h,$$

because each side equals $f(g(h(x)))$ for all x in the domain. We now recognize identity, inverses, and associativity as the defining properties of a **group** (see the group axioms below), but in the beginning all groups were permutation groups. Indeed, there is no loss of generality in assuming all groups to be permutation groups. So, instead of saying "permutation group" we will simply say "group" from now on.

When the domain consists of five things, as with the quintic equation, its permutations comprise the symmetric group S_5 under the operation of composition. A simpler example, already seen in the case of the quadratic equation, is the symmetric group S_2 of permutations of two things. S_2 consists of just two permutations: the identity $\mathbf{1}$ and the permutation f that exchanges x_1 and x_2: $f(x_1) = x_2$ and $f(x_2) = x_1$. Since $f \circ f = \mathbf{1}$, S_2 consists of the powers of the single element f.

Definitions. A group consisting of powers of a single element is called **cyclic**. A group G is said to have a **cyclic quotient** C if there is a map φ of G onto a cyclic group C such that $\varphi(xy) = \varphi(x) \circ \varphi(y)$.

Galois discovered, essentially (though in different language), that breaking symmetry by an mth root operation amounts to mapping the symmetry group onto the cyclic group with m elements. The elements of the group sent to **1** by this map is a smaller group whose symmetry we seek to break further, and so on, until only the identity symmetry remains (indicating that we have reached the completely asymmetric functions that are the roots).

It so happens that S_5 *cannot be broken down in this way*, which is why *the general quintic equation is not solvable by radicals*. Obviously, we have glossed over many details that lie behind Galois's discoveries, particularly the relationship between mth roots and cyclic groups and what makes S_5 "unbreakable." That Galois was able to find his way through these details and to organize them in a new system of concepts is one of the great achievements in the history of mathematics.

Axioms for Groups

Since the groups G considered by Galois consisted of bijective functions, under the operation $\circ$ of function composition, certain properties of their elements $f, g, h, \ldots$ were obvious (and generally used without comment):

1. The group operation is *associative*; that is, $f \circ (g \circ h) = (f \circ g) \circ h$.
2. G includes an *identity* element **1**, such that $\mathbf{1} \circ g = g \circ \mathbf{1} = g$ for any g.
3. Each $g \in G$ has an *inverse* g^{-1}, such that $g \circ g^{-1} = g^{-1} \circ g = \mathbf{1}$.

Later, when a plethora of examples had been noticed, these three properties became the defining properties of a group, or the **group axioms**. In the axioms, first stated in this way by Weber (1896), the elements need not be functions, and the operation $\circ$ need not be function composition but can be any binary operation whatever (though generally called "multiplication").

One is free to think of group elements as functions, and the group operation as function composition, because of the observation of Cayley (1878) that members of group G correspond to bijective functions on G. Namely, any element $g \in G$ corresponds to the permutation $g\circ$ of G induced by multiplying each member of G on the left by g.

4.6 MODERN ALGEBRA: FIELDS AND RINGS

Section 4.3 raised the question, If equations are theorems, what are the axioms? Some, but not all, such axioms had been noticed in ancient times. In particular, Euclid's "axioms of equality" was mentioned in section 1.1:

1. Identical figures are equal.
2. Things equal to the same thing are equal to each other.
3. If equals are added to equals the sums are equal.
4. If equals are subtracted from equals the differences are equal.

By the seventeenth century, "figures" could be understood as "algebraic expressions," at which time it must have been understood that:

5. If equals are multiplied by equals the products are equal.
6. If equals are divided by (nonzero) equals the quotients are equal.

But what happens *inside* one side of an equation when addition or multiplication occurs? This was not completely spelled out, though Euclid gave some thought to its geometric equivalent. Proposition 1 of book 2 of the *Elements* is the geometric equivalent of the general **distributive law**

$$a(b + c + d + \cdots) = ab + ac + ad + \cdots.$$

Here, $b, c, d, \ldots$ stand for what Euclid calls "any number of parts," which is understood to be a finite number. So Euclid's distributive law is really a consequence of *our* distributive law

$$a(b + c) = ab + ac$$

together with the **associative law for addition**

$$b + (c + d) = (b + c) + d,$$

which says that the two interpretations of the expression $b + c + d$ are equal.

It seems that associative laws went unnoticed until the **associative law for multiplication**,

$$b(cd) = (bc)d,$$

was found to *fail* for an exotic kind of multiplication. The word "associative" was introduced by William Rowan Hamilton in a paper dated November 13, 1843, at which time he and his friend John Graves were studying generalizations of the complex numbers called the **quaternions** and **octonions**. Hamilton found that his system of quaternions has associative multiplication, whereas Graves's system of octonions does not. Neither system, however, satisfies the **commutative law of multiplication**,

$$bc = cb,$$

so commutative multiplication (which of course holds for complex numbers) had been brought to their attention and was known to be capable of failing.

Complex and Hypercomplex Numbers

The passage from real to complex numbers, and then to quaternions and octonions, is an interesting story in itself but also important for bringing axioms of algebra to light. It began when Hamilton in 1834 defined the complex numbers as ordered pairs $\langle a, b \rangle$ of real numbers.

If $z_1 = \langle a_1, b_1 \rangle$ and $z_2 = \langle a_2, b_2 \rangle$ are complex numbers, then Hamilton *defined* their sum and product in terms of the sum and product of real numbers as follows:

$$z_1 + z_2 = \langle a_1 + a_2, b_1 + b_2 \rangle, \quad z_1 \cdot z_2 = \langle a_1 a_2 - b_1 b_2, a_1 b_2 + a_2 b_1 \rangle.$$

It is easy to check that this sum and product agree with the sum and product of $a_1 + b_1 i$ and $a_2 + b_2 i$ calculated using the rule $i^2 = -1$. The point of Hamilton's definition is that it reduces any statement about sums and products of complex numbers to a statement about sums and products of real numbers, so *no contradiction can arise in the theory of complex numbers* (unless there is a contradiction in the theory of real numbers). As Hamilton himself said:

> Were these definitions even altogether arbitrary, they would at least not contradict each other, nor the earlier principles of Algebra, and it would be possible to draw legitimate conclusions, by rigorous mathematical reasoning.

> (Hamilton 1837: 403)

Among the conclusions Hamilton drew from his definitions were the commutative laws for addition and multiplication and the distributive law.

Hamilton's next step was to attempt to define sums and products for ordered triples of real numbers. Sums are naturally defined the same way as for pairs,

$$\langle a_1, b_1, c_1 \rangle + \langle a_2, b_2, c_2 \rangle = \langle a_1 + a_2, b_1 + b_2, c_1 + c_2 \rangle,$$

but Hamilton was unable to find a definition of product with good algebraic properties, even after giving up commutative multiplication. At last, in October 1843, he hit on a system of *quadruples* with all the algebraic properties of real numbers *except* commutative multiplication. These are the **quaternions**, which he described concisely as objects of the form

$$q = a + bi + cj + dk,$$

added in the obvious way and multiplied using the distributive law and the rules

$$i^2 = j^2 = k^2 = ijk = -1.$$

As with the complex numbers, it is possible to reduce statements about sums and products of quaternions to statements about sums and products of real numbers and to rigorously prove properties of the quaternions, such as the associative laws of addition and multiplication (which Hamilton did). The commutative law for multiplication fails because $ij = k = -ji$.

Finally, in December 1843, Hamilton's friend John Graves devised his system of octuples of real numbers, now known the **octonions**. These "hypercomplex numbers," like the quaternions, have most but not all the algebraic properties of real numbers. They lack only commutativity and associativity of multiplication. As mentioned above, it appears that Hamilton introduced the concept of associativity to highlight this "defect" of the octonions.

At any rate, it seems likely that the commutative and associative properties of the real and complex numbers were first appreciated—rather than overlooked or taken for granted—because they failed to hold in the quaternions and octonions.

Axioms for Fields and Rings

Another mathematician to recognize the importance of commutative and associative properties was Hermann Grassmann (1861). In his *Lehrbuch der Arithmetik*—a textbook for high school students!—he proved the commutative, associative, and distributive properties of addition and multiplication for the system $\mathbb{N}$ of natural numbers, and then extended them to the integers $\mathbb{Z}$ and rational numbers $\mathbb{Q}$ by making appropriate definitions of negative integers and fractions. Altogether, Grassmann proved all the *field properties* that we saw in section 3.1:

$$A + B = B + A \qquad A \cdot B = B \cdot A$$
$$A + (B + C) = (A + B) + C \qquad A \cdot (B \cdot C) = (A \cdot B) \cdot C$$
$$A + 0 = A \qquad A \cdot 1 = A$$
$$A + (-A) = 0 \qquad A \cdot A^{-1} = 1 \text{ when } A \neq 0$$
$$A \cdot (B + C) = A \cdot B + A \cdot C.$$

In proving the properties of $\mathbb{N}$ (namely, those not involving $-A$ and A^{-1}) Grassmann made an even greater contribution to the history of proof in arithmetic than he did in algebra, by showing that all of them follow from the principle of **induction**. As we saw in section 2.6, induction is visible in Euclid's *Elements*, but for a long time it made only sporadic appearances in mathematics. By showing for $\mathbb{Q}$ that induction is more fundamental than the field properties, Grassman ensured the place of induction in any future axiom system for $\mathbb{N}$ or $\mathbb{Q}$. We will take up this story again in chapter 13.

As it happened, Grassmann's *Lehrbuch* went almost unnoticed, but some of Grassmann's theorems eventually became *axioms*—a common occurrence in modern algebra. The ones above, as noted, are the defining properties of a **field**. Thus a field is any system of objects with sum and product operations satisfying the field properties. The classical examples are $\mathbb{Q}$, $\mathbb{R}$, and $\mathbb{C}$, but there are also finite fields, as we will see later.

If we omit the axiom $AA^{-1} = 1$ we get the **ring** axioms:

$$A + B = B + A \qquad A \cdot B = B \cdot A$$
$$A + (B + C) = (A + B) + C \qquad A \cdot (B \cdot C) = (A \cdot B) \cdot C$$
$$A + 0 = A \qquad A \cdot 1 = A$$
$$A + (-A) = 0$$
$$A \cdot (B + C) = A \cdot B + A \cdot C.$$

The classical example of a ring is $\mathbb{Z}$, but there are also finite rings, which we will discuss when we discuss finite fields. The theory of fields and rings developed beside the theory of **algebraic numbers** and **algebraic integers**, explored mainly by Dedekind and Kronecker in the 1870s and later. Since algebraic numbers and integers lie in $\mathbb{C}$, their sum and product "inherit" properties from $\mathbb{C}$, so at first it was not thought necessary to state them.

The field axioms were first explicitly stated by Weber (1893), and the ring axioms by Fraenkel (1914). The axiomatic approach to algebra in general took hold only in the 1920s, under the influence of Emmy Noether.

4.7 LINEAR ALGEBRA

Linear algebra came both early and late in the development of mathematics. It appeared in ancient times with linear equations, before "algebra" was a recognized discipline. Then for a long time linear algebra was a poor cousin of the "higher" algebra of polynomial equations and the abstractions arising from them, considered too simple to be an independent branch of mathematics. Now that linear algebra has reasserted itself and is known to all undergraduates, we can identify different stages in its historical development.

The early stage was the solution of systems of linear equations, achieved in China about 2,000 years ago by essentially the method now known as "Gaussian elimination." In our notation, one has a system of n equations in n unknowns $x_1, x_2, \ldots, x_n$:

$$a_{11}x_1 + a_{12}x_2 + \cdots + a_{1n}x_n = b_1$$
$$a_{21}x_1 + a_{22}x_2 + \cdots + a_{2n}x_n = b_2$$
$$\vdots$$
$$a_{n1}x_1 + a_{n2}x_2 + \cdots + a_{nn}x_n = b_n.$$

By subtracting suitable multiples of the first equation from the second, third, and so on, one eliminates the x_1 term in all but the first equation. Then one similarly eliminates the x_2 term in all but the first two equations. Repeating this process gives a system in "triangular form":

$$a'_{11}x_1 + a'_{12}x_2 + \cdots + a'_{1n}x_n = b'_1$$
$$a'_{22}x_2 + \cdots + a'_{2n}x_n = b'_2$$
$$\vdots$$
$$a'_{nn}x_n = b'_n,$$

from which one solves for $x_n, x_{n-1}, \ldots, x_1$ in that order, starting with the last equation and substituting successive results in the equations above it.

Just as we now do with matrices, the Chinese operated only on the array of coefficients a_{ij}, which they entered on a device called a counting board. Underlying the process are the six "axioms of equality" that were listed in section 4.6, which allow multiples of one row to be subtracted from another. With this very simple logic, the solution of linear equations is reduced to basic arithmetic.[5]

The intermediate stage of linear algebra began with the introduction of **determinants** to express the solution of a linear equation system as a function of the coefficients. Determinants were discovered independently by Leibniz and Seki around 1680. The explicit formula for the solution in terms of determinants ("Cramer's rule") appeared in Cramer (1750), ushering in the era of determinant theory that dominated linear algebra until the twentieth century. This era is celebrated in the four-volume history of determinants by Muir (1960). The determinant function is a very powerful function, with some important applications that we will see later. However, much of linear algebra can be done without it, and today we prefer to base the subject on simpler foundations.

The late stage of linear algebra, based on the concept of vector space, began with Grassmann (1844).

4.8 MODERN ALGEBRA: VECTOR SPACES

In section 4.6 we saw that Hamilton knew how to add ordered n-tuples but that he found it hard to multiply them. It *is* hard to multiply n-tuples so that the product is another n-tuple, but Grassmann (1844) realized that there is much to be said about addition and *scalar multiples* of n-tuples. Essentially, he had arrived at the concept of a (real) **vector space**. In this context, the ordered n-tuples are called **vectors**, say,

$$\boldsymbol{u} = \langle u_1, u_2, \ldots, u_n \rangle, \quad \boldsymbol{v} = \langle v_1, v_2, \ldots, v_n \rangle,$$

---

5. In later centuries the Chinese used the same logic to eliminate unknowns from systems of polynomial equations. By eliminating all but one unknown, x, a system of polynomial equations is reduced to a single polynomial equation in x.

with **sum** defined by

$$u + v = \langle u_1 + v_1, u_2 + v_2, \ldots, u_n + v_n \rangle,$$

and **scalar multiple** by r, for any $r \in \mathbb{R}$, defined by

$$ru = \langle ru_1, ru_2, \ldots, ru_n \rangle.$$

Grassmann's creation, of which this is only the simplest part, was intended as a new approach to geometry—which indeed it is today. But sadly it went unnoticed by contemporary mathematicians, even after Grassmann made fresh attempts to popularize it in 1847 and 1862. It was resurrected only when Peano (1888) wrote down an axiomatic description of the vector sum and scalar multiple operations:

$$u + v = v + u$$
$$u + (v + w) = (u + v) + w$$
$$u + 0 = u$$
$$u + (-u) = 0$$
$$r(u + v) = ru + rv$$
$$(r + s)u = ru + su$$
$$r(su) = (rs)u,$$

where r, s are any real numbers, $0 = \langle 0, 0, \ldots, 0 \rangle$, and $-u = \langle -u_1, -u_2, \ldots, -u_n \rangle$. Among the fundamental results proved by Grassmann (1862), which are now staples of linear algebra courses, are **existence of a basis** and **invariance of dimension**. That is, any real vector space V of finite dimension includes vectors $e_1, e_2, \ldots, e_m$, called a **basis** of V, such that

- For each $v \in V$ there are numbers $r_1, r_2, \ldots, r_m \in \mathbb{R}$ such that

$$v = r_1 e_1 + \cdots + r_m e_m$$

  (which says that $e_1, e_2, \ldots, e_m$ **span** V).
- If $r_1 e_1 + \cdots + r_m e_m = 0$ then $r_1 = \cdots = r_m = 0$ (which says that $e_1, \ldots, e_m$ are **linearly independent**).

Moreover, m (the **dimension** of V over $\mathbb{R}$) is the same for any basis.

It follows from the existence of basis and dimension that any vector $v = r_1 e_1 + \cdots + r_m e_m$ in a real vector space of dimension m is uniquely determined by the m-tuple $\langle r_1, r_2, \ldots, r_m \rangle$, so the space is essentially $\mathbb{R}^m$, the space of ordered m-tuples of real numbers.

In the meantime, a generalization of this vector space concept had been used implicitly in the 1870s, by Dedekind and Kronecker in algebraic number theory. In effect, they were working with vector spaces over other fields $\mathbb{F}$, for which the axioms are exactly those of Peano except that $\mathbb{R}$ is replaced by a field $\mathbb{F}$. When this is done, one speaks of a basis for V *over* $\mathbb{F}$, linear independence *over* $\mathbb{F}$, and dimension *over* $\mathbb{F}$. Dedekind and Kronecker did not consciously rely on axioms, though they did make frequent use of linear independence and dimension over various fields. Also, it was important for them to allow $\mathbb{F}$ to vary.

Vector Spaces in Algebraic Number Theory

Dedekind and Kronecker began with the field $\mathbb{Q}$ of rational numbers and then extended $\mathbb{Q}$ to larger fields that include irrational numbers, such as $\sqrt{2}$. We find, for example, that the smallest field containing $\mathbb{Q}$ and $\sqrt{2}$ is

$$\mathbb{Q}(\sqrt{2}) = \{a + b\sqrt{2} : a, b \in \mathbb{Q}\}.$$

It is obvious that a field containing $\mathbb{Q}$ and $\sqrt{2}$ must include all numbers of the form $a + b\sqrt{2}$ for $a, b \in \mathbb{Q}$. Conversely, it can be checked that the sum, difference, product, and quotient of numbers of this form are again of this form; hence $\mathbb{Q}(\sqrt{2})$ is a field. It is also pretty obvious that the numbers 1 and $\sqrt{2}$ are a basis for $\mathbb{Q}(\sqrt{2})$ over $\mathbb{Q}$, so this field is a *vector space over* $\mathbb{Q}$, of dimension 2 (over $\mathbb{Q}$).

The dimension 2 is also the **degree** of the number $\sqrt{2}$ over $\mathbb{Q}$, that is, the degree of the polynomial equation with rational coefficients satisfied by $\sqrt{2}$: $x^2 - 2 = 0$. Since $\sqrt{2}$ is irrational, 2 is the *minimal* degree of a polynomial equation, with rational coefficients, satisfied by $\sqrt{2}$. More generally, if a number α satisfies a polynomial equation with rational coefficients, there is such an equation of minimal degree, and the latter degree equals the dimension of the smallest field containing $\mathbb{Q}$ and α. This agreement between degree and dimension accounts for the applicability of vector spaces in the study of algebraic numbers.

In fact, the dimension concept makes short work of some seemingly hard problems. For example, is $\sqrt[3]{2}$ expressible in terms of rational numbers and square roots? This is a modern formulation of the ancient Greek problem called *duplicating the cube*, which asked for a straightedge-and-compass construction of (the side of) a cube with volume 2. The dimension concept cuts this problem down to size because of the following theorem of Dedekind (1894):

Dedekind dimension theorem. *If $\mathbb{E} \supseteq \mathbb{F} \supseteq \mathbb{G}$ are fields, with $\mathbb{E}$ of dimension e over $\mathbb{F}$ and $\mathbb{F}$ of dimension f over $\mathbb{G}$, then ef is the dimension of $\mathbb{E}$ over $\mathbb{G}$.*

The proof is merely a matter of checking that if elements e_i form a basis of $\mathbb{E}$ over $\mathbb{F}$, and if elements f_j form a basis of $\mathbb{F}$ over $\mathbb{G}$, then the elements $e_i f_j$ form a basis for $\mathbb{E}$ over $\mathbb{G}$.

We apply this dimension theorem to the problem of duplicating the cube by supposing that α is a number formed from rational numbers by square roots. This means that α lies in a field

$$\mathbb{F}_n \supseteq \mathbb{F}_{n-1} \supseteq \cdots \supseteq \mathbb{F}_1 \supseteq \mathbb{Q},$$

where each field contains the square root of an element from the one before. For example, if $\alpha = \sqrt{5 + \sqrt{3}}$, then we have

$$3 \in \mathbb{Q},$$
$$\text{then } \sqrt{3} \in \mathbb{Q}(\sqrt{3}) = \mathbb{F}_1,$$
$$\text{so } 5 + \sqrt{3} \in \mathbb{F}_1,$$
$$\text{then } \sqrt{5 + \sqrt{3}} \in \mathbb{F}_1(\sqrt{5 + \sqrt{3}}) = \mathbb{F}_2.$$

Since each extension is by a square root, it is of degree 2, and hence of dimension 2, over the field before. Therefore, by the Dedekind dimension theorem, the dimension of $\mathbb{F}_n$ over $\mathbb{Q}$ is 2^n. So the *degree of α is 2^n.* But the degree of $\sqrt[3]{2}$ is 3, as one would expect and can quite easily prove, so $\sqrt[3]{2}$ *cannot be obtained from rational numbers by square roots.*

4.9 REMARKS

Before we pass on to the most famous and extensive application of algebra to geometry—the *algebraic geometry* of Fermat and Descartes—it is worth mentioning two cameo appearances of algebra in geometry.

Finite Permutation Groups in Geometry

Section 4.4 talked about symmetric functions and the permutation groups that represent their symmetry, but it did not mention the most ancient form of symmetry: the symmetry of geometric objects, particularly the **regular polyhedra**. Figure 4.4 shows all five of them: the tetrahedron, cube, octahedron, dodecahedron, and icosahedron. These figures have been known since prehistoric times, and they are the climax of Euclid's *Elements*, where it is shown that they exist and are the only regular polyhedra in three-dimensional space. Each of them is "regular" in the sense that its faces are identical regular polygons and its vertices are the meeting points of identical numbers of faces. Also, each is **symmetric** in the sense that it can be mapped onto itself by rotations sending any vertex, edge, or face to any other.

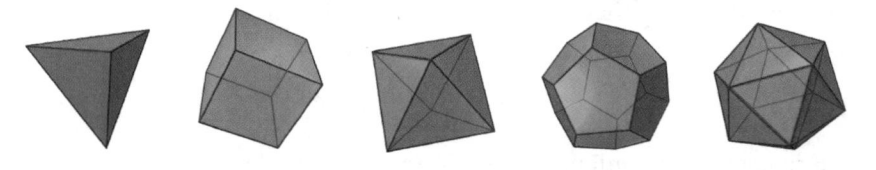

Figure 4.4 : Tetrahedron, cube, octahedron, dodecahedron, icosahedron

For example, if we fix a position of the tetrahedron, then any one of the four faces can be rotated to the position of the "front" face; moreover, any of the three edges of that face may be rotated to the position of the "top" edge. This gives $4 \times 3 = 12$ rotations that leave the tetrahedron looking exactly the same. These twelve rotations are the **symmetry group** of the tetrahedron, because the product of any two of them is another, the inverse of any one of them is another, and between them they give all positions of the tetrahedron that "look the same." There are twelve positions that look the same because the position of the tetrahedron is completely determined by which of the four faces, F, is in front, and which of the three edges of F is on top.

By similar arguments, multiplying the number of faces by the number of edges in a face, we see that:

- The symmetry group of the cube has $6 \times 4 = 24$ elements.
- The symmetry group of the octahedron has $8 \times 3 = 24$ elements.
- The symmetry group of the dodecahedron has $12 \times 5 = 60$ elements.
- The symmetry group of the icosahedron has $20 \times 3 = 60$ elements.

Now 24 is the size of S_4, because there are $4 \times 3 \times 2 \times 1$ permutations of four things, and indeed, the symmetry group of the cube is **isomorphic** to S_4. That is, there is a bijection between the permutations and rotations under which products also correspond. This is not a coincidence; in fact there are four parts of the cube that are permuted by rotations, and the twenty-four rotations give twenty-four distinct permutations. The four parts of the cube are its *diagonals*, shown in four different colors in figure 4.5.

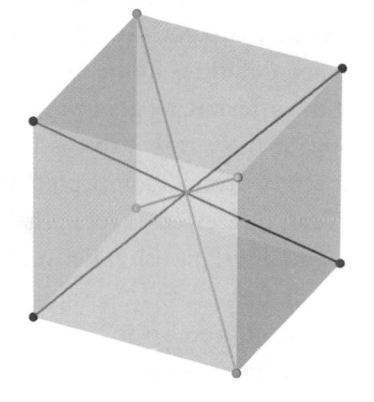

Figure 4.5 : The cube and its diagonals

The number 12, occurring as the size of the symmetry group of the tetrahedron, is of course half of 24, and indeed, rotations of the tetrahedron correspond to half the rotations of the cube. This is because the tetrahedron can be placed in a cube as shown in figure 4.6, and then all rotations of the tetrahedron result from rotations of the cube. But only *half* the rotations of the cube rotate the tetrahedron to a position that

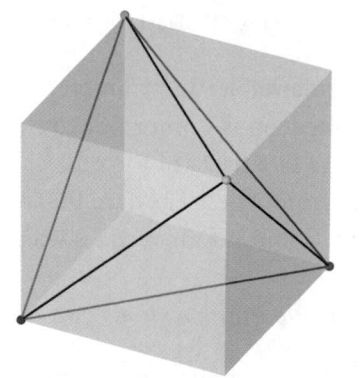

Figure 4.6 : Tetrahedron in a cube

looks the same—the other half move the vertices of the tetrahedron to the other four vertices of the cube.

These rotations correspond to what are called the **even permutations** of the four vertices. A permutation of n objects, call them $1, 2, \ldots, n$, is called **even** if it puts an even number of pairs out of order; otherwise it is called **odd**. For example, the permutation 2134 of 1234 is odd, because only one pair $\langle 2, 1 \rangle$ is out of order. On the other hand, the permutation 2143 is even because two pairs $\langle 2, 1 \rangle$ and $\langle 4, 3 \rangle$ are out of order. It happens that half the permutations in the group S_n are even, and they form a subgroup called the **alternating group** A_n.

The symmetry group of the tetrahedron is isomorphic to A_4. The symmetry group of the octahedron is isomorphic to S_4, because all rotations of the octahedron can be viewed as rotations of the cube, and vice versa. This is due to the "dual" relationship between the octahedron and cube shown on the left in figure 4.7. On the right we see that the dodecahedron and the icosahedron are also in a dual relationship, so their symmetry groups are isomorphic as well.

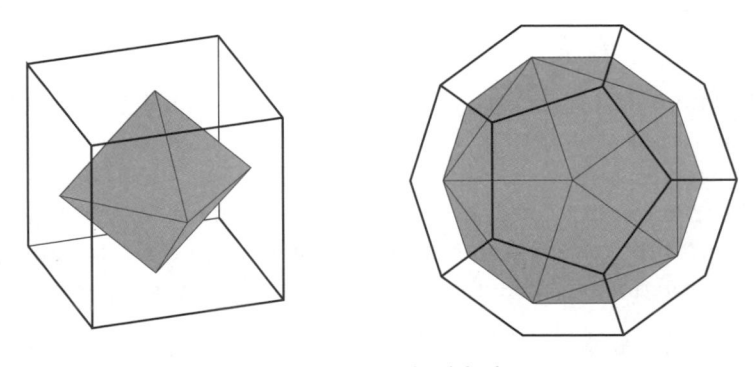

Figure 4.7 : Dual polyhedra

Finally, and most remarkable, the symmetry group of both the dodecahedron and the icosahedron is isomorphic to A_5.

We can even see the five parts of the dodecahedron whose permutations represent the symmetries. They are the five tetrahedra shown in figure 4.8, each of which shares four vertices with the dodecahedron.

Projective Plane Axioms and Algebra

Section 3.8 mentioned the Pappus and Desargues theorems of projective geometry and how Hilbert used them to define sum and product of

Figure 4.8 : Five tetrahedra in the dodecahedron

points on a line satisfying the field properties:

$$A + B = B + A \qquad A \cdot B = B \cdot A$$
$$A + (B + C) = (A + B) + C \qquad A \cdot (B \cdot C) = (A \cdot B) \cdot C$$
$$A + 0 = A \qquad A \cdot 1 = A$$
$$A + (-A) = 0 \qquad A \cdot A^{-1} = 1 \text{ when } A \neq 0$$
$$A \cdot (B + C) = A \cdot B + A \cdot C.$$

More recently, as mentioned in section 4.6, the discovery in 1843 of quaternions by Hamilton and octonions by Graves threw the spotlight on the commutative and associative properties of multiplication—

$$A \cdot B = B \cdot A \qquad \text{(commutativity)}$$
$$A \cdot (B \cdot C) = (A \cdot B) \cdot C \qquad \text{(associativity)}$$

—because commutativity fails for the quaternions and both commutativity and associativity fail for the octonions.

But, despite being "failed fields," the quaternions $\mathbb{H}$ and the octonions $\mathbb{O}$ can serve as coordinates for projective planes, and their algebraic "failings" correspond precisely to geometric "failings" of the corresponding planes. It is relatively straightforward to construct a **quaternion projective plane** $\mathbb{HP}^2$ by analogy with our construction of the real projective plane $\mathbb{RP}^2$ in section 3.8. Its "points" are quaternion lines (given by parametric equations) through the origin in the space $\mathbb{H}^3$ or ordered triples of quaternions, and its "lines" are quaternion planes through the origin (given by homogeneous linear equations). This was pointed out by Ruth Moufang, who also achieved the more difficult construction of the **octonion projective plane** $\mathbb{OP}^2$ in Moufang (1933). One of the obstacles to the construction of an octonion projective plane is that *octonion projective space does not exist*, whereas there are quaternion projective spaces $\mathbb{HP}^n$ of all dimensions n.

What are the geometric "failings" of the quaternion and octonion projective planes? Here is a summary, together with the positive features of these planes and of other planes satisfying the same geometric axioms:

1. $\mathbb{HP}^2$ satisfies the Desargues theorem but not the Pappus theorem. (If it satisfied Pappus theorem then the quaternions would have commutative multiplication, which they do not.)
2. Conversely, Hilbert (1899) proved that any projective plane in which the Desargues theorem holds has a sum and product with all field properties except commutative multiplication.
3. Any projective *space* satisfies the Desargues theorem, because of the "spatial" proof of Desargues suggested by figure 4.9, in which the line $\mathscr{L}$ arises as the line of intersection of the planes of the two triangles.

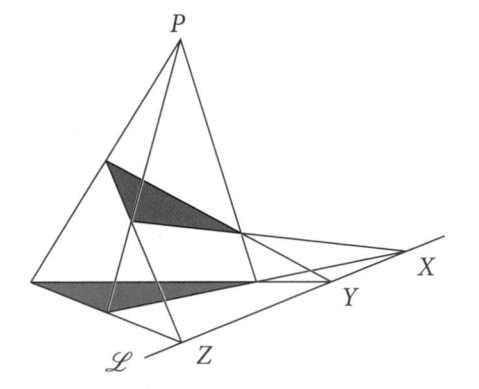

Figure 4.9 : Spatial view of Desargues's theorem

4. If there were an octonion projective space the octonions would have associative multiplication, which they do not.
5. $\mathbb{OP}^2$ does not satisfy the Desargues theorem; otherwise the octonions would have associative multiplication, which they do not.

6. However, $\mathbb{OP}^2$ satisfies the **little Desargues theorem**: the special case of the Desargues theorem where the two triangles have pairs of corresponding sides that meet on a line $\mathscr{L}$ through the center P of perspective (figure 4.10). In this case the conclusion of the theorem is that the third pair of corresponding sides also meet on $\mathscr{L}$.

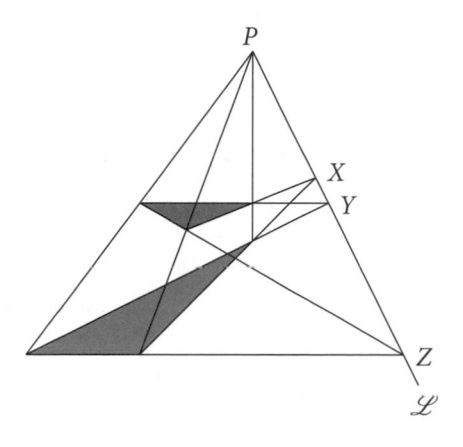

Figure 4.10 : The little Desargues theorem

7. Conversely, Moufang (1933) showed that any projective plane in which the little Desargues theorem holds has a sum and product with all field properties except commutative and associative multiplication.

There are names for the structures satisfying the field axioms minus the commutative and associative properties, though rather ill-chosen ones. A structure such as $\mathbb{H}$, satisfying all but commutative multiplication, is called a **skew field**. A structure such as $\mathbb{O}$, satisfying all but commutative and associative multiplication, is called an **alternative field**.

CHAPTER 5

■ ■ ■ ■ ■

Algebraic Geometry

The development of symbolic algebra in the late sixteenth century led to a turning point in the history of proof. This is when algebra became more efficient than geometric reasoning, so algebra and geometry could exchange roles: instead of using geometry to prove results in algebra, one could use algebra to prove results in geometry. Thus, for the first time since Euclid, the concept of proof was significantly expanded.

In effect, instead of taking concepts such as "point," "line," "length," and "angle" as primitive terms, they could be defined *in terms of* numbers and equations. Then, instead of appealing to axioms about points, lines, and so on, one could simply calculate with equations. Today, this seems clearly more efficient, but the authority of Euclid was such that algebra was applied only cautiously at first. The early practitioners of the algebraic approach to geometry, such as Descartes, continued to view Euclid as "real geometry" and algebra only as a shortcut to results about Euclid's geometric objects. Wallis was the first to see it as a genuine alternative ("algebraic geometry"). We would now say that coordinates provided a **model** of Euclid's geometry, though this became explicit only with Hilbert (1899).

In a way, caution was justified, since the underlying concept of "number" remained undefined, so the algebraic geometers had to fall back on Euclid's concepts of point and line to support it. But in practice, algebra quickly outpaced geometric reasoning, since it could be applied to **algebraic curves** that ancient Greek geometry could barely touch.

Later, linear algebra provided an approach to geometry that was less than the full algebraic geometry but better matched to Euclid. In linear algebra the **inner product** plays the same role as the Pythagorean theorem.

5.1 CONIC SECTIONS

The conic sections, as their name suggests, are curves obtained by cutting a cone by a plane. Those shown, left to right, in figure 5.1 are called the **ellipse**, **parabola**, and **hyperbola**, respectively. Strictly speaking, these are *nondegenerate* sections of the cone; there are also degenerate sections, such as a single point or a pair of lines, that result from exceptional positions of the cutting plane.

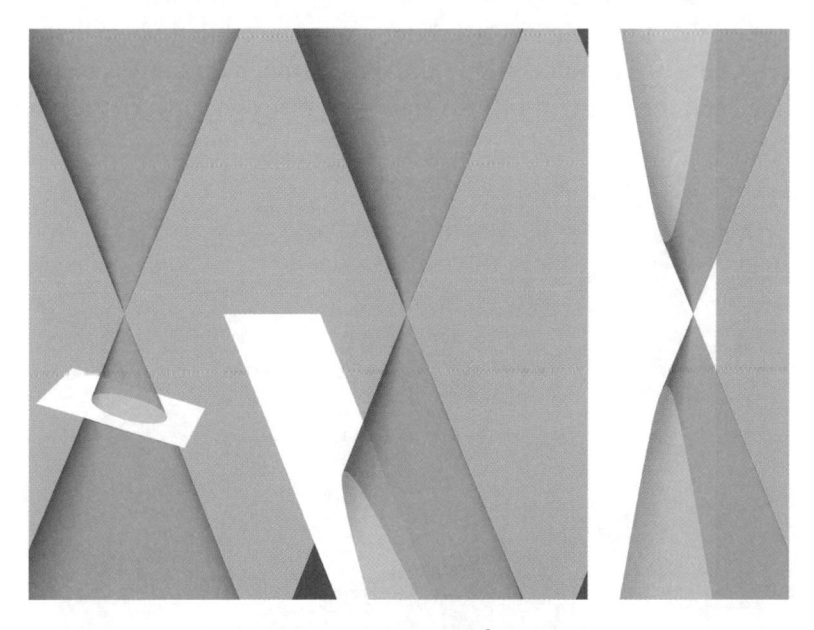

Figure 5.1 : Sections of a cone

Most of the interest is in the ellipse, parabola, and hyperbola, and many of their properties have been known since ancient times. Hundreds of theorems about them may be found in the *Conics* of Apollonius of Perga, a Greek mathematician from the period between Euclid and Archimedes. The proofs in the *Conics* are geometric, in the style of Euclid. They often describe relationships between lengths or areas that we would write as equations but that Apollonius always states, and reasons about, in words.

For example, after cutting the cone by a plane parallel to one side, as in the middle image in figure 5.1, Apollonius deduces in his proposition 11 what we would call the equation $y = x^2$ of the parabola. But his x and y are lines, laboriously described by their position in the cone, and he states the equality of the "square" on x with a "rectangle" on y and a unit length.

Before claiming that algebra makes it much easier to prove properties of conic sections (which is generally the case), I must admit that proofs involving the cone can be surprisingly elegant. One such was discovered by the French mathematician Germinal Pierre Dandelin in 1822. It shows that there are two points inside any ellipse, called its **foci**, such that *the sum of the distances from any point on the ellipse to the foci is constant.*

The proof is almost obvious from figure 5.2, which shows two spheres inside the cone just touching it and the cutting plane that creates the ellipse. The points where the spheres touch the cutting plane are the foci, and the constant is the distance on the cone between the circles

Figure 5.2 : Dandelin spheres for the ellipse

(drawn in black) where the spheres touch it. To see why this constant equals the sum of the distances from any point on the ellipse to the foci, pick a point on the ellipse, notice that its lines to the foci are tangents to the spheres, and bear in mind that tangents to a sphere from any external point have equal length.

5.2 FERMAT AND DESCARTES

Pierre de Fermat and René Descartes independently discovered the coordinate approach to geometry in the 1620s. Because it was first published by Descartes (1637), he has received most of the credit for it; in particular, the coordinate system is called "Cartesian" after him. Section 3.5 has already presented the modern version of the coordinate system (presented again in figure 5.3), as a model of the Euclidean plane in terms of the real numbers.

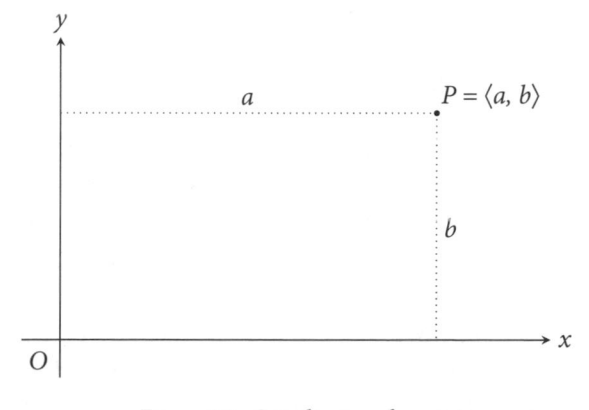

Figure 5.3 : Coordinates of a point

Descartes had a similar idea but was closer to Euclid. He did not admit negative coordinates, since he viewed coordinates as lengths, and he did not have an independent concept of real number. In fact, Descartes thought that "real" quantities were geometric and that coordinates were merely a convenient way to approach them. What made coordinates convenient, of course, were the great advances in algebra over the preceding decades, which made it possible to describe many curves by *polynomial equations* and to discover their properties by algebraic manipulation.

Both Fermat and Descartes realized that polynomial equations $p(x, y) = 0$ give a classification of curves by degree. Equations of degree 1, $ax + by + c = 0$, represent straight lines, which is why we now call such equations "linear." More interesting, they both discovered that equations of degree 2 represent conic sections. As every mathematics student knows, parabolas, ellipses, and hyperbolas are represented by particular types of quadratic equation:

$$y = x^2 \qquad \text{(parabola)}$$

$$\frac{x^2}{a^2} + \frac{y^2}{b^2} = 1 \qquad \text{(ellipse)}$$

$$\frac{x^2}{a^2} - \frac{y^2}{b^2} = 1. \qquad \text{(hyperbola)}$$

The most general quadratic equation,

$$ax^2 + bxy + cy^2 + dx + ey + f = 0,$$

can be converted into one of the above forms by changing variables, unless it is a "degenerate" form, such as

$x^2 + y^2 = 0$, which represents the single point $\langle 0, 0 \rangle$;
$x^2 - y^2 = 0$, which represents the line pair $y = \pm x$; or
$(x - y)^2 = 0$, which represents the line $y = x$, "counted twice."

The degenerate forms in fact correspond to "degenerate" conic sections, where the cutting plane is in an exceptional position. In the non-degenerate cases the proof can be viewed geometrically as translation and rotation of axes: the terms dx and ey can be removed by translation, and the term bxy can be removed by rotation.

A simple case of removing the xy term is the hyperbola $xy = 1$, which becomes $X^2 - Y^2 = 2$ when the x- and y-axes are rotated by $45°$ (figure 5.4). This is because the new coordinates are $X = \frac{x}{\sqrt{2}} + \frac{y}{\sqrt{2}}$, $Y = \frac{x}{\sqrt{2}} - \frac{y}{\sqrt{2}}$.

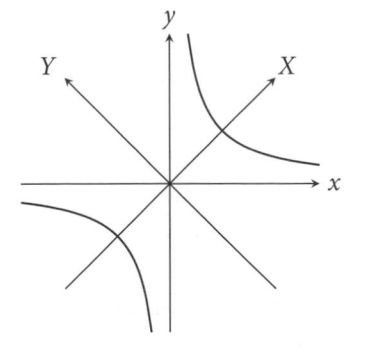

Figure 5.4 : Rotating axes for the hyperbola

5.3 ALGEBRAIC CURVES

Restricting coordinate geometry to curves with polynomial equations was natural when the known methods for dealing with equations were algebraic. This situation was to change in a few decades with the arrival of calculus, but Descartes doubted that nonalgebraic, or **transcendental**, curves—which he called "mechanical" as opposed to "geometric"—could ever be well understood. His doubts stemmed from the problem of *arc length*, which seemed out of reach even for algebraic curves:

> Geometry should not include lines that are like strings, in that they are sometimes straight and sometimes curved, since the ratios between straight and curved lines are not known, and I believe cannot be discovered by human minds.
>
> (Descartes 1637: 91)

Ironically, the first curve to have its arc length determined was a "mechanical" curve, the so-called **equiangular spiral** (figure 5.5), whose arc length was found by Thomas Harriot around 1590 (see Lohne 1979). His result was unpublished but rediscovered by Torricelli in 1645. The spiral is called "equiangular" because it meets lines from its center at a constant angle. Harriot's discovery is described in section 6.1.

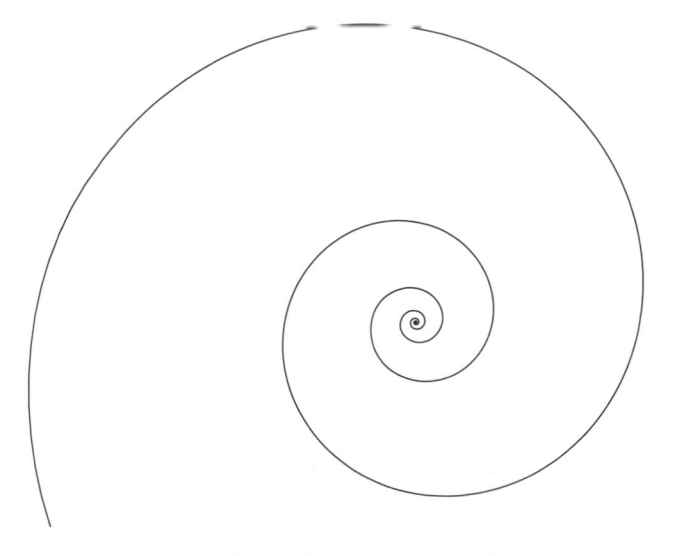

Figure 5.5 : The equiangular spiral, a nonalgebraic curve

It is worth explaining here why a spiral is not an algebraic curve. This was done by Newton (1687: bk. 1, sec. 6, lemma 28) because a spiral meets certain lines infinitely often. On the other hand, an algebraic curve $p(x, y) = 0$ of degree n meets the line $y = mx + c$ only *finitely* often: at the points where $p(x, mx + c) = 0$. This is a polynomial equation of degree n, so it has at most n solutions, as we saw in section 4.5.

Tangents

Descartes was shortsighted to dismiss nonalgebraic curves because their arc length was hard to determine. But there is another problem, normally solved by calculus today, that he and Fermat were able to solve for algebraic curves by pure algebra: the problem of finding tangents.

A simple example is given by the parabola $y = x^2$ and the line $y = 2x - 1$. These intersect where x satisfies the equation $x^2 = 2x - 1$, or $(x - 1)^2 = 0$. This equation has a "double solution" $x = 1$, because of the squared factor $(x - 1)$. This is an *algebraic* indication of tangency. Intuitively speaking, a line $\mathscr{L}$ through the point $P = \langle 1, 1 \rangle$ on the parabola and "near" the tangent $y = 2x - 1$ meets the parabola in *two* points P, P', which "come together" as $\mathscr{L}$ approaches the tangent $\mathscr{T}$ (see figure 5.6).

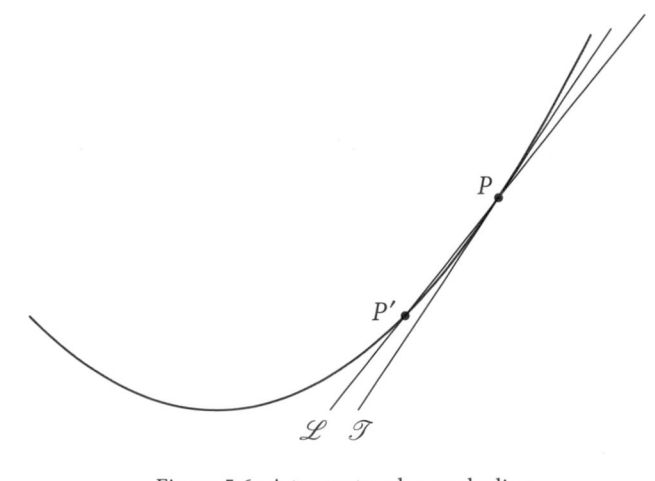

Figure 5.6 : A tangent and a nearby line

In general, the line $y = mx + c$ is a tangent to the curve $p(x, y) = 0$ at $x = a$ if the factor $(x - a)$ occurs at least twice in the polynomial $p(x, mx + c)$.

A more interesting curve, which Descartes actually posed to Fermat in an attempt to stump him, is $x^3 + y^3 = 3xy$. It is called the **folium of Descartes**[1] and is shown in figure 5.7.

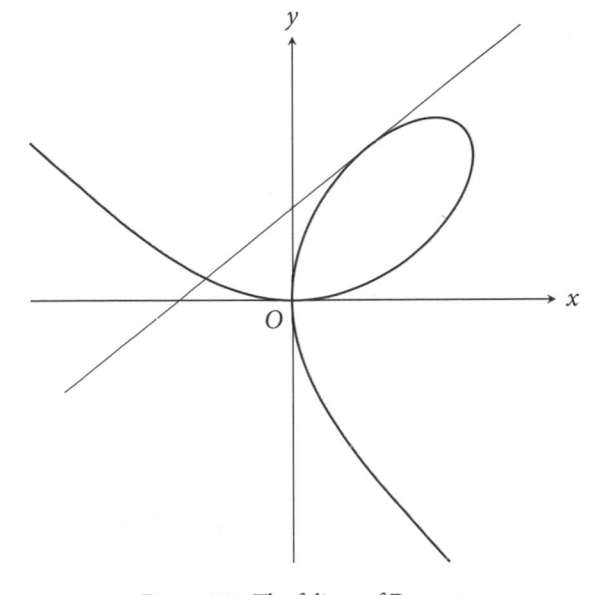

Figure 5.7 : The folium of Descartes

Consider the blue line $y = \frac{4}{5}(x+1)$, which passes through the point $\left\langle \frac{2}{3}, \frac{4}{3} \right\rangle$ on the folium. It is a tangent because substituting for y gives the equation

$$x^3 + \left(\frac{4}{5}\right)^3 (x+1)^3 = 3x \cdot \frac{4}{5}(x+1),$$

which simplifies to

$$189x^3 - 108x^3 - 108x + 64 = 0, \quad \text{or} \quad (3x-2)^2(21x+16) = 0.$$

The repeated root $x = 2/3$ shows that the line has double contact with the folium at $x = 2/3$, and hence it is a tangent. Of course, such calculations with polynomials are not what one wants to do any more—calculus is easier!—but they show that finding tangents to an algebraic curve is essentially an algebraic problem.

---

1. "Folium" is Latin for "leaf" and the "leaf," in case you are wondering, is the portion of the curve in the positive quadrant, where $x, y \geq 0$. Descartes noticed only this portion because he did not admit negative coordinates.

Points at Infinity

We saw in section 3.8 that the behavior of straight lines in the plane is simplified by adding a line at infinity to form the **real projective plane**. In this extended plane any two lines have exactly one point in common. The only difference between an ordinary line pair and a pair of parallels is that the latter meet at infinity.

The behavior of algebraic curves is also simplified by including points at infinity. The only difference between the ellipse, parabola, and hyperbola, for example, is in the number of their points at infinity: the ellipse has none, the parabola has one, and the hyperbola has two. This distinction was pointed out in Desargues (1639), the first book on projective geometry. Figure 5.8, for example, shows how the parabola becomes an ellipse when viewed projectively: it touches the line at infinity at a single point.

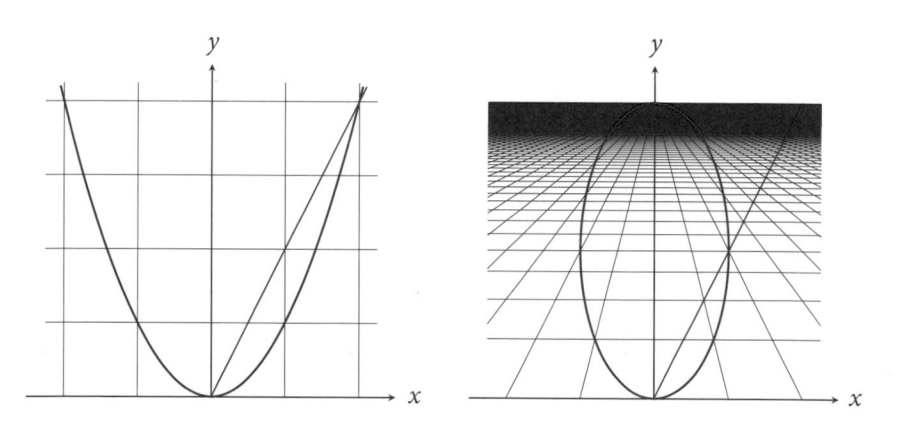

Figure 5.8 : Direct and projective views of the parabola

Indeed, if we use the model of the real projective plane described in section 3.8, whose "points" are lines through the origin in $\mathbb{R}^3$, then a "conic section" literally *is* a cone! In figure 5.9, the eye "sees" the ellipse by looking from the pointy end of the cone. Thus all (nondegenerate) conic sections are the same when viewed projectively.

5.4 CUBIC CURVES

The algebraic view of conic sections, while convenient for calculations, was not considered enlightening by everyone. Thomas Hobbes, whose

Figure 5.9 : Projective model of a conic section

first encounter with Euclid was described in section 2.4, called the algebraic treatment of conics "a scab of symbols" (Hobbes 1656: 316) and
denounced "the whole herd of them that apply their algebra to geometry" (Hobbes 1672: 447). After all, it was hard to find a theorem about
conic sections that had not already been proved by Apollonius.

The real test of algebra came with *cubic* curves: those with the general
equation

$$ax^3 + bx^2y + cxy^2 + dy^3 + ex^2 + fxy + gy^2 + hx + jy + k = 0.$$

One or two special cubic curves were considered by the Greeks, though of
course not in terms of equations, so the Greeks had no general concept
of cubic curve. The first to tackle them was Newton (1667), who succeeded in reducing the general equation to the following special types by
translation and rotation of axes:

$$Exy^2 + Fy = Ax^3 + Bx^2 + Cx + D$$
$$xy = Ax^3 + Bx^2 + Cx + D$$
$$y^2 = Ax^3 + Bx^2 + Cx + D$$
$$y = Ax^3 + Bx^2 + Cx + D.$$

Newton then divided the curves into species according to the nature of the roots of the right-hand side. He came up with 72 species, and missed six. Stirling (1717) supplied more details but still missed two species, so the situation was far less satisfactory than the classification of quadratic curves. However, Newton also made an aside "on the genesis of curves by shadows" that dramatically simplifies the situation. Grouping curves according to the nature of their *projections* reduces the number of types of cubic curves to five! They all have equations of the form

$$y^2 = Ax^3 + Bx^2 + Cx + D. \qquad\qquad (*)$$

The greater diversity of cubic curves can be attributed to **singularities**, which quadratic curves do not have. The cubic singularities are illustrated by the curves $y^2 = x^3$, which has a **cusp** at the origin; $y^2 = x^2(x+1)$, which has a **crossing** at the origin; and $y^2 = x^2(x-1)$, which has an **isolated point** at the origin (see figure 5.10).

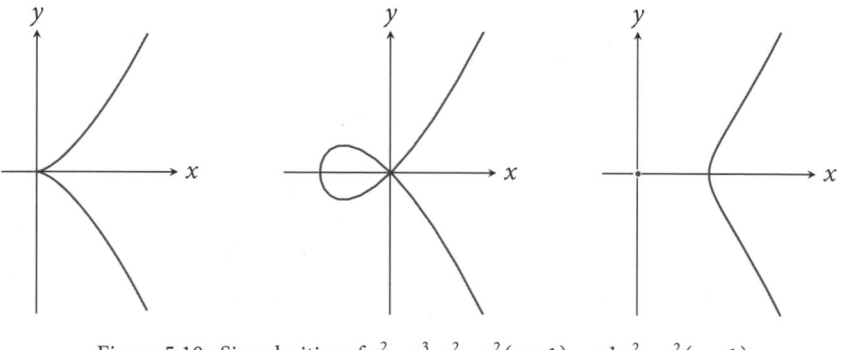

Figure 5.10 : Singularities of $y^2 = x^3$, $y^2 = x^2(x+1)$, and $y^2 = x^2(x-1)$

The nonsingular cubics are of two types, depending on the real roots of $Ax^3 + Bx^2 + Cx + D$ on the right-hand side of (*). If there is a single real root the curve is a single infinite piece. If there are three distinct real roots then the curve consists of two pieces: an infinite piece and an oval. These two possibilities are illustrated by the curves

$$y^2 = (x+1.5)(x^2+1) \qquad\qquad \text{(real root } x = -1.5),$$
$$y^2 = x(x^2-1) \qquad\qquad \text{(real roots } x = -1, 0, 1),$$

shown in blue and red, respectively, in figure 5.11.

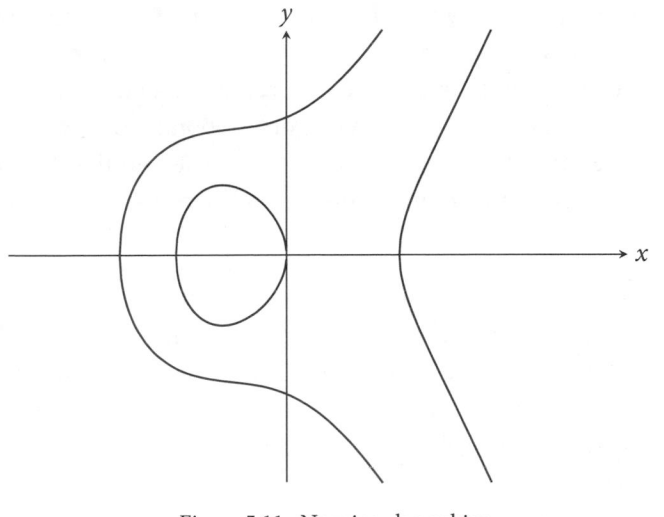

Figure 5.11 : Nonsingular cubics

The two nonsingular types become one if we allow *complex* values of x and y, because in this case $Ax^3 + Bx^2 + Cx + D = 0$ has three distinct roots and it does not matter whether they are real or not. Other reasons for admitting complex coordinates are explained in the next section.

5.5 BÉZOUT'S THEOREM

The story of Bézout's theorem begins with a remark of Newton from 1665:

> For y^e number of points in wch two lines may intersect can never bee greater y^n ye rectangle of ye numbers of their dimensions. And they always intersect in soe many points, excepting those wch are imaginarie onely.
>
> (Newton 1665: 498)

In modern language what Newton said is that, in the plane, a curve of degree m meets a curve of degree n in mn points.[2] At first glance, there

---

2. Notice that Newton calls mn the "rectangle of" m and n—an interesting relic of Euclid's concept of product of lengths. He also calls the degree of a polynomial its "dimension," accidentally anticipating the confluence of "degree" and "dimension" in algebraic number theory, which we saw in section 4.8.

seem to be many counterexamples to this statement, even when we count imaginary points, for example, parallel lines, which have degree 1 but zero intersections, and tangents, which can meet a curve of degree 2 in one point or a curve of degree 3 in one or two points. Despite this, mathematicians felt that the theorem *ought* to be true and that the concepts of "plane," "curve," and "intersection" should be reshaped to *make* it true. This means:

1. Admit complex coordinates, so that a polynomial equation of degree d can have d solutions. Also, the **fundamental theorem of algebra** must be proved, to guarantee that each polynomial equation has a solution in the complex numbers. The fundamental theorem of algebra is a story in itself, which we take up in chapter 8.

2. Count solutions, and hence intersections, with **multiplicity**; that is, count a solution $x = a$ of a polynomial equation $p(x) = 0$ as often as $x - a$ occurs as a factor of the polynomial $p(x)$. We saw in section 5.3 that multiplicity is useful for detecting tangents.

3. Place curves in a *projective* plane, so that "intersections at infinity" can be included.

When these steps are taken, it can be shown that the intersections of the curves are determined by a polynomial equation $p(x) = 0$ of degree mn. The polynomial $p(x)$ splits into mn factors by the fundamental theorem of algebra, and the mn factors give mn intersections, when counted according to multiplicity.

Although foreseen by Newton (1665) and discussed by Bézout (1779) (among many others), the theorem was not completely proved until the nineteenth century. By that time, it was of interest mainly for its contribution to the reshaping of algebraic geometry.

The demand for complex coordinates, and for points at infinity, forced mathematicians to see curves differently: as *surfaces*. The simplest example is the **complex projective line**, $\mathbb{CP}^1$, the result of extending the plane $\mathbb{C}$ of complex numbers by a point at infinity. $\mathbb{CP}^1$ is a *sphere*, for the simple reason that $\mathbb{C}$ corresponds to a sphere minus one point, and the missing point corresponds very naturally to a "point at infinity" for $\mathbb{C}$ (see figure 5.12).

General algebraic curves are also surfaces, with some interesting complications that we will study in chapter 10.

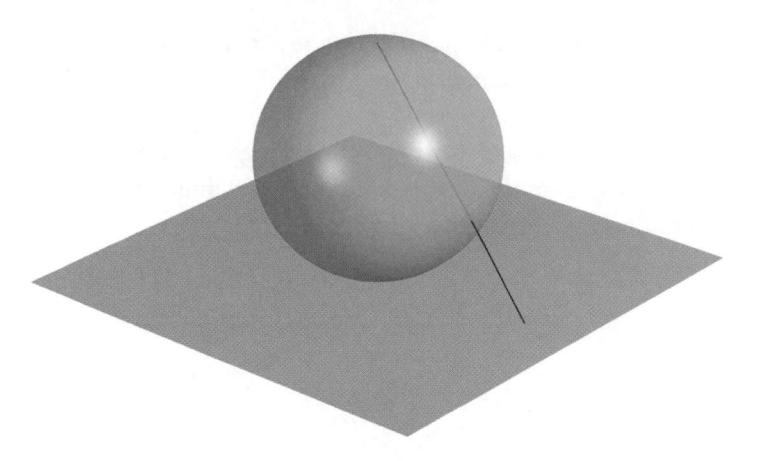

Figure 5.12 : The complex projective line

5.6 LINEAR ALGEBRA AND GEOMETRY

The linear part of algebraic geometry is not generally reckoned to be part of algebraic geometry, since linear algebra has a different flavor from the algebra of general polynomials. However, the linear part is a better match for the geometry in Euclid, so it has been studied independently.

It is based on the theory of real vector spaces, originating in Grassmann (1844), discussed in section 4.8. In fact, some important theorems of geometry depend *only* on the real vector space structure of the plane $\mathbb{R}^2$. The vector difference gives a notion of *relative position* or *direction*, and the real multiples give a notion of *relative distance* in a given direction.

Definitions. If $u, v \in \mathbb{R}^2$ then $u - v$ is the position of u *relative to* v, or the *direction from* v *to* u. Two nonzero vectors s and t have the same direction, or are *parallel*, if $t = as$ for some real number a. In the latter case we also say that t has a times the *length* of s.

With these definitions we can now give a quick proof of a classical theorem, which is proposition 2 of book 6 in the *Elements* and is believed to be much older than that. It is attributed to the semilegendary Thales, who lived about 600 BCE, some 300 years before Euclid. The vector space proof relies on the fact that if vectors do *not* have the same direction, then they are linearly independent.

Theorem of Thales. *A line parallel to one side of a triangle divides the other two sides proportionately.*

Proof. Choose **0** to be one vertex of the triangle, and let the other two be **t** and **v**. Draw a parallel to the side opposite **0**, meeting the other two sides at **s** = a**t** and **u** = b**v** (figure 5.13).

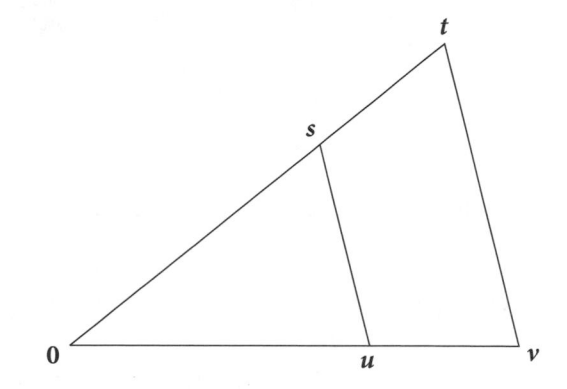

Figure 5.13 : The theorem of Thales

Since **s** − **u** is parallel to **t** − **v** we have

$$\mathbf{t}-\mathbf{v}=c(\mathbf{s}-\mathbf{u}) \qquad\qquad \text{(for some } c\text{)}$$
$$=c(a\mathbf{t}-b\mathbf{v}) \qquad\qquad \text{(by assumption)}$$
$$=ca\mathbf{t}-cb\mathbf{v} \qquad\qquad \text{(by axioms for scalar multiples).}$$

Then, since **t** and **v** are linearly independent we can equate their coefficients, obtaining

$$1=ca=cb, \quad \text{hence} \quad a=b,$$

so the two sides of the triangle are divided proportionally by **s** and **u**. □

Euclidean Spaces

Despite having a concept of parallels, vector space geometry alone does not model the geometry of Euclid because it does not have a concept of length that is independent of direction. Nor does it have a concept of angle. To give a real vector space these attributes, Grassmann introduced the **inner product** **u** · **v** of vectors **u** and **v**. To state the definition of inner product we can assume, as we saw in section 4.8, that the space is $\mathbb{R}^m$.

Definition. If $u = \langle u_1, u_2, \ldots, u_m \rangle$ and $v = \langle v_1, v_2, \ldots, v_m \rangle$ then

$$u \cdot v = u_1 v_1 + u_2 v_2 + \cdots + u_m v_m.$$

It follows from this definition that

$$u \cdot u = u_1^2 + u_2^2 + \cdots + u_m^2 = |u|^2,$$

where $|u|$ is the length of u, by the Pythagorean theorem. Grassmann (1847) emphasized that the inner product plays the role of the Pythagorean theorem in his approach to geometry.

The concept of angle is implicit in the concept of length, since an angle is determined by the sides of a triangle that contains it, and in fact there is an elegant formula,

$$u \cdot v = |u||v| \cos \theta,$$

where θ is the angle between u and v. Among other things, this formula implies the important **orthogonality condition**: *the vectors u and v are orthogonal, or perpendicular, if and only if $u \cdot v = 0$.*

As an example that applies the inner product, we prove a theorem not found in ancient Greek mathematics. Its first known proof is due to the tenth-century Persian mathematician Abū Sahl Wayjan ibn Rustam al-Qūhī. The inner product proof makes highly efficient use of the inner product criterion for orthogonality.

Concurrence of altitudes. *In any triangle, the lines through each vertex orthogonal to the opposite side (its "altitudes") meet at a common point.*

Proof. Let u, v, and w be the vertices of the triangle, and choose $\mathbf{0}$ to lie at the intersection of the lines through u and v that are orthogonal to the opposite sides (figure 5.14). Since the side opposite to u has direction

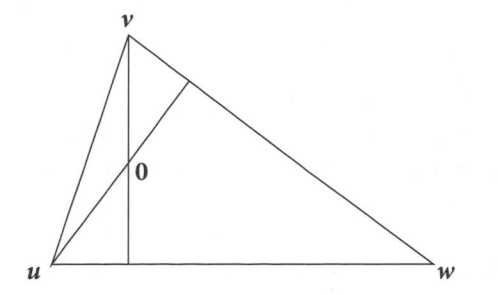

Figure 5.14 : Altitudes of a triangle

$v - w$, and the line through u and $\mathbf{0}$ has direction u, the orthogonality condition gives

$$u \cdot (v - w) = 0, \quad \text{or} \quad u \cdot v - u \cdot w = 0. \qquad (*)$$

Similarly, the other pair of orthogonal lines gives

$$v \cdot (w - u) = 0, \quad \text{or} \quad v \cdot w - v \cdot u = 0. \qquad (**)$$

Adding the equations (*) and (**), and bearing in mind that $u \cdot v = v \cdot u$, and the like, we get

$$w \cdot (u - v) = 0,$$

which says that the line through w and $\mathbf{0}$ is orthogonal to the side through u and v. In other words, the point $\mathbf{0}$ is on all three altitudes. □

5.7 REMARKS

The decision to place algebraic curves in complex projective space so as to obtain Bézout's theorem is an example of a common phenomenon in mathematics: choosing context and definitions so as to simplify the statement of theorems. Other examples we have seen, or will see later, are:

1. Including $\sqrt{2}$ among the real numbers so as to arithmetize geometry
2. Not including the number 1 among the primes so as to ensure unique prime factorization of natural numbers
3. Introducing "ideal numbers" so as to extend unique prime factorization to algebraic number fields
4. Making the real numbers complete to ensure that continuous functions behave as expected and that polynomial equations have solutions
5. Assuming there are no infinitesimals to make the theory of real numbers as simple as possible
6. Including the axiom of choice in set theory so as to simplify the theory of cardinal numbers, and also parts of algebra and analysis.

Classical algebraic geometry assumed real coordinates and then later complex coordinates to ensure the existence of solutions for polynomial equations. Linear algebra also was based on the real field $\mathbb{R}$ at first. This situation began to change in the second half of the nineteenth century, when vector spaces over other fields were found to be useful in number theory. We will see how this came about in chapter 7. It led in the end to a vast expansion of algebraic geometry, in which ideas from algebra, number theory, geometry, and topology are intermingled.

Even the linear part of geometry, which as noted in the previous section is not generally considered part of algebraic geometry, experienced growth in several directions. In particular, the field $\mathbb{R}$ can be replaced by other fields, such as $\mathbb{C}$, and the inner product can be replaced by other quadratic forms. A famous example is the **Minkowski inner product** on $\mathbb{R}^4$, which comes up again in section 13.2:

$$\langle x_1, y_1, z_1, t_1 \rangle \cdot \langle x_2, y_2, z_2, t_2 \rangle = x_1 x_2 + y_1 y_2 + z_1 z_2 - t_1 t_2.$$

This inner product gives squared "distances" that can be negative, which happens to be just the right thing for Einstein's special theory of relativity. In Einstein's theory, the "time" coordinate t needs to behave differently from the "space" coordinates x, y, and z in order to explain the constancy of the speed of light for all observers.

Given an inner product on a vector space V and the corresponding concept of distance, we can study the linear transformations of V that preserve distance, called **isometries**. This generalization of Euclidean geometry is sometimes called **geometric algebra**, after the book of that name by Artin (1957). A more leisurely treatment of geometric algebra may be found in Snapper and Troyer (1971).

CHAPTER 6

Calculus

Calculus, as its name suggests, is a system of calculation. It calculates quantities in geometry, such as slopes, lengths, and areas, and quantities in mechanics, such as velocity, acceleration, and energy. Its calculations are similar to, and made possible by, the symbolic calculations in algebra that flourished in the early seventeenth century and had already been successful with classical questions in geometry.

In fact, early calculus was very much an extension of algebraic geometry to include nonalgebraic curves, which Descartes called *mechanical curves*. And, as in algebraic geometry, the new methods of calculation took on a life of their own. Algebra powered calculus in three specific ways:

- By providing a setting of coordinates and equations for the study of geometric objects
- By suggesting an extension of algebra to "infinite polynomials" (Newton's calculus of **power series**, inspired, so he said, by calculation with infinite decimals)
- By suggesting an algebra and geometry of *infinitesimals* (Leibniz).

Infinitesimals provoked criticism from the philosophers Hobbes and Berkeley, and indeed, their criticisms were not properly answered until the **limit** concept was formulated in the nineteenth century. But the seventeenth century was a time, perhaps like now, when most mathematicians thought the foundations were sound (then, thanks to the method of exhaustion), so they did not wish to spend time examining them. This set in motion a slow-moving "crisis of foundations" that did not come to a head until the late nineteenth century.

6.1 FROM LEONARDO TO HARRIOT

This section is a prelude to calculus, involving infinite geometric processes and their limits. It is not calculus proper because it involves no general method of calculation, but it shows two ingenious solutions of what are now considered calculus problems. The first, the area and circumference of a disk, is by Leonardo da Vinci, around 1500, and the second, the equiangular spiral mentioned in section 5.3, is by Thomas Harriot from around 1590.

Area and Circumference of the Disk

As everyone now knows, a circular disk of radius r has circumference $2\pi r$ and area πr^2. The relationship between the two was shown by Leonardo da Vinci, using a simple argument that is summed up by figure 6.1. (Da Vinci's sketch of the idea can be seen online at codex-atlanticus.it, page 518 recto.)

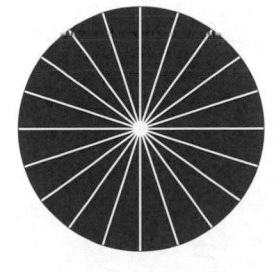

Figure 6.1 : Relating circumference to area

Leonardo imagined slicing the disk into small identical triangles. When these triangles are placed side by side as shown, their height is close to r and their total width is close to the circumference of the disk, $2\pi r$. Admittedly, the "triangles" are not quite true triangles, since their bases are parts of the circle. Nevertheless, these approximations become more nearly true as the triangles become thinner. So, "in the limit," the total area of the triangles, and hence of the disk, becomes

$$\frac{1}{2} \text{ total base } \times \text{ height } = \frac{1}{2} 2\pi r \times r = \pi r^2.$$

The Equiangular Spiral

The equiangular spiral was introduced in section 5.3 as an example of a curve that is not algebraic. Nevertheless, it has a very simple *geometric* description—that the radius vector makes a constant angle with the curve—and this description leads to an equally simple determination of its arc length. This unpublished discovery of Harriot, from around 1590, was uncovered by Lohne (1979). Harriot's idea is illustrated by figure 6.2.

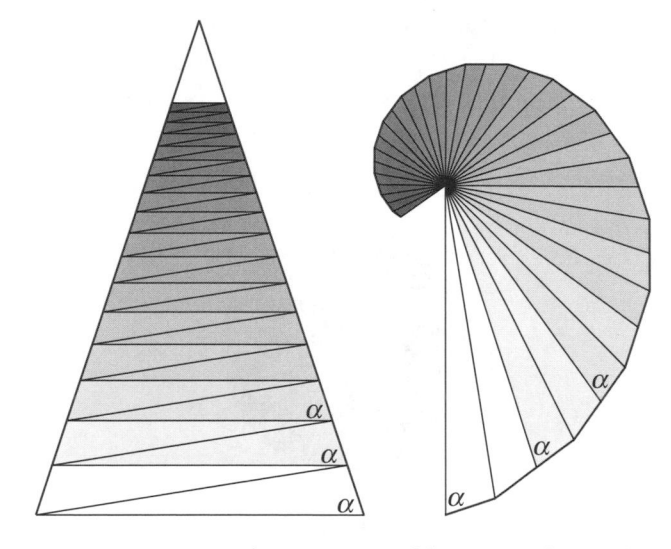

Figure 6.2 : Harriot's construction of the equiangular spiral

Harriot begins with an isosceles triangle, with base angles α, and slices it into similar trapezoids as shown on the left in figure 6.2. Since the trapezoids are similar, we get two sets of similar triangles when each trapezoid is cut in two by its diagonal. The resulting triangles are then reassembled into a polygonal spiral shown on the right side of the figure.

The boundary of the spiral, shown in red, consists of pieces of the two equal sides of the triangle. This red spiral is equiangular in the sense that every other line from the common apex of the triangles meets it at angle α, and its length equals the total length of the two red sides of the triangle. This fact is *independent of the height of the trapezoids*.

Now, if we let the height of the trapezoids approach zero, the polygonal spiral approaches a smooth equiangular spiral curve *whose arc length equals the total length of the red sides of the triangle*.

6.2 INFINITE SUMS

We saw in section 2.7 that Euclid had a clear understanding of both finite and infinite geometric series and that he found the volume of the tetrahedron as the sum of an infinite geometric series. It was not until the middle ages that a significantly new infinite series was studied—the harmonic series—and not until the seventeenth century that infinite series (and other infinite processes) really began to proliferate.

In this chapter we are mainly concerned with seventeenth century methods of proof, which are not rigorous by later standards but can be *made* rigorous, just as Euclid's axioms for geometry were made rigorous by Hilbert. But making seventeenth-century calculus rigorous was not simply a matter of new axioms—it was about new *concepts*, which are best left until later chapters. In the meantime, we should appreciate how amazingly fruitful the seventeenth-century methods were. Those mathematicians knew they were doing something right, even if they could not explain exactly what it was!

First, let us examine some discoveries made before the seventeenth century that were ahead of their time, as will be clear when they are seen against the background of calculus: the harmonic series and the power series for circular functions.

The Harmonic Series

The series

$$1 + \frac{1}{2} + \frac{1}{3} + \frac{1}{4} + \frac{1}{5} + \cdots$$

is called the **harmonic series** because of its relation to the *modes of vibration* of a stretched string (figure 6.3). Modes of vibration are in turn related to the tones emitted by the string as it vibrates, and the first

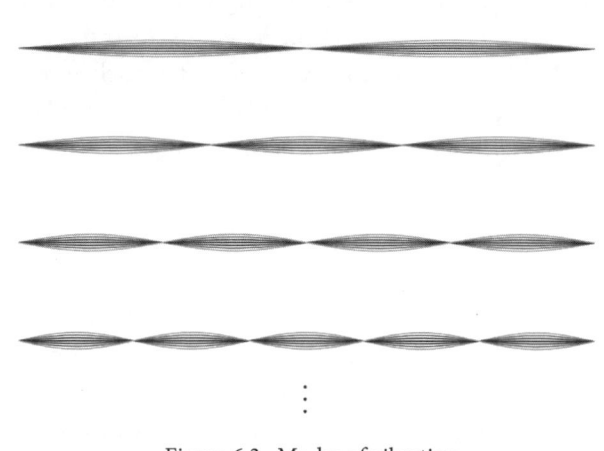

Figure 6.3 : Modes of vibration

few—when the string vibrates in halves, thirds, quarters, and fifths—harmonize with the fundamental tone.

Like the series $1 + \frac{1}{2} + \frac{1}{4} + \frac{1}{8} + \cdots$, the terms of the harmonic series become arbitrarily small, so one might expect that the infinite sum $1 + \frac{1}{2} + \frac{1}{3} + \frac{1}{4} + \frac{1}{5} + \cdots$ exists. Indeed, calculation shows that the sum grows extremely slowly; it takes 12,367 terms for the sum to exceed 10. Nevertheless, the sum grows beyond all bounds!

Oresme (1350) proved this by an incisive elementary argument that we still use today: collect the terms of the series in groups, each of which has sum greater than 1/2. We can let the first group be

$$1 + \frac{1}{2} > \frac{1}{2}.$$

Then take

$$\frac{1}{3} + \frac{1}{4} > \frac{1}{4} + \frac{1}{4} = \frac{1}{2},$$

$$\frac{1}{5} + \frac{1}{6} + \frac{1}{7} + \frac{1}{8} > \frac{1}{8} + \frac{1}{8} + \frac{1}{8} + \frac{1}{8} = \frac{1}{2},$$

$$\frac{1}{9} + \frac{1}{10} + \frac{1}{11} + \frac{1}{12} + \cdots + \frac{1}{16} > \frac{1}{16} + \frac{1}{16} + \frac{1}{16} + \frac{1}{16} + \cdots + \frac{1}{16} = \frac{1}{2},$$

and so on, where each group has twice as many terms as the one before.

Thus the partial sum of the harmonic series can be made to exceed any multiple of 1/2. It grows beyond all bounds and hence *the "sum of all terms" does not exist.* Oresme's proof is impeccably rigorous and a warning that the existence of infinite sums should not be taken for granted. Nevertheless, until the nineteenth century, mathematicians were generally willing to trust their intuition about infinite sums, and they treated them like finite sums.

Power Series for Circular Functions

The most spectacular discoveries before the calculus explosion in the seventeenth century were made in Kerala, southern India, in the fifteenth century. They were the infinite series for the sine, cosine, and inverse tangent functions:

$$\sin x = x - \frac{x^3}{3!} + \frac{x^5}{5!} - \frac{x^7}{7!} + \cdots,$$

$$\cos x = 1 - \frac{x^2}{2!} + \frac{x^4}{4!} + \frac{x^6}{6!} + \cdots,$$

$$\tan^{-1} x = x - \frac{x^3}{3} + \frac{x^5}{5} - \frac{x^7}{7} + \cdots, \quad \text{for } -1 < x \leq 1,$$

where $n! = n(n-1)(n-2)\cdots 3 \cdot 2 \cdot 1$. The formula for $\tan^{-1} x$ has the stunning special case (with $x = 1$)

$$\frac{\pi}{4} = 1 - \frac{1}{3} + \frac{1}{5} - \frac{1}{7} + \cdots,$$

providing an exact and simple expression for π for the first time.

These results did not become known outside India until the seventeenth century, when they were rediscovered by various European mathematicians using calculus. Calculus now seems the most efficient and natural way to prove them—so much so that we are amazed they could be discovered without it. Here we briefly explore the Indian approach to the sine function, due to the mathematician Mādhava who lived from approximately 1350 to 1425 CE. More details may be found in Plofker (2009: 235–246).

To keep the story short we will use modern notation for algebra and take advantage of the familiarity with trigonometric formulas we now have. We will see *why* Mādhava's approach worked, rather than exactly *how* it worked. It worked because of two basic facts that Mādhava knew:

- From basic geometry/trigonometry he knew the **addition formula** for the sine function:

$$\sin(\theta + \varphi) = \sin\theta \cos\varphi + \cos\theta \sin\varphi.$$

From this formula, together with $\cos^2\theta = 1 - \sin^2\theta$, one can find formulas for $\sin n\theta$ in terms of $\sin\theta$ for odd values of n. For example,

$$\sin 3\theta = 3\sin\theta - 4\sin^3\theta,$$
$$\sin 5\theta = 5\sin\theta - 20\sin^3\theta + 16\sin^5 x.$$

- As θ approaches 0, $\sin\theta/\theta$ approaches 1. This is clear when we view θ as arc length on the unit circle, as in figure 6.4.

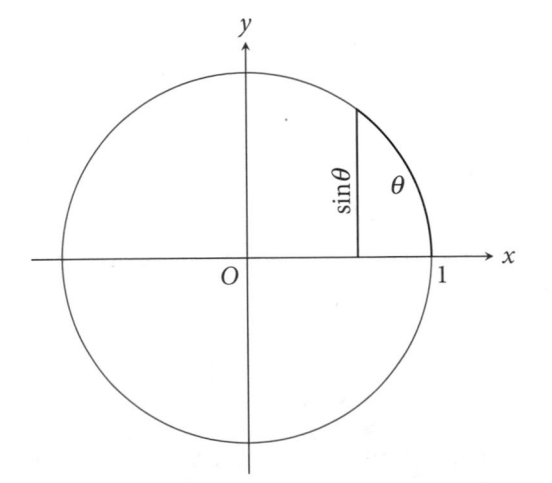

Figure 6.4 : The sine function as a function of arc length

The second fact is certainly a *limit argument*, but it is not *calculus*, since it applies only to the sine function—it is not a general method for calculating limits. To use it, Mādhava divided the angle x into n small parts, θ.

The first fact gives a formula for $\sin n\theta$ in terms of $\sin\theta$, which Mādhava knew in some form. A convenient form is the following, due to Newton (1676a): This form ends with the $\sin^n\theta$ term when n is odd:

$$\sin n\theta = n\sin\theta - \frac{n(n^2 - 1^2)}{3!}\sin^3\theta + \frac{n(n^2 - 1^2)(n^2 - 3^2)}{5!}\sin^5\theta - \cdots.$$

Now we rewrite the right-hand side in powers of $n \sin \theta$ by pulling factors of n out of the coefficients:

$$\sin n\theta = n \sin \theta - \frac{(1 - 1^2/n^2)}{3!}(n \sin \theta)^3$$

$$+ \frac{(1 - 1^2/n^2)(1 - 3^2/n^2)}{5!}(n \sin \theta)^5 - \cdots.$$

Next, for a given x we set $x = n\theta$ and let n grow arbitrarily large. Then $1/n^2$ approaches 0 and the right-hand side approaches

$$n \sin \theta - \frac{(n \sin \theta)^3}{3!} + \frac{(n \sin \theta)^5}{5!} - \cdots.$$

At the same time, $\sin \theta / \theta$ approaches 1, so each $n \sin \theta$ approaches $n\theta = x$, giving us finally

$$\sin x = x - \frac{x^3}{3!} + \frac{x^5}{5!} - \cdots. \qquad \square$$

6.3 NEWTON'S BINOMIAL SERIES

The binomial theorem was known to Chinese and Indian mathematicians in medieval times but was first treated rigorously (with proofs by induction) by Pascal (1654). For simplicity we will take the theorem in the form with binomial $1 + x$, so the **binomial theorem** expresses $(1 + x)^n$ as a sum of powers of x. For the first few values of n we get

$$
\begin{aligned}
(1+x)^1 &= & 1+x \\
(1+x)^2 &= & 1+2x+x^2 \\
(1+x)^3 &= & 1+3x+3x^2+x^3 \\
(1+x)^4 &= & 1+4x+6x^2+4x^3+x^4 \\
(1+x)^5 &= & 1+5x+10x^2+10x^3+5x^4+x^5 \\
(1+x)^6 &= & 1+6x+15x^2+20x^3+15x^4+6x^5+x^6 \\
(1+x)^7 &= & 1+7x+21x^2+35x^3+35x^4+21x^5+7x^6+x^7,
\end{aligned}
$$

and so on. When the binomial coefficients are tabulated as follows (with a trivial row 1 added at the top, corresponding to the power 0 of $1 + x$), we get the so-called **Pascal's triangle**, in which each entry $\neq 1$ is the sum of the two above it:

$$
\begin{array}{ccccccccccccccc}
 & & & & & & & 1 & & & & & & & \\
 & & & & & & 1 & & 1 & & & & & & \\
 & & & & & 1 & & 2 & & 1 & & & & & \\
 & & & & 1 & & 3 & & 3 & & 1 & & & & \\
 & & & 1 & & 4 & & 6 & & 4 & & 1 & & & \\
 & & 1 & & 5 & & 10 & & 10 & & 5 & & 1 & & \\
 & 1 & & 6 & & 15 & & 20 & & 15 & & 6 & & 1 & \\
1 & & 7 & & 21 & & 35 & & 35 & & 21 & & 7 & & 1
\end{array}
$$

The triangle is convenient for calculating the coefficients of $(1+x)^n$ and was used for this purpose by Chinese mathematicians centuries before Pascal. However, it is also useful to have a formula in which the coefficients are explicit functions of n. This formula is one of Pascal's contributions, and today it is written

$$
(1+x)^n = \binom{n}{0} + \binom{n}{1}x + \binom{n}{2}x^2 + \cdots + \binom{n}{n-1}x^{n-1} + \binom{n}{n}x^n,
$$

where the **binomial coefficient** $\binom{n}{k}$ is called "n choose k" and

$$
\binom{n}{k} = \frac{n(n-1)(n-2)\cdots(n-k+1)}{k(k-1)\cdots 3 \cdot 2 \cdot 1}.
$$

The reason for the name is that a term x^k arises from the product $(1+x)^n$ by choosing x from k of the factors and 1 from the rest. Therefore, the coefficient $\binom{n}{k}$ of x^k is the *number of ways* of choosing k things from n. The reason for the formula[1] can be explained as follows:

- Suppose first that we are choosing the k things in order—"first thing," "second thing," and so on—from n things. Then there are n ways to choose the first thing, after which $n-1$ things remain, so there are $n-1$ ways to choose the second thing, and so on. Finally, there are $n-k+1$ ways to choose the kth thing, so the number of sequences of choices is $n(n-1)(n-2)\cdots(n-k+1)$.

---

1. It is clear from its meaning as a "number of ways" that $\binom{n}{k}$ is a positive integer. It follows that its denominator $k(k-1)\cdots 3 \cdot 2 \cdot 1$ must divide its numerator $n(n-1)(n-2)\cdots(n-k+1)$. Gauss (1801: article 41) was unhappy about this type of reasoning and sought a proof of this fact about divisibility using basic facts about products and divisors. He succeeded, but only with considerable difficulty.

- However, we do *not* care about the order in which terms x are chosen from the n factors $1 + x$, so we need to divide by the number of ways of arranging k things in a sequence. This number is $k(k-1)\cdots3\cdot2\cdot1$ because there are k choices for first place in the sequence, then $k-1$ choices remain for the second place, and so on.

Special cases of the formula are $\binom{n}{1} = n$ and $\binom{n}{n} = 1$. By convention we take $\binom{n}{0} = 1$.

Newton, around 1670, boldly allowed fractional and negative values of n in the binomial theorem, in which case $(1+x)^n$ is an infinite power series instead of a polynomial. For example, if $n = -1$ we get the "binomial coefficients"

$$\binom{-1}{k} = \frac{(-1)(-2)\cdots(-k)}{k(k-1)\cdots3\cdot2\cdot1} = (-1)^k,$$

for all positive integer values of k. This gives

$$(1+x)^{-1} = 1 - x + x^2 - x^3 + \cdots,$$

which agrees with the known sum of the geometric series $1 - x + x^2 - x^3 + \cdots$. However, we know that this sum makes sense only for $|x| < 1$, so binomial series should be used with care. In fact, it was not until around 1800 that the binomial series was properly understood. But this did not stop Newton from using it to make some spectacular discoveries, which we will see in section 6.8. One instance of the binomial series that he used was

$$(1+x)^{-1/2} = 1 + \frac{-1/2}{1}x + \frac{-1/2(-1/2-1)}{1\cdot2}x^2$$
$$+ \frac{-1/2(-1/2-1)(-1/2-2)}{1\cdot2\cdot3}x^3 + \cdots$$
$$= 1 - \frac{1}{2}x + \frac{(1/2)(3/2)}{1\cdot2}x^2 - \frac{(1/2)(3/2)(5/2)}{1\cdot2\cdot3}x^3 + \cdots$$
$$= 1 - \frac{1}{2}x + \frac{1\cdot3}{2\cdot4}x^2 - \frac{1\cdot3\cdot5}{2\cdot4\cdot6}x^3 + \cdots.$$

6.4 EULER'S SOLUTION OF THE BASEL PROBLEM

Series of the form $a_0 + a_1x + a_2x^2 + \cdots$ are known as **power series** because they are combinations of powers of x. As we have just seen, Newton found power series for functions such as $(1+x)^{-1/2}$ by analogy with the binomial theorem for positive integer powers of $1 + x$, where one has a *finite* power series that is, a polynomial. The idea that power series behave like polynomials, though not rigorously justified, was very fruitful. Perhaps the most spectacular success of this idea was the Euler (1734) summation of the series of reciprocals of the positive integer squares:

$$\frac{1}{1^2} + \frac{1}{2^2} + \frac{1}{3^2} + \frac{1}{4^2} + \frac{1}{5^2} + \cdots.$$

This problem was known as the *Basel problem* because it was first attacked, without success, by the brothers Jakob and Johann Bernoulli in Basel (which was also Euler's hometown).

Euler found the solution indirectly, by assuming a "factor theorem" for a function with infinitely many roots, analogous to Descartes's factor theorem for polynomials (section 4.5). Under this assumption, he found an *infinite product formula* for the sine function. The starting point of the argument is the definition of $\sin\theta$ as a function of arc length θ on the unit circle (figure 6.4), which we will say more about in section 6.7. For now, it suffices to know that $\sin\theta$ is the y-coordinate of the point on the circle at arc length θ from the point $(1,0)$. It follows, since 2π is the length of one turn around the circle, that

$$\sin\theta = 0 \quad \text{precisely when} \quad \theta = 0, \pm\pi, \pm 2\pi, \pm 3\pi, \ldots.$$

It is also clear that, as θ approaches 0, $\sin\theta/\theta$ approaches 1, so we can assign the value 1 to the function $\sin\theta/\theta$ when $\theta = 0$. Thus $f(\theta) = \sin\theta/\theta$ is a function such that $f(0) = 1$ and $f(\theta) = 0$ for $\theta = \pm\pi, \pm 2\pi, \pm 3\pi, \ldots$.

Now if $f(x)$ is a *polynomial* function such that $f(0) = 1$ and $f(x) = 0$ for $x = x_1, x_2, \ldots, x_n$, then we can conclude, from Descartes's factor theorem, that

$$f(x) = \left(1 - \frac{x}{x_1}\right)\left(1 - \frac{x}{x_2}\right)\cdots\left(1 - \frac{x}{x_n}\right).$$

Assuming the analogous "factor theorem" for the function $f(\theta) = \sin\theta/\theta$ gives

$$\frac{\sin\theta}{\theta} = \left(1 - \frac{\theta}{\pi}\right)\left(1 + \frac{\theta}{\pi}\right)\left(1 - \frac{\theta}{2\pi}\right)\left(1 + \frac{\theta}{2\pi}\right)\left(1 - \frac{\theta}{3\pi}\right)\left(1 + \frac{\theta}{3\pi}\right)\cdots$$

$$= \left(1 - \frac{\theta^2}{\pi^2}\right)\left(1 - \frac{\theta^2}{2^2\pi^2}\right)\left(1 - \frac{\theta^2}{3^2\pi^2}\right)\cdots.$$

At this point we also need an infinite sum formula for $\sin\theta/\theta$, so as to compare the θ^2 terms. We will assume the power series for $\sin\theta$ from section 6.2, but see also section 6.8 for another approach by Newton.

Since

$$\sin\theta = \theta - \frac{\theta^3}{3!} + \frac{\theta^5}{5!} - \cdots, \text{ then } \frac{\sin\theta}{\theta} = 1 - \frac{\theta^2}{3!} + \frac{\theta^4}{5!} - \cdots,$$

and therefore the coefficient of θ^2 is $-1/3! = -1/6$. On the other hand,

$$\frac{\sin\theta}{\theta} = \left(1 - \frac{\theta^2}{\pi^2}\right)\left(1 - \frac{\theta^2}{2^2\pi^2}\right)\left(1 - \frac{\theta^2}{3^2\pi^2}\right)\cdots$$

implies that the coefficient of θ^2 is

$$-\left(\frac{1}{\pi^2} + \frac{1}{2^2\pi^2} + \frac{1}{3^2\pi^2} + \cdots\right),$$

since we get a θ^2 term by taking $-\theta^2/n^2\pi^2$ from one factor and 1 from all the rest. Equating this coefficient to $-1/6$ gives

$$\frac{1}{6} = \frac{1}{\pi^2}\left(\frac{1}{1^2} + \frac{1}{2^2} + \frac{1}{3^2} + \cdots\right),$$

and therefore

$$\frac{1}{1^2} + \frac{1}{2^2} + \frac{1}{3^2} + \cdots = \frac{\pi^2}{6}.$$

Impressive, no?

While we have the infinite product formula,

$$\frac{\sin\theta}{\theta} = \left(1 - \frac{\theta^2}{\pi^2}\right)\left(1 - \frac{\theta^2}{2^2\pi^2}\right)\left(1 - \frac{\theta^2}{3^2\pi^2}\right)\cdots,$$

it is worth pointing out the special case with $\theta = \pi/2$. Since $\sin\frac{\pi}{2} = 1$ we get

$$\frac{2}{\pi} = \left(1 - \frac{1}{2^2}\right)\left(1 - \frac{1}{4^2}\right)\left(1 - \frac{1}{6^2}\right)\cdots$$

$$= \frac{2^2 - 1}{2\cdot 2}\cdot\frac{4^2 - 1}{4\cdot 4}\cdot\frac{6^2 - 1}{6\cdot 6}\cdots$$

$$= \frac{1\cdot 3}{2\cdot 2}\cdot\frac{3\cdot 5}{4\cdot 4}\cdot\frac{5\cdot 7}{6\cdot 6}\cdots,$$

which is known as **Wallis's product** for π. It was discovered by Wallis (1655), rather mysteriously, at the dawn of the calculus era.

6.5 RATES OF CHANGE

As all calculus students learn, we often want to know the *rate of change* of one quantity, y, with respect to a quantity x that y depends on. For example, if a car travels along a straight road, reaching position y at time x, then the rate of change of y with respect to x is the *speed* of the car.

Possibly the first interesting calculation of a rate of change was that of the function $\sin\theta$ with respect to θ, due to the Indian mathematician Aryabhata in 499 CE. For a nice account of the background to this problem, which had to do with calculating values of the sine function, see Bressoud (2019: 50–56).

Aryabhata based his calculation on the two properties of the sine function listed in section 6.2: the addition formula and the limit property. He considered the change $\sin(\theta+\Delta\theta)-\sin\theta$ in the sine function due to a small change $\Delta\theta$ in θ. This gives the average rate of change

$$\frac{\sin(\theta+\Delta\theta)-\sin\theta}{\Delta\theta} = \frac{\sin\theta\cos\Delta\theta + \cos\theta\sin\Delta\theta - \sin\theta}{\Delta\theta}$$

$$= \cos\theta\,\frac{\sin\Delta\theta}{\Delta\theta} - \sin\theta\,\frac{1-\cos\Delta\theta}{\Delta\theta},$$

with the help of the addition formula. Now figure 6.5 not only shows why

$$\frac{\sin\Delta\theta}{\Delta\theta} = \frac{PR}{QR} \quad \text{approaches 1 as } \Delta\theta \text{ approaches 0;}$$

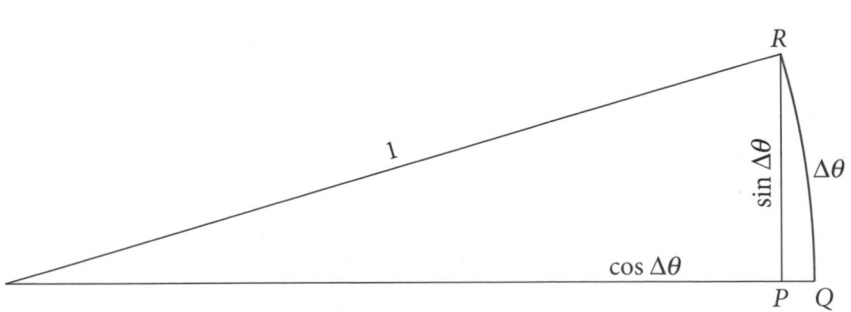

Figure 6.5 : Sine and cosine of a small angle $\Delta\theta$

it also shows that

$$\frac{1-\cos\Delta\theta}{\Delta\theta} = \frac{PQ}{QR} \quad \text{approaches 0 as } \Delta\theta \text{ approaches 0.}$$

Thus the average rate of change, $\frac{\sin(\theta+\Delta\theta)-\sin\theta}{\Delta\theta}$, approaches $\cos\theta$ as $\Delta\theta$ approaches 0. In other words, *the rate of change of* $\sin\theta$ *equals* $\cos\theta$. The proof we have just seen appeals to our geometric intuition of the circle, but it is even better to appeal to our *physical* intuition of circular motion. Here is what I mean.

Suppose a point is moving around the unit circle at constant speed equal to 1. Then its position at time t is $\langle\cos t, \sin t\rangle$ and its velocity is obviously tangential to the circle, hence perpendicular to the radius. That means the direction $\langle-\sin t, \cos t\rangle$, so the horizontal and vertical speeds of the point are proportional to $-\sin t$ and $\cos t$. In fact these speeds must be precisely $-\sin t$ and $\cos t$, since together they give speed 1. However,

horizontal speed $-\sin t$ is the rate of change of the horizontal distance $\cos t$, and vertical speed $\cos t$ is the rate of change of the vertical distance $\sin t$.

This example shows that physical intuition may sometimes have more to offer than geometric intuition—as was the case for Newton and others we will discuss in section 6.10—but it also underlines the importance of the geometric concept of *tangent* in calculus.

Tangents

A few examples of tangents to curves were known in ancient times; for example, Euclid proved, in the *Elements*, book 3, proposition 18, that the tangent to a circle is perpendicular to the radius. More surprising, around 100 CE Heron proved that *the tangent $\mathcal{T}$ to an ellipse at a point P makes equal angles, $\angle F_1 PQ$ and $\angle F_2 PS$, with the lines from the foci F_1, F_2 to P* (figure 6.6). That is, the line from F_1 to P "reflects back" through F_2.

Suppose, in search of a contradiction, that the lines from the foci to the point of contact with the tangent do *not* meet the tangent at equal angles. Then F_2PF_1 is *not* the path of reflection from F_1 to F_2 via $\mathcal{T}$. It follows, by the shortest path property of reflection (section 3.6), that the path F_1QF_2 of reflection from F_1 to F_2 via $\mathcal{T}$ is *shorter* than F_1PF_2. (This hypothetical path of reflection of course does not *look* like reflection, since it is only hypothetical.) But since $\mathcal{T}$ touches the ellipse only at P, Q is outside the ellipse. Then the path F_1RF_2 will be even shorter than F_1QF_2, by the triangle inequality (section 3.6). This contradicts the

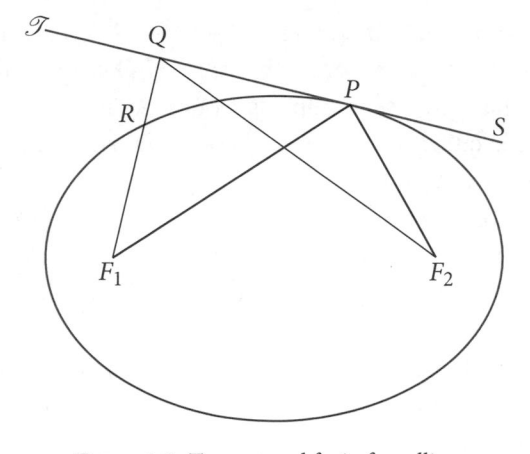

Figure 6.6 : Tangent and foci of an ellipse

"constant sum" property of the ellipse mentioned in section 5.1. So in fact F_1PF_2 is the path of reflection, and it *does* meet the tangent at equal angles.

This example and the previous one depend on special properties of the circle and the ellipse. Wholesale calculation of tangents to curves became possible only with the development of algebra in the late sixteenth century. To illustrate, we look at the example of the parabola, whose tangent at any point was calculated by Fermat in 1629, by essentially the method used today.

Modernizing notation only slightly, we consider the parabola $y = x^2$ and seek the tangent at the point P, where $x = X$. The main problem is to find its slope, which we approximate by the slope of the chord PQ from P to the nearby point Q where $x = X + E$ (figure 6.7).

We see from the diagram that the height of the curve rises from X^2 to $(X + E)^2$ as x runs from X to $X + E$. Therefore,

$$\text{slope of } PQ = \frac{\text{rise}}{\text{run}} = \frac{(X+E)^2 - X^2}{E}$$
$$= \frac{2XE + E^2}{E}$$
$$= 2X + E.$$

At this stage we want to let E approach 0, because that is what happens as Q approaches P. However, as Fermat realized, we cannot honestly say

$$\text{slope of tangent} = (\text{slope of } PQ \text{ when } E = 0),$$

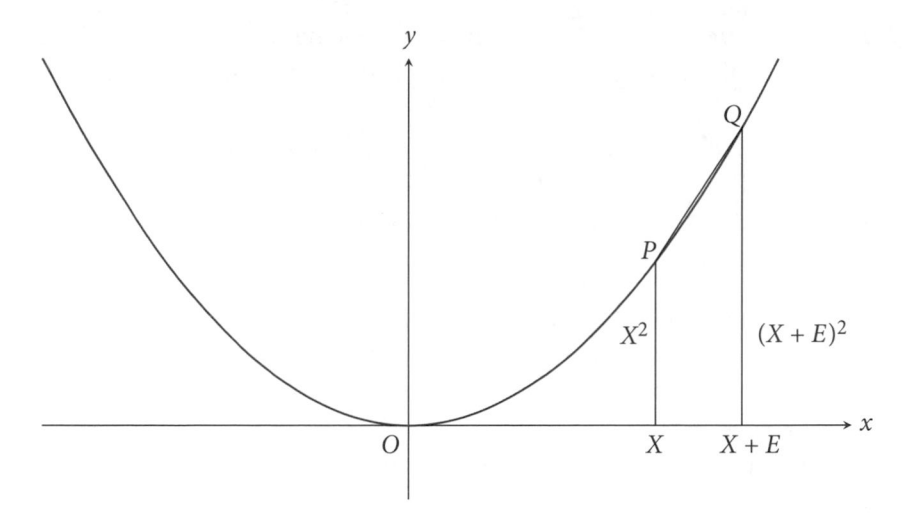

Figure 6.7 : Approximating the tangent by a chord

because we assumed E *not* equal to zero in order to divide by it in the above calculation. Instead, Fermat introduced a new concept he called *adequality*, meaning something like "virtual equality" or what we might now call "equality in the limit," except that he did not hit on the limit concept itself. Rather, he said the slope of the tangent is $2X$ because $2X$ is *adequal to* $2X + E$.

Fermat was flirting with the idea that E was an *infinitesimal*: a quantity smaller than any positive rational number, yet nonzero. He realized he could not say that $2X + E$ and $2X$ are equal, but he could say that they are adequal. Later in the seventeenth century mathematicians would throw logic to the winds and even assert that

> a quantity which is increased or decreased by another quantity that is infinitely smaller than it is, may be considered as remaining the same.

which was precisely what l'Hôpital (1696) said on pages 2 and 3 of the first calculus textbook,[2] *Analyse des infiniment petits, pour l'intelligence des lignes courbes (Analysis of the Infinitely Small, for the Understanding*

---

2. A 1768 edition of this book, now in the Boston Public Library, was owned by John Quincy Adams, the sixth president of the United States. He studied mathematics at the urging of his father, John Adams, the second president. Adams senior himself taught Adams junior the basics of arithmetic, algebra, and geometry. In a letter to Benjamin Waterhouse, April 23, 1785, John Adams wrote

of Curved Lines). We will see that much of seventeenth-century calculus operated in this strange, but strangely effective, world of infinitesimals.

6.6 AREA AND VOLUME

Some notable exact formulas for areas and volumes were found by Euclid and Archimedes:

- Volume of tetrahedron = $\frac{1}{3}$ base area × height was found by Euclid using infinite geometric series, as we saw in section 2.7.
- Area of parabolic segment = $\frac{4}{3}$ area of inscribed triangle was found by Archimedes.
- Volume of a sphere = $\frac{2}{3}$ volume of circumscribing cylinder was also found by Archimedes.

Archimedes found the area of a parabolic segment by cutting it into infinitely many triangles and grouping their areas, which miraculously form a geometric series. Figure 6.8 shows the first three groups of triangles in black (the inscribed triangle), dark gray, and light gray.

Figure 6.8 : Filling the parabolic segment with triangles

The areas of the three groups turn out to be the first three terms of the series

$$1 + \frac{1}{4} + \frac{1}{4^2} + \cdots,$$

I then attempted a sublime flight, and endeavoured to give him some idea of the differential method of the Marquis de l'Hôpital, and the method of fluxions and infinite series of Sir Isaac Newton; but alas! it is thirty years since I thought of mathematics, and I found I had lost the little I once knew, especially of these higher branches of geometry, so that he is as yet but a smatterer, like his father.

See https://founders.archives.gov/documents/Adams/06-17-02-0020.

and subsequent groups of triangles—each falling into the gaps between the previous group and the curve—correspond to the subsequent terms, so the sum of their areas is 4/3.

To find the volume of the sphere, Archimedes compared it with that of the cylinder circumscribing it as shown in figure 6.9. Archimedes found the volume of the sphere by considering infinitesimal slices, an idea that was revived by the Italian mathematician Bonaventura Cavalieri in the 1630s. In fact, Cavalieri's proof is more elegant: he compared the sphere with a cone, inscribing each in a cylinder, and observed that each horizontal slice *inside* the sphere has the same area as the corresponding slice of the cylinder *outside* the cone (figure 6.10). Supposing that volumes are sums of infinitesimal slices, Cavalieri concluded that the volume of the sphere equals the volume outside the cone, that is, 2/3 the volume of

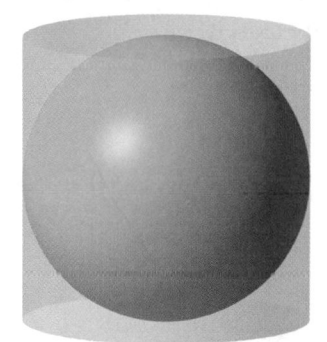

Figure 6.9 : Sphere with circumscribed cylinder

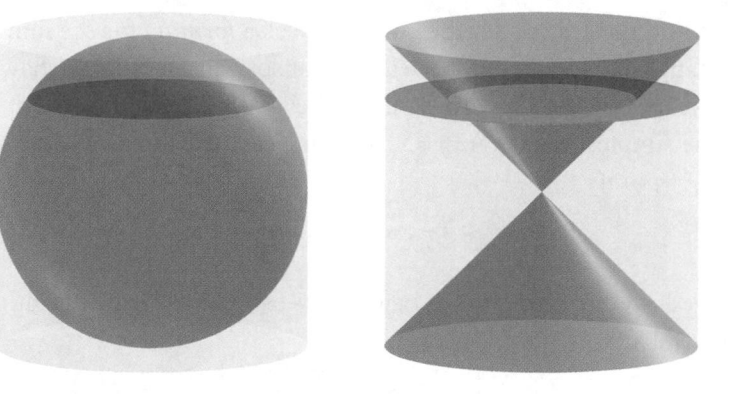

Figure 6.10 : Cavalieri's proof

the cylinder (because the volume of a cone equals 1/3 base area × height, as one sees by slicing it into infinitesimal tetrahedra).

Approximating Areas by Rectangles

When curves are described by equations, as became common after the introduction of algebraic geometry, it is possible to calculate the area of curved regions by algebra too. Here we concentrate on a simple but typical case: the region bounded by the x-axis and a curve $y = f(x)$.

A natural and uniform way of estimating the area under a curve $y = f(x)$ is by *rectangular approximation*, as shown in figure 6.11 in the three cases $f(x) = x, f(x) = x^2$, and $f(x) = x^3$, between $x = 0$ and $x = 1$. The interval from $x = 0$ to $x = 1$ is divided into $n = 20$ equal parts, each of which is the base of a rectangle just tall enough to touch the curve. It is clear that, as n increases, the total area of the rectangles approaches the area under the curve more and more closely.

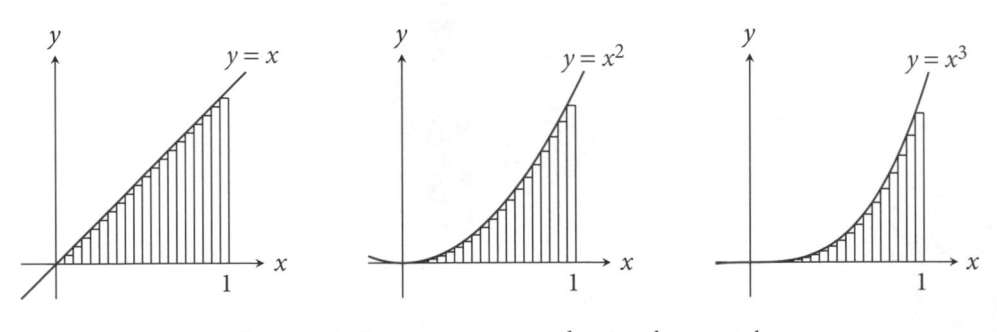

Figure 6.11 : Approximating curved regions by rectangles

To find the area under the curve we seek a *formula* for the sum of the n rectangles, which we hope will be transparent enough to show what value it approaches as n grows indefinitely.

The simplest case is $f(x) = x$, in which case the interval is divided by the $n - 1$ points

$$x = \frac{1}{n}, \frac{2}{n}, \frac{3}{n}, \ldots, \frac{n-1}{n},$$

and we have n rectangles, of heights $0, \frac{1}{n}, \frac{2}{n}, \frac{3}{n}, \ldots, \frac{n-1}{n}$ and width $\frac{1}{n}$, so their total area is

$$\frac{1}{n^2} \left(0 + 1 + 2 + 3 + \cdots + (n-1) \right).$$

Finding a formula for the area of the rectangles therefore reduces to finding the sum $1 + 2 + 3 + \cdots + (n-1)$, which is most easily done by the trick of writing it twice, forward and backward:

1	+	2	+	3	+	$\cdots$	+	$(n-1)$
$(n-1)$	+	$(n-2)$	+	$(n-3)$	+	$\cdots$	+	1

which sums to

n	+	n	+	n	+	$\cdots$	+	$n,$

Thus $2\big(1 + 2 + 3 + \cdots + (n-1)\big) = n(n-1)$, and therefore

$$1 + 2 + 3 + \cdots + (n-1) = \frac{n(n-1)}{2} = \frac{n^2 - n}{2}.$$

To find the total area of the rectangles we divide this by n^2, obtaining $\frac{1}{2} - \frac{1}{2n}$, which clearly approaches $\frac{1}{2}$ as n increases indefinitely. This is as we should expect, because the region under $y = x$ is a triangle with base and height each equal to 1.

When $f(x) = x^2$ we similarly find that the total area of the rectangles is

$$\frac{1}{n^3}\left(0^2 + 1^2 + 2^2 + 3^2 + \cdots + (n-1)^2\right),$$

so the main problem is to evaluate $1^2 + 2^2 + 3^2 + \cdots + (n-1)^2$. The answer, not so easily found (but, once found, easily proved by induction), is

$$\frac{n(n+1)(2n+1)}{6} = \frac{n\left(n + \frac{1}{2}\right)(n+1)}{3} = \frac{n^3}{3} + \frac{n^2}{2} + \frac{n}{6},$$

so the area of the rectangles is $\frac{1}{3} + \frac{1}{2n} + \frac{1}{6n^2}$. This clearly approaches the value $\frac{1}{3}$ as n increases. This gives another way to find the area bounded by a parabolic segment, and indeed, Archimedes did it by this method, as well as by the method mentioned at the beginning of this section.

Increasingly complicated arguments give $1^p + 2^p + 3^p + \cdots + (n-1)^p$ for larger and larger values of the power p. Around 1000 CE al-Haytham found the formulas for $p = 1, 2, 3, 4$ and used the result for $p = 4$ to find the volume of the solid formed by rotating the parabola $y = x^2$ around the x-axis.

There is a method to obtain the formula for $p + 1$ from the formula for p, so in principle one can go up to any value. However, this is a lot of work, and the formula does not depend in a transparent way on p. Fortunately, it can be shown that the term involving the highest power of n is always $\frac{n^{p+1}}{p+1}$, which is all we need to know to prove that *the area under* $y = x^p$ *between* $x = 0$ *and* $x = 1$ *is simply* $\frac{1}{p+1}$. Essentially, this fact was used by Mādhava in his proof (mentioned in section 6.2) that

$$\frac{\pi}{4} = 1 - \frac{1}{3} + \frac{1}{5} - \frac{1}{7} + \cdots.$$

We are now close to seeing how calculus explains this wonderful result. But first we need an easier way to find the area under the curve $y = x^p$.

6.7 INFINITESIMAL ALGEBRA AND GEOMETRY

Calculations like Fermat's for the tangent to the parabola (section 6.5) were systematized by Leibniz in the 1680s, in a fully fledged **infinitesimal calculus**. Leibniz postulated the existence of infinitesimals and subjected them to algebraic operations.

In particular, when y is a function of x, he let dy denote the infinitesimal change in y produced by an infinitesimal change dx in x, so the quotient dy/dx represents the *rate of change* of y. As in ordinary algebra, we require $dx \neq 0$ in order to divide by it, and this tends to leave behind terms like dx, $(dx)^2$, and so on, at the end of calculations. Leibniz believed such terms could be omitted, since they are "incomparably small" (Struik 1969: 280).

For example, if $y = x^2$ then

$$dy = (x + dx)^2 - x^2 = 2x\,dx + (dx)^2,$$

so

$$\frac{dy}{dx} = 2x + dx,$$

and Leibniz would conclude $\frac{dy}{dx} = 2x$. He and his followers were aware that $2x$ is *approached by* $2x + dx$ rather than equal to it—and this is what we really want to know—but their failure to say so led to criticism and ridicule by philosophers. Berkeley (1734: 59) memorably and aptly called infinitesimals "Ghosts of departed Quantities."

Nevertheless, calculating with infinitesimals is an easy shortcut to correct results. To find the tangent to $y = x^p$, for example, we let

$$dy = (x + dx)^p - x^p$$

$$= \left[x^p + px^{p-1}dx + \frac{p(p-1)}{2}x^{p-2}(dx)^2 + \cdots + (dx)^p \right] - x^p$$

by the binomial theorem of section 6.3

$$= px^{p-1}dx + \frac{p(p-1)}{2}x^{p-2}(dx)^2 + \cdots + (dx)^p,$$

so $\quad \dfrac{dy}{dx} = px^{p-1}$, omitting terms in $dx, (dx)^2, \ldots$.

Rules for Differentiation

When Leibniz (1684) introduced the notation dx he intended the d to stand for "difference." Thus dx was the difference between two "infinitesimally close" values of x. This idea is echoed in our term **differentiation** for the operation $\frac{d}{dx}$, which produces $\frac{dy}{dx}$ from y. However, while dx and dy exist for us only as parts of the symbol $\frac{dy}{dx}$, it is still helpful to pretend that they are parts of a fraction, because certain rules for differentiation look like rules for fractions (which is where they came from).

Inverse function rule. If $y = f(x)$ and f has an inverse, so $x = f^{-1}(y)$, then

$$\frac{dx}{dy} = \frac{1}{dy/dx}.$$

Chain rule. If z is a function of y and y is a function of x, then

$$\frac{dz}{dx} = \frac{dz}{dy}\frac{dy}{dx}.$$

The other rules for differentiation also come from easy calculations, ending with omission of infinitesimal terms. For example,

Product rule. If $y = uv$ then $\frac{dy}{dx} = u\frac{dv}{dx} + v\frac{du}{dx}$.

This is because

$$dy = (u + du)(v + dv) - uv$$

$$= u\,dv + v\,du + du\,dv$$

so $\quad \dfrac{dy}{dx} = u\dfrac{dv}{dx} + v\dfrac{du}{dx}$, omitting the infinitesimal $du\dfrac{dv}{dx}$.

The Fundamental Theorem of Calculus

Leibniz also introduced a corresponding concept of "sum," denoted by the long S symbol $\int$, which is none other than the **integral** symbol we use today.[3]

And, for Leibniz, $\int$ really *was* a sum, of infinitesimal values— typically, a sum of values of the form $f(x)\, dx$, each of which represents the area of a rectangle of height $f(x)$ and infinitesimal width dx. When such values are added for x from a to b we get

$$\int_a^b f(x)\, dx,$$

which is what we call the **definite integral** of f for x in $[a, b]$. It represents the area beneath the graph of $y = f(x)$ between $x = a$ and $x = b$, and its affinity with the idea of rectangular approximation in section 6.6 is obvious. In some sense, the sum of infinitesimal rectangles represents the *limit of* rectangular approximations, but Leibniz skipped the step of passing to the limit by postulating a division of the interval $[a, b]$ into infinitesimal parts of width dx.

By associating differentiation with differences and integration with sums, Leibniz made it clear that integration and differentiation are inverse operations. It is basically because sum and difference are inverse operations. A more formal explanation goes as follows. Consider the area

$$A = \int_a^u f(x)\, dx, \text{ beneath the graph of } y = f(x) \text{ between } x = a \text{ and } x = u,$$

as a function of u. Let dA be the increase in A as x increases from u to $u + du$ (figure 6.12).

Then we have, up to an infinitesimal error in $f(u)$,

$$dA = f(u)\, du,$$

since dA is the area of a rectangle with height $f(u)$ and infinitesimal width du. It follows that

$$\frac{dA}{du} = f(u),$$

---

3. Unfortunately, the word "integral" is not as close to "sum" as "differentiation" is to "difference." There does not seem to be a good word in English to reflect the sum analogy without causing confusion with ordinary summation.

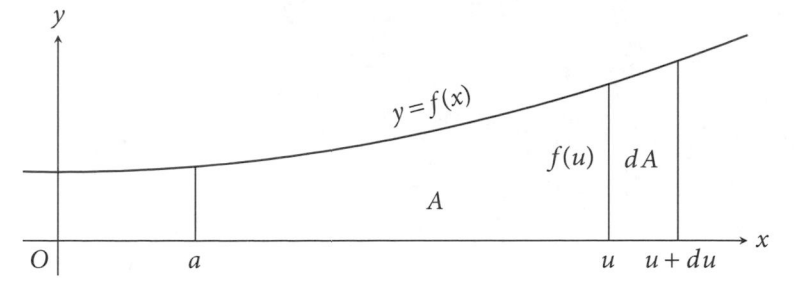

Figure 6.12 : The fundamental theorem of calculus

so the differentiation operation reverses the integration operation that produced the function A from the function f.

This is the **fundamental theorem of calculus**, which in principle reduces all problems of integration to problems of differentiation. To find $\int_0^1 x^p\,dx$, for example, it suffices to find an **antiderivative** of x^p, that is, a function that gives x^p by differentiation. We know from section 6.3 that

$$\frac{d}{dx}x^{p+1} = (p+1)x^p,$$

and from this it easily follows that

$$\frac{d}{dx}\frac{x^{p+1}}{p+1} = x^p.$$

Therefore

$$\int_0^u x^p\,dx = \frac{u^{p+1}}{p+1}, \quad \text{and in particular} \quad \int_0^1 x^p\,dx = \frac{1}{p+1},$$

a result obtained with much greater difficulty in section 6.6.

Arc Length

Another geometric problem solvable by integration is finding the length of a curve. This problem, too, is most easily handled by infinitesimal geometry. On a curve in the plane, we look at two infinitesimally close points $P = \langle x, y \rangle$ and $Q = \langle x + dx, y + dy \rangle$, as shown in figure 6.13.

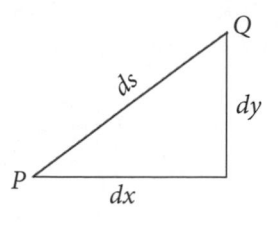

Figure 6.13 : Infinitesimal arc of a curve

Since dx and dy are infinitesimal, the arc between P and Q can be considered straight,[4] so its length ds is given by the Pythagorean theorem:

$$|PQ| = ds = \sqrt{(dx)^2 + (dy)^2}.$$

Now, depending on the relation between x and y that defines the curve, we transform the expression for ds into a function whose integral is the length of the curve.

Case 1. Given y as a function of x, write

$$ds = \sqrt{1 + \left(\frac{dy}{dx}\right)^2}\, dx.$$

Then the arc length s between $x = a$ and $x = b$ is given by

$$s = \int_a^b \sqrt{1 + \left(\frac{dy}{dx}\right)^2}\, dx.$$

Essentially this formula was found by Gregory (1668).

Case 2. If both x and y are given as functions of a parameter t, write

$$ds = \sqrt{\left(\frac{dx}{dt}\right)^2 + \left(\frac{dy}{dt}\right)^2}\, dt.$$

Then the arc length s between the values c and d of t is given by

$$s = \int_c^d \sqrt{\left(\frac{dx}{dt}\right)^2 + \left(\frac{dy}{dt}\right)^2}\, dt.$$

---

4. In fact Leibniz (1684)—see Struik (1969: 276)—went so far as to postulate that the tangent can be considered "the continued side of a polygon with an infinite number of angles, which for us takes the place of the curve."

In both cases the function to be integrated involves a square root, which often makes finding an antiderivative difficult. Perhaps this is why Descartes (1637: 91) believed that the "ratios between straight and curved lines . . . cannot be discovered by human minds." Another reason may be that the simplest curves, the conic sections, do *not* have simple arc length functions. In particular, the arc length of the circle depends on the elusive number π. One might therefore despair of dealing with curves of higher degree.

However, for certain *cubic* curves, $\left(\frac{dx}{dt}\right)^2 + \left(\frac{dy}{dt}\right)^2$ is a perfect square and we get an easy integral. For example, if we start with the equations

$$\frac{dx}{dt} = 2t, \quad \frac{dy}{dt} = t^2 - 1,$$

then

$$\sqrt{\left(\frac{dx}{dt}\right)^2 + \left(\frac{dy}{dt}\right)^2} = \sqrt{4t^2 + t^4 - 2t^2 + 1} = \sqrt{t^4 + 2t^2 + 1} = \sqrt{(t^2 + 1)^2} = t^2 + 1.$$

Then we can easily calculate the arc length of this curve between any two values of t. What is the curve? Well, taking obvious antiderivatives of $\frac{dx}{dt}$ and $\frac{dy}{dt}$, we get the parametric equations

$$x = t^2, \quad y = \frac{t^3}{3} - t = t\left(\frac{t^2}{3} - 1\right),$$

which give $t = \pm\sqrt{x}$ and therefore $y = \pm\sqrt{x}\left(\frac{x}{3} - 1\right)$. Finally, squaring gives the Cartesian equation

$$y^2 = x\left(\frac{x}{3} - 1\right)^2,$$

which represents the curve shown in figure 6.14.

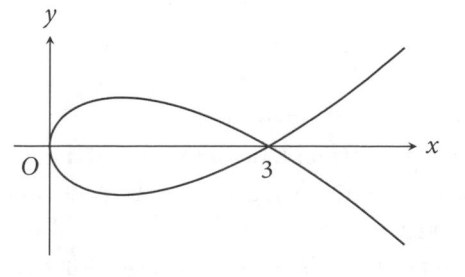

Figure 6.14 : A cubic curve with easily calculable arc length

6.8 THE CALCULUS OF SERIES

The notation we use in calculus today originated in the work of Leibniz in the 1680s. However, the basic *theorems* of calculus, including the fundamental theorem and the integrals of powers of x, were known to Newton in the 1660s. For that matter, they were also known to James Gregory in Scotland soon after. But calculus for Newton and Gregory was a somewhat different subject, with emphasis on infinite series rather than infinitesimals. As we have already seen in section 6.3, Newton broke new ground with his discovery of the binomial series. He and Gregory also made great discoveries from the humble starting point of the geometric series.

For Newton, the basic objects of calculus were power series, that is, infinite sums of the form

$$a_0 + a_1 x + a_2 x^2 + a_3 x^3 + \cdots, \quad \text{where } a_0, a_1, a_2, a_3, \ldots \text{ are real numbers.}$$

Newton found it easy to handle infinite series: he could add, subtract, multiply, and divide them, and even compute fractional powers such as square roots, thanks to his general binomial theorem. He could also differentiate and integrate them simply by differentiating and integrating powers of x. His first discovery was a power series for what we now call the **natural logarithm** function:

$$\ln(1 + x) = x - \frac{x^2}{2} + \frac{x^3}{3} - \frac{x^4}{4} + \cdots, \quad \text{valid for } |x| \le 1.$$

The logarithm was defined by the integral

$$\ln x = \int_0^x \frac{dt}{1 + t},$$

which Newton evaluated by expanding the integrand as a geometric series

$$\frac{1}{1 + t} = 1 - t + t^2 - t^3 + \cdots$$

and integrating each term from 0 to x.

To his great disappointment, he then learned that the same result had already been found, and published, by Nicolas Mercator (1668) in a book called *Logarithmotechnia*. Newton then redoubled his efforts, applying his series manipulation skills to obtain series for the sine, cosine, and exponential functions. The exponential function did not even have a name at that time, but Newton found its series by his big weapon, *inversion*, applied to the logarithm series.

To invert a series for a function y in powers of x, say,

$$y = x - \frac{x^2}{2} + \frac{x^3}{3} - \frac{x^4}{4} + \cdots,$$

one supposes that there is a series for x in powers of y,

$$x = a_0 + a_1 y + a_2 y^2 + \cdots,$$

and proceeds to find the coefficients $a_0, a_1, a_2, \ldots$ one by one, in that order. In this example we can see that $a_0 = 0$, because $x = 0$ when $y = 0$. To find a_1 we substitute

$$x = a_1 y + a_2 y^2 + \cdots$$

in the series for y, obtaining

$$y = \left(a_1 y + a_2 y^2 + \cdots\right) - \frac{1}{2}\left(a_1 y + a_2 y^2 + \cdots\right)^2 + \frac{1}{3}\left(a_1 y + a_2 y^2 + \cdots\right)^3 - \cdots.$$

Comparing coefficients of y on both sides gives $a_1 = 1$. By repeating this process Newton found $a_2 = 1/2$, $a_3 = 1/6$, $a_4 = 1/24$, and $a_5 = 1/120$, at which point he was confident that $a_n = 1/n!$. In this way he discovered (though did not rigorously prove) that the inverse function x, which we call $e^y - 1$, is

$$x = y + \frac{y^2}{2!} + \frac{y^3}{3!} + \frac{y^4}{4!} + \frac{y^5}{5!} + \cdots.$$

Newton's discovery of the sine series was even more spectacular, though he did it the hard way compared with the argument in section 6.2. His starting point was an integral expression for the inverse sine function, which involves a square root. He expanded this by his binomial theorem and integrated term by term to obtain the series

$$y = \sin^{-1} x = x + \frac{1}{2}\frac{x^3}{3} + \frac{1 \cdot 3}{2 \cdot 4}\frac{x^5}{5} + \frac{1 \cdot 3 \cdot 5}{2 \cdot 4 \cdot 6}\frac{x^7}{7} + \cdots.$$

Then he inverted this series to obtain

$$x = \sin y = y - \frac{1}{6}y^3 + \frac{1}{120}y^5 - \frac{1}{5040}y^7 + \frac{1}{362880}y^9 - \cdots,$$

which was enough to convince him that y^{2n+1} has coefficient $(-1)^n / (2n+1)!$.

Rediscovery of the Series for π

As mentioned in section 6.2, in the fifteenth century Mādhava discovered the series for the inverse tangent function:

$$\tan^{-1} x = x - \frac{x^3}{3} + \frac{x^5}{5} - \frac{x^7}{7} + \cdots, \quad \text{for } -1 < x \le 1,$$

with the consequence (for $x = 1$) that

$$\frac{\pi}{4} = 1 - \frac{1}{3} + \frac{1}{5} - \frac{1}{7} + \cdots.$$

These results were rediscovered in the seventeenth century by Gregory and Leibniz. They used an argument distantly related to Mādhava's, though streamlined by calculus. By some infinitesimal geometry one first expresses the inverse tangent function as an integral:

$$\tan^{-1} x = \int_0^x \frac{dt}{1 + t^2}.$$

Then one expands the integrand as a geometric series,

$$\frac{1}{1 + t^2} = 1 - t^4 + t^6 - t^8 + \cdots,$$

and integrates term by term to get the above power series for $\tan^{-1}$.

We can streamline this argument even further, bypassing the infinitesimal geometry. The idea is to use the following parametric equations for the circle $x^2 + y^2 = 1$:

$$x = \frac{1 - t^2}{1 + t^2}, \quad y = \frac{2t}{1 + t^2}.$$

With these parametric equations the arc length integral $\int \sqrt{\left(\frac{dx}{dt}\right)^2 + \left(\frac{dy}{dt}\right)^2} \, dt$ turns out to be none other than $\int_0^x \frac{2dt}{1+t^2}$. A similar derivation of this integral was found by Jakob Bernoulli (1696). The magic parametric equations for the circle used here had their origin in number theory and are discussed further in section 7.2.

6.9 ALGEBRAIC FUNCTIONS AND THEIR INTEGRALS

Newton's calculus of power series was simple because it did integration and differentiation only on powers of x. But this approach erased distinctions between functions, such as algebraic versus nonalgebraic, and indeed, it avoided thinking about the **function** concept much at all. Leibniz, on the other hand, thought in terms of functions, with some functions considered more "elementary" than others. He and his followers sought to express integrals, where possible, by functions that were as elementary as the functions being integrated.

Some nonalgebraic functions arise very simply as integrals, for example,

$$\ln u = \int_1^u \frac{dx}{x} \quad \text{and} \quad \tan^{-1} u = \int_0^u \frac{dx}{1+x^2},$$

so the class of **elementary functions** is taken to include the natural logarithm and the inverse tangent functions. Then functions obtainable from these by inversion and algebraic operations should also be included,[5] so we end up with a class that also includes the exponential and trigonometric functions. The initial goal of Leibniz and his followers was to find classes of functions with elementary integrals.

The biggest natural class with this property is the **rational functions**, which are those functions that can be written as quotients of polynomials. The rational functions are of course algebraic, but integrals of other algebraic functions, such as $\sqrt{1-x^3}$, are not elementary. Even showing that the integrals of rational functions are elementary poses formidable problems. One example that gave Leibniz a headache is the integral of $1/(x^4+1)$. He did not at first see how to split x^4+1 into real quadratic factors. Once this is done, however, by observing

$$x^4 + 1 = x^4 + 2x^2 + 1 - 2x^2 = (x^2+1)^2 - (\sqrt{2}x)^2$$
$$= (x^2 + 1 + \sqrt{2}x)(x^2 + 1 - \sqrt{2}x),$$

we can split $1/(x^4+1)$ into partial fractions of the form

$$\frac{ax+b}{x^2+1+\sqrt{2}x} + \frac{cx+d}{x^2+1-\sqrt{2}x}.$$

Then the integral reduces to a sum of simpler integrals of the ln and $\tan^{-1}$ type by suitable changes of variable. This example contains the

---

5. The functions thus obtainable from $\tan^{-1}$ include the tan function, which is clearly not algebraic since its graph meets a straight line in infinitely many points, so we get a proof that $\tan^{-1}$ is not algebraic. We can also prove that the exponential function, and hence the natural logarithm, is nonalgebraic by obtaining the sine function from it. This is possible with the help of complex numbers; in fact, $\sin x = \frac{1}{2i}\left(e^{ix} - e^{-ix}\right)$.

essence of the problem for rational functions in general, namely, splitting an arbitrary polynomial into real linear or quadratic factors. This algebraic problem is essentially equivalent to the **fundamental theorem of algebra**, which we return to in chapter 8.

Elliptic Integrals

Certain integrals of irrational algebraic functions are also elementary, for example,

$$\int_0^u \frac{dx}{\sqrt{1-x^2}} = \sin^{-1} u.$$

The same goes for any integrand that is rational except for occurrences of $\sqrt{1-x^2}$. Such an integral can be "rationalized" by the change of variable

$$x = \frac{2t}{1+t^2},$$

because $\frac{dx}{dt}$ and

$$\sqrt{1-x^2} = \sqrt{1 - \frac{4t^2}{(1+t^2)^2}} = \sqrt{\frac{1+2t^2+t^4-4t^2}{(1+t^2)^2}} = \sqrt{\frac{(1-t^2)^2}{(1+t^2)^2}} = \frac{1-t^2}{1+t^2}$$

are both rational functions of t. Thus each x in the integrand is replaced by a rational function of t, as is $\sqrt{1-x^2}$, and dx is replaced by $\frac{1-t^2}{1+t^2} dt$, giving the integral of a rational function of t. The same goes for any integrand that is rational except for occurrences of a term of the form $\sqrt{\text{quadratic in } x}$.

The trouble starts when the integrand contains a term such as $\sqrt{\text{cubic in } x}$ or $\sqrt{\text{quartic in } x}$. These are called **elliptic integrals** because an example is the integral for the arc length of an ellipse. In fact, the same is true for the arc length of the hyperbola, and the simplest example is $y = 1/x$. For this curve we have

$$\frac{dy}{dx} = -x^{-2},$$

and hence the first arc length integral from section 6.7 is

$$\int_a^b \sqrt{1 + \left(\frac{dy}{dx}\right)^2}\, dx = \int_a^b \sqrt{1 + x^{-4}}\, dx = \int_a^b \frac{1}{x^2} \sqrt{x^4 + 1}\, dx.$$

The troublemaker in this integral is the term $\sqrt{x^4 + 1}$, which cannot be "rationalized" by replacing x by a rational function of t as we did above for $\sqrt{1 - x^2}$. The same goes for the so-called **lemniscatic integral**,

$$\int_a^b \frac{dx}{\sqrt{1 - x^4}},$$

which is the most famous elliptic integral. It too cannot be "rationalized," and we will see why in the next chapter. The reason for its name, by the way, is that Jakob Bernoulli (1694) found that it expresses the arc length of the **lemniscate** (from the Greek for "ribbon"), which is a figure-eight curve with Cartesian equation

$$(x^2 + y^2)^2 = x^2 - y^2,$$

shown in figure 6.15.

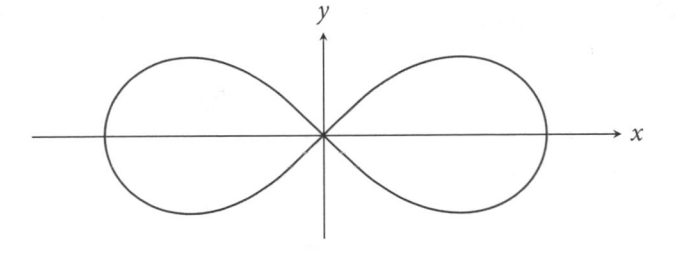

Figure 6.15 : The lemniscate of Bernoulli

These examples hint at a connection between rationalization of integrals and finding rational parametric equations for curves. The example of $\sqrt{1 - x^2}$ goes with the curve $y = \sqrt{1 - x^2}$, which is unit circle $x^2 + y^2 = 1$, and the change of variable we used,

$$x = \frac{2t}{1 + t^2},$$

is part of the parameterization of the circle by rational functions:

$$x = \frac{2t}{1 + t^2}, \quad y = \frac{1 - t^2}{1 + t^2}.$$

On the other hand, the curve $y = \sqrt{1 - x^4}$, or $y^2 = 1 - x^4$, does not have a parameterization by rational functions, as we will see in the next chapter. The proof is connected with number theory, which was suspected, but not quite understood, by Leibniz when he wrote

> I ... remember having suggested (what could seem strange to some) that the progress in our integral calculus depended in good part upon the development of that type of arithmetic which, as far as we know, Diophantus has been the first to treat systematically.
>
> (Leibniz 1702, translated in Weil 1984: 121)

Another to notice a connection between integral calculus and number theory was Jakob Bernoulli (1696), who explicitly credited the Greek number theorist Diophantus with a change of variable that rationalizes the expression $\sqrt{2x - x^2}$. We learn more about Diophantus in the next chapter.

6.10 REMARKS

Many books have been written on calculus and its role in expanding the concept of proof. An excellent recent example is Bressoud (2019), which I recommend for a more detailed account of the history of calculus and its foundations. I prefer to view calculus and its difficulties as part of a larger struggle, *reasoning about infinity*, which began before the discovery of calculus and is not over yet. Accordingly, this chapter has picked up some threads from earlier chapters, such as the method of exhaustion, and has left some loose ends to be picked up later, such as the nature of the real numbers and whether it is possible to give a consistent definition of infinitesimals.

Newton likened calculation with power series to calculation with infinite decimals:

> Since there is a great conformity between the Operations in Species, and the same Operations in common numbers; nor do they seem to differ, except in the Characters by which they are represented, the first being general and indefinite, and the other definite and particular: I cannot but wonder that no body has thought of accommodating the lately-discover'd Doctrine of Decimal Fractions in like manner to Species, (unless you except the Quadrature of the Hyperbola by Mr. *Nicolas Mercator*); especially since it might have open'd a way to more abstruse discoveries. But since this Doctrine of Species has the same relation to Algebra as the Doctrine of Decimal Numbers has to common Arithmetic; the Operations of Addition, Subtraction, Multiplication, Division and Extraction of Roots, may easily be learned from thence.
>
> (Newton 1736: 1–2, translated in Newton 1671: 38–39)

Stevin (1585a) had briefly introduced infinite decimals in his book *De Thiende* (*The Tenth*), although the book is mainly about calculation with finite decimals. I do not know what "Doctrine" on infinite decimals Newton could refer to here. He seems to rely on familiarity with numbers and arithmetic in general, rather than familiarity with infinite decimals, in order to "sell" power series. In reality, calculation with infinite decimals is probably *less* understood than calculation with power series, especially with regard to "Extraction of Roots." We know less about the infinite decimal for $\sqrt{2}$, for example, than we do about the power series for $\sqrt{1+x}$. Specifically, the binomial series gives a formula for the nth term of the power series for $\sqrt{1+x}$, but there is no known formula for the nth decimal digit of $\sqrt{2}$.

Mechanics

In ancient times only a small part of mechanics was well understood: the basic statics involved in the law of the lever. But even this small idea was enough to inspire some of the most spectacular discoveries of Archimedes, in a book called *The Method*. His "method" was used for discovery rather than proof, but as he tells us,

> it is of course easier to supply the proof when we have previously acquired some knowledge of the questions by the method, than it is to find it without any previous knowledge.
>
> (Heath 1897: 13 of Supplement.)

The mechanics of continuous motion and continuous media was the subject of a series of calculus triumphs in the eighteenth century, so much so that physics and mathematics began to look like two sides of the same coin. "Rational mechanics" became a body of theorems and a style of proof, and mathematical discoveries could be prompted by physical phenomena (for example, Fourier series by the modes of vibration of a taut string, and later by the theory of heat). Some later results, such as the mapping theorem of Riemann (1851), were even claimed on physical grounds before they could be proved in pure mathematics. Indeed, until the late nineteenth century many mathematicians did not draw a clear distinction between geometry and physics: both were thought to be based on intuitive ideas about the natural world.

Conversely, when Hilbert (1899) put geometry on a completely axiomatic basis it raised the possibility of doing the same for physics.

Indeed, Hilbert in 1900 posed the problem of finding axioms for physics, which by that time included probability theory because of the growing belief in atoms and their role in statistical mechanics and thermodynamics. Corry (2004) gives an excellent account of Hilbert's own attempts to find axioms for physics, which were partially successful—they led to an anticipation of general relativity theory in 1915 at about the same time that Einstein was formulating his own version of the theory. For physics as a whole this problem is still open, but axioms for probability theory were given by Kolmogorov (1933).

CHAPTER 7

Number Theory

The natural numbers $0, 1, 2, 3, 4, \ldots$ are the most fundamental mathematical objects, understood to some extent by everybody. They are also the subject of the oldest unsolved problems in mathematics. For example, are there any odd perfect numbers? Infinitely many twin primes? Yet, until recent times number theory was often derided as a "bag of tricks," of little interest or use to most mathematicians.

Attitudes changed in recent decades, when the world went digital and numbers became its lifeblood, needing the protection of encryption, which ultimately relies on number theory. So the problem today is not to justify number theory but to understand it.

As we will see in this chapter, number theory is hard because its methods of proof draw on virtually all areas of mathematics. They include geometry, algebra, calculus, and some we haven't discussed yet, such as topology. This is surprising because number theory has extremely simple ingredients: 0, the successor function that takes us from one natural number to the next, and the principle of **induction**, which essentially says that all natural numbers come from 0 by repeated application of the successor function.

The fact is, simple ingredients can create extreme complexity, which is why all the resources of mathematics have been recruited to help number theory. In this chapter we will discuss the influence of geometry, algebra, and calculus on proofs in number theory—and vice versa, particularly in the case of algebra. Later, when we come to the mathematical study of proof itself, we will see what makes number theory capable of such complexity.

7.1 ELEMENTARY NUMBER THEORY

As we saw in section 2.6, Euclid developed the fundamentals of number theory in books 7–9 of the *Elements*. To do so, he used **induction**, the fundamental method of proof in number theory and other areas of mathematics that study finite objects, such as combinatorics. In particular, Euclid used induction to prove that there are infinitely many prime numbers, the existence of a prime factorization of any integer greater than 1, and (essentially) the uniqueness of that factorization.

The latter result emerged from an investigation of what we now call the **Euclidean algorithm** for finding $\gcd(a, b)$, the greatest common divisor of positive integers a and b. A basic consequence of the Euclidean algorithm, derived in section 2.6 and used again below, is that

$$\gcd(a, b) = ma + nb, \quad \text{for some integers } m \text{ and } n.$$

Congruences

The relation of congruence modulo a given integer was introduced by Gauss (1801: article 1), and it gives a convenient language for stating and proving many facts about divisibility. Today we use the following notation for the relation of **congruence modulo a given integer** c:

$$a \equiv b \,(\text{mod } c).$$

It simply means c *divides* $a - b$. The symbol $\equiv$ is chosen because it is similar to the = symbol, and congruence has many of the properties of equality. Informally, one can say that $a \equiv b$ (mod c) means that a and b are "equal up to multiples of c."

Just as one can add, subtract, multiply, and (sometimes) divide equations, one can add, subtract, multiply, and (sometimes) divide congruences modulo a given c. That is, if $a_1 \equiv b_1$ (mod c) and $a_2 \equiv b_2$ (mod c) then

$$a_1 + a_2 \equiv b_1 + b_2 \,(\text{mod } c)$$
$$a_1 - a_2 \equiv b_1 - b_2 \,(\text{mod } c)$$
$$a_1 a_2 \equiv b_1 b_2 \,(\text{mod } c)$$
$$a_1 a_2^{-1} \equiv b_1 b_2^{-1} \,(\text{mod } c) \quad \text{if } a_2 \not\equiv 0 \,(\text{mod } c).$$

The first two of these follow easily from the definition of congruence. To prove the third, we first convert the congruences into ordinary equations:

$$a_1 \equiv b_1 \pmod{c} \quad \text{and} \quad a_2 \equiv b_2 \pmod{c} \quad \text{mean}$$
$$a_1 = b_1 + m_1 c \quad \text{and} \quad a_2 = b_2 + m_2 c \quad \text{for some } m_1, m_2 \in \mathbb{Z},$$

so multiplication gives

$$a_1 a_2 = (b_1 + m_1 c)(b_2 + m_2 c)$$
$$= b_1 b_2 + c(m_1 b_2 + m_2 b_1 + c m_1 m_2)$$
$$\equiv b_1 b_2 \pmod{c}.$$

Thus addition, subtraction, and multiplication of congruences mod c are always valid. To accomplish division we need to know when the inverse a^{-1} of a exists, mod c. Well, m is an inverse of a, mod c, if

$$ma \equiv 1 \pmod{c},$$

or, equivalently, $1 = ma + nc$ for some $n \in \mathbb{Z}$.

Since any common divisor of a and c divides $ma + bc$, this last equation implies that $\gcd(a, c) = 1$. Conversely, if $\gcd(a, c) = 1$ then $1 = ma + nc$ for some $m, n \in \mathbb{Z}$, by the basic consequence of the Euclidean algorithm. So *a has an inverse, mod c if and only if* $\gcd(a, c) = 1$. It is precisely by these values of a that we can divide congruences mod c.

In particular, *if p is prime then* a^{-1} *exists for each* $a \not\equiv 0 \pmod{p}$, because in this case $\gcd(a, p) = 1$ for any a not a multiple of p. To sum up: *if p is prime then it is valid to add, subtract, multiply, and divide (by nonzero elements) congruences mod p.*

Example: Casting out nines (or threes). The base 10 system of numerals, used in India for more than 1,000 years and introduced to Europe by Fibonacci (1202), has a useful property known as **casting out nines**. In its simplest form, the property states that the remainder when a number is divided by 9 equals the remainder when the sum of its digits is divided by 9. The earliest known statement of the property is by the Indian mathematician Aryabhata II around 950 CE. Fibonacci (1202: chap. 2) uses the method to check correctness of multiplications.

As an example, to find the remainder when 4877 is divided by 9 one computes

$$4 + 8 + 7 + 7 = 26,$$

which has remainder 8 when divided by 9. It then follows, by casting out nines, that 4877 has remainder 8 when divided by 9. The reason this property holds is that, by definition of decimal numerals,

$$4877 = 4 \cdot 10^3 + 8 \cdot 10^2 + 7 \cdot 10 + 7.$$

And of course $10 \equiv 1 \pmod 9$. Since it is valid to add and multiply congruences, we can replace each occurrence of 10 in the above equation by 1, which gives

$$4 \cdot 10^3 + 8 \cdot 10^2 + 7 \cdot 10 + 7 \equiv 4 + 8 + 7 + 7 \pmod 9.$$

This says precisely that 4877 and $4 + 8 + 7 + 7$ have the same remainder on division by 9.

The same is true for division by 3, because $10 \equiv 1 \pmod 3$ also.

Some Finite Fields and Rings

The concept of congruence also leads to an algebra of fields and rings like those studied in section 4.6, except that the new fields and rings are *finite*.

The pseudo-equality of the congruence relation can be replaced by actual equality if, instead of operating on numbers, we operate on their **congruence classes** mod c. Today this is a simple step, but it was a dramatic innovation in the 1850s, when introduced by Dedekind, because it amounts to *treating infinite sets as mathematical objects*.[1] For any integer a we define the **congruence class of a, mod** c to be

$$[a] = \{a' : a' \equiv a \pmod c\} = \{\ldots, a - 2c, a - c, a, a + c, a + 2c, \ldots\}$$
$$= \{a + mc : m \in \mathbb{Z}\}.$$

Then we define sum and product of congruence classes in the obvious way:

$$[a] + [b] = [a + b], \quad [a][b] = [ab].$$

The subtle part is to check that these definitions are *meaningful*, that is, independent of the element chosen to represent the congruence class. If we choose $a' = a + nc$ instead of a we get

$$[a'] + [b] = [a' + b] = [a + nc + b] = \{a + b + mc : m \in \mathbb{Z}\},$$

---

1. In retrospect we can see another more ancient example of this: treating a class of fractions, such as $\{1/2, 2/4, 3/6, \ldots\}$, as a single rational number. In a sense, this is a more sophisticated example, since defining the sum of fractions is quite complicated.

but this is exactly the same set as $[a] + [b]$. Similarly, if we change the representative of $[b]$. Likewise,

$$[a'][b] = [(a + nc)b] = [ab + nbc] = \{ab + mc : m \in \mathbb{Z}\},$$

which is exactly the same set as $[a][b]$.

Thus it is meaningful to add and multiply congruence classes. Now comes the easy part: *the sum and product of congruence classes inherit the ring properties from those of* $\mathbb{Z}$. (Recall the defining properties of rings and fields from section 4.6.) For example, here is why $[a] + [b] = [b] + [a]$:

$$[a] + [b] = [a + b] \qquad \text{(by definition of sum of congruence classes)}$$
$$= [b + a] \qquad \qquad \text{(because } a + b = b + a \text{ in } \mathbb{Z})$$
$$= [b] + [a]. \qquad \text{(by definition of sum of congruence classes)}$$

All the other ring properties hold for similar reasons. Thus *the congruence classes of integers mod c form a ring under their addition and multiplication operations.* The ring is denoted by $\mathbb{Z}/c\mathbb{Z}$.

Moreover, *when p is prime, the ring $\mathbb{Z}/p\mathbb{Z}$ is a field.* The reason is that, in this case, each congruence class $[a] \neq [0]$ has an inverse, as we saw in the previous subsection.

To illustrate how algebraic structure can simplify proofs in number theory, we give a new statement and proof of a theorem first proved by Fermat using properties of binomial coefficients.

Fermat's little theorem. *If p is prime and a is not divisible by p, then*

$$a^{p-1} \equiv 1 \pmod{p}.$$

Proof. Consider the nonzero congruence classes mod p: $[1], [2], \ldots,$ $[p - 1]$. Multiplying these $p - 1$ nonzero classes by $[a] \neq [0]$ gives nonzero classes $[a][1], [a][2], \ldots, [a][p - 1]$. These, too, are distinct nonzero classes, since we can recover the classes $[1], [2], \ldots, [p - 1]$ from them by multiplying by $[a]^{-1}$.

Thus $[a][1], [a][2], \ldots, [a][p - 1]$ are the *same classes* as $[1], [2], \ldots,$ $[p - 1]$ (though possibly in a different order), and hence their product is the same:

$$[a][1][a][2] \cdots [a][p - 1] = [1][2] \cdots [p - 1].$$

Multiplying each side by the inverses of $[1], [2], \ldots, [p - 1]$ then gives

$$[a]^{p-1} = [1],$$

which is equivalent to $a^{p-1} \equiv 1 \pmod{p}$. $\qquad \square$

7.2 PYTHAGOREAN TRIPLES

The Pythagorean theorem, as we saw in section 1.2, goes hand in hand with the number theory of Pythagorean triples: the integer triples $\langle a, b, c \rangle$ such that $a^2 + b^2 = c^2$. We saw that, as long ago as 1800 BCE, the Babylonians could apparently generate an unlimited supply of Pythagorean triples. Whether they could generate *all* Pythagorean triples is unclear: the first complete description of all the triples is in Euclid's *Elements*, book 10, in lemma 1 following proposition 28.

Euclid's description amounts to the formula

$$a = (p^2 - q^2)r, \quad b = 2pqr, \quad c = (p^2 + q^2)r,$$

where p, q, r are positive integers. If we omit the factor r, then the formula

$$a = p^2 - q^2, \quad b = 2pq, \quad c = p^2 + q^2$$

includes all triples that are *essentially different*, in the sense that they give triangles of different shapes. (Though there are still some repeated shapes unless we further restrict p and q.)

Euclid's statement and proof are in geometric terms, involving rectangles and squares. They benefit from translation into algebraic terms, since

$$a^2 + b^2 = c^2 \quad \text{implies} \quad b^2 = c^2 - a^2 = (c - a)(c + a),$$

suggesting b should be a product. But some difficulties remain, involving rather tedious consideration of even and odd numbers. Here is a more precise statement and a proof by elementary number theory.

Primitive Pythagorean triples. *If a, b, c are positive integers with greatest common divisor 1, and $a^2 + b^2 = c^2$, we call $\langle a, b, c \rangle$ a primitive Pythagorean triple. Then exactly one of a, b is even and there are positive integers p, q, with greatest common divisor 1, such that*

$$a = p^2 - q^2, \quad b = 2pq, \quad c = p^2 + q^2.$$

Proof. Since a, b, c have gcd 1, a and b cannot both be even; otherwise, 2 would divide a, b, c. They also cannot both be odd, since

$$(\text{odd})^2 = (2m + 1)^2 = m^2 + 4m + 1 \equiv 1 \ (\text{mod} \ 4),$$

in which case $a^2 + b^2 \equiv 2 \ (\text{mod} \ 4)$, whereas $c^2 \equiv 0 \ (\text{mod} \ 4)$ since c is even in this case.

Thus we can assume, without loss of generality, that a is odd and b is even. Since $\langle a, b, c \rangle$ is a Pythagorean triple we have

$$b^2 = c^2 - a^2 = (c-a)(c+a) \quad \text{so} \quad \left(\frac{b}{2}\right)^2 = \frac{c-a}{2} \cdot \frac{c+a}{2}.$$

Now $\frac{c-a}{2}$ and $\frac{c+a}{2}$ are integers with gcd 1, because any common prime divisor will divide their sum c and difference a, contrary to the hypothesis. It then follows, from unique prime factorization, that these factors of $\left(\frac{b}{2}\right)^2$ are both squares, say,

$$\frac{c-a}{2} = q^2, \quad \frac{c+a}{2} = p^2.$$

Subtracting and adding these equations, we get

$$a = p^2 - q^2, \quad c = p^2 + q^2.$$

They also give $\left(\frac{b}{2}\right)^2 = p^2 q^2$, so $b = 2pq$.

Finally, $\gcd(p, q) = 1$, because any common prime divisor of p, q would also divide a, b, c. □

In the next subsection we give an algebro-geometric proof that avoids consideration of common divisors by finding the ratios a/c and b/c instead. These ratios encapsulate the "shape" of the $\langle a, b, c \rangle$ triangle.

Rational Points on the Circle

If a, b, c are integers with $a^2 + b^2 = c^2$ then $\left(\frac{a}{c}\right)^2 + \left(\frac{b}{c}\right)^2 = 1$, which says that

$$\left(\frac{a}{c}, \frac{b}{c}\right) \text{ is a rational point on the circle } x^2 + y^2 = 1.$$

Algebraic geometry suggests a simple method for finding all rational points on this circle.

- There are several obvious rational points, for example, $P = \langle -1, 0 \rangle$.
- The line through the rational point P and any other rational point Q has rational slope.
- Take the line $y = t(x + 1)$ through the point P with rational slope t, and find its second point, R, of intersection with the circle (figure 7.1). If R is rational, then we have found *all* rational points on the circle.

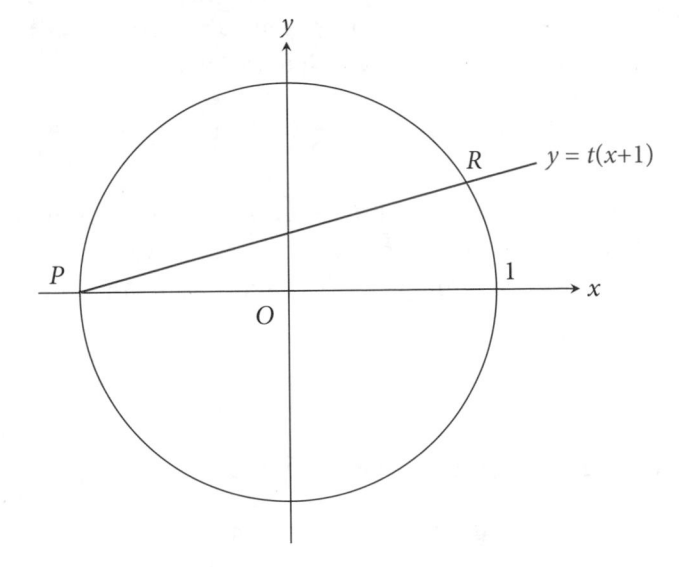

Figure 7.1 : Chord construction of rational points

The intersections of the line $y = t(x+1)$ with the circle $x^2 + y^2 = 1$ have x-values that are solutions of the quadratic equation

$$x^2 + t^2(x+1)^2 = 1, \quad \text{or} \quad (1+t^2)x^2 + 2t^2x + t^2 - 1 = 0,$$

$$\text{or} \quad x^2 + \frac{2t^2x}{1+t^2} + \frac{t^2-1}{1+t^2} = 0.$$

We know that $x = -1$ is one solution of this equation (giving the point P), so one factor of the quadratic is $x + 1$. This means the factorization is necessarily

$$x^2 + \frac{2t^2x}{1+t^2} + \frac{t^2-1}{1+t^2} = (x+1)\left(x - \frac{1-t^2}{1+t^2}\right),$$

so the other solution of the quadratic equation is $x = \frac{1-t^2}{1+t^2}$.

Since $y = t(x+1)$ it follows that the y-value at R is $y = \frac{2t}{1+t^2}$. This gives the point

$$R = \left(\frac{1-t^2}{1+t^2}, \frac{2t}{1+t^2}\right), \qquad (*)$$

which is indeed rational, since t is. Thus all rational points $R \neq P$ on the unit circle are given by $(*)$, as t runs through all rational numbers.

Finally, if we let $t = q/p$ in lowest terms and recall that $x = a/c$ and $y = b/c$ for some Pythagorean triple $\langle a, b, c \rangle$, we have

$$x = \frac{a}{c} = \frac{p^2 - q^2}{p^2 + q^2}, \quad y = \frac{b}{c} = \frac{2pq}{p^2 + q^2}.$$

Thus instead of finding a, b, c outright we find the ratios a/c and b/c, which correspond to the essentially different triples $\langle a, b, c \rangle$.

The above style of proof is credited to Diophantus by Weil (1984: 28). Diophantus lived during the late period of Greek mathematics, somewhere between 150 CE and 300 CE. He used neither the algebraic nor the geometric language of the proof above and worked out only special cases with particular numbers. Nevertheless, when European mathematicians read Diophantus in the light of algebra they could tell what he was doing and saw how to generalize it. Viète (1593: 65–67) saw how to find all rational points on the circle $x^2 + y^2 = a^2 + b^2$, and Fermat, Wallis, and others found the rational points on other quadratic curves.

Parametric Equations for Quadratic Curves

The proof above involves more algebra and geometry than some readers may have wished for, but it tells us more about the circle than just its rational points. The chord construction is valid for any real value of t, so the equations

$$x = \frac{1 - t^2}{1 + t^2}, \quad y = \frac{2t}{1 + t^2}$$

are *parametric equations* for the circle, which give all of its points (other than $\langle -1, 0 \rangle$) as t runs through the real numbers. These are precisely the parametric equations we used to "rationalize" the arc length integral

$$\int \sqrt{\left(\frac{dx}{dt}\right)^2 + \left(\frac{dy}{dt}\right)^2}\, dt \quad \text{to} \quad \int \frac{2\, dt}{1 + t^2}$$

in section 6.8. More generally, the substitution $x = \frac{1 - t^2}{1 + t^2}$ will rationalize any integral $\int f(x, \sqrt{1 - x^2})\, dx$ where f is a rational function of x and $\sqrt{1 - x^2}$. Thus number theory throws light on integral calculus by suggesting a parallel between rational numbers and rational functions—bearing out the speculation of Leibniz mentioned in section 6.9.

The chord construction leads to parametric equations for any quadratic curve. For example, if the curve is the hyperbola $x^2 - y^2 = 1$:

- Choose a point on the curve, say, $P = \langle -1, 0 \rangle$ (though it need not be a rational point).
- Consider the line $\mathscr{L}$ of slope t through P, in this case $y = t(x+1)$.
- Find the second point R at which $\mathscr{L}$ meets the curve, in this case

$$R = \left\langle \frac{1+t^2}{1-t^2}, \frac{2t}{1-t^2} \right\rangle.$$

- Then, in this case,

$$x = \frac{1+t^2}{1-t^2}, \quad y = \frac{2t}{1-t^2}$$

are rational functions that parameterize the curve.

Of course, if we want only the rational points on the curve we must choose P rational and t rational. Not every quadratic curve contains a rational point, for example, the circle $x^2 + y^2 = 3$. But this circle can be parameterized by rational *functions*, for example,

$$x = \sqrt{3}\frac{1-t^2}{1+t^2}, \quad y = \sqrt{3}\frac{2t}{1+t^2},$$

because a rational function need not have rational number coefficients.

7.3 FERMAT'S LAST THEOREM

The formula for Pythagorean triples shows that there are infinitely many solutions of the equation $x^2 + y^2 = z^2$ in positive integers x, y, z. However, the same does not apply to $x^n + y^n = z^n$ when n is an integer greater than 2. In fact, there are *no* positive integer solutions in this case. This statement is the famous **Fermat's last theorem**, claimed by Fermat in the 1630s but not proved until 1994. Fermat himself seems to have proved the theorem only for $n = 4$, using a very interesting elementary method.

Here we present a version of Fermat's proof, somewhat streamlined but with essentially the same ideas, namely, the formula for Pythagorean triples and an ingenious induction argument in "infinite descent" form.

Fermat's last theorem for fourth powers. *If x, y, z are positive integers, then $x^4 + y^4 \neq z^4$.*

Proof. Suppose, on the contrary, that x, y, z are positive integers such that $x^4 + y^4 = z^4$. By dividing out any common divisors, we can assume that the greatest common divisor of x, y, z is 1. The equation $x^4 + y^4 = z^4$ implies that $\langle x^2, y^2, z^2 \rangle$ is a Pythagorean triple, and the triple is primitive because the greatest common divisor of x^2, y^2, z^2 is also 1.

Then it follows, from the discussion of primitive Pythagorean triples in the previous section, that one of x^2, y^2 is even and the other is odd. Without loss of generality we can assume that y^2 is even. We now view $x^2 = u$, $y = v$, and $z = w$ as positive integer solutions of the equation

$$u^2 + v^4 = w^4, \tag{*}$$

in which case $\langle u, v^2, w^2 \rangle$ is also a primitive Pythagorean triple, with v^2 even.

By the theorem on primitive Pythagorean triples in the previous section, we get positive integers p, q such that

$$u = p^2 - q^2, \quad v^2 = 2pq, \quad w^2 = p^2 + q^2.$$

The last equation here shows that $\langle p, q, w \rangle$ is a Pythagorean triple, which is also primitive since a common prime divisor of p, q would give a common prime divisor of u, v^2, w^2.

By the formula for primitive Pythagorean triples again, we get positive integers s, t with greatest common divisor 1 such that either

$$p = s^2 - t^2 \text{ and } q = 2st, \quad \text{or} \quad p = 2st \text{ and } q = s^2 - t^2.$$

In either case we have

$$v^2 = 2pq = 4st(s^2 - t^2), \quad \text{with } \gcd(s, t) = 1.$$

Unique prime factorization implies that s, t, and $s^2 - t^2$ are all squares, say,

$$s = a^2, \quad t = b^2, \quad s^2 - t^2 = a^4 - b^4 = c^2.$$

Now the last equation, $c^2 + b^4 = a^4$, *has the same form as equation* (*). Also, tracing back, we find $a \leq s < p < w$ or $a \leq s < q < w$. Thus, from any positive integer solution w of (*) we can find a *smaller* positive integer solution a. This "infinite descent" is impossible, so there is no positive integer solution of (*), or of $x^4 + y^4 = z^4$. $\qquad \square$

Translating Fermat from Numbers to Functions

The above proof about integers translates to a proof about polynomials, with the remarkable conclusion that *the curve $y^2 = 1 - x^4$ cannot be parameterized by rational functions*. Here is an outline of the argument.

If $y^2 = 1 - x^4$ can be parameterized by rational functions, let the functions be

$$x = \frac{p(t)}{q(t)}, \quad y = \frac{r(t)}{s(t)}, \quad \text{where } p(t), q(t), r(t), s(t) \text{ are polynomials in } t.$$

Substituting these in the equation $y^2 = 1 - x^4$ gives

$$\frac{p(t)^2}{q(t)^2} = 1 - \frac{r(t)^4}{s(t)^4}$$

and therefore, multiplying through by $q(t)^4 s(t)^4$,

$$p(t)^2 q(t)^2 s(t)^4 = q(t)^4 s(t)^4 - q(t)^4 r(t)^4.$$

This means that

$$u = p(t)q(t)s(t)^2, \quad v = q(t)r(t), \quad w = q(t)s(t)$$

is a *polynomial* solution of the equation $u^2 + v^4 = w^4$.

Jakob Bernoulli (1704) speculated that Fermat's proof that $u^2 + v^4 = w^4$ has no integer solutions might explain why $\sqrt{1 - x^4}$ cannot be rationalized by substituting a rational function for x, though he did not work out the details. It is now not difficult to do so, by showing that $u^2 + v^4 = w^4$ has no polynomial solution.

Stevin (1585b) already pointed out that there is a Euclidean algorithm for polynomials and that it can be used to calculate the gcd of two polynomials. (For more about this Euclidean algorithm, see section 7.5.) At the time, there was no well-trodden path from the Euclidean algorithm to unique prime factorization, but now that we know such a path (section 2.6), it is easy to retrace it with polynomials in mind and to carry on to the results on Pythagorean triples that depend on it. We find the following:

- For any polynomials $a(t)$ and $b(t)$ there are polynomials $m(t)$ and $n(t)$ such that

$$\gcd(a(t), b(t)) = m(t)a(t) + n(t)b(t).$$

- Any polynomial has a factorization into **irreducible** polynomials—that is, polynomials not the product of polynomials of lower degree—and the factorization is unique up to nonzero constant factors. This is the polynomial version of **unique prime factorization**.
- If polynomials $a(t), b(t), c(t)$ are such that

$$a(t)^2 + b(t)^2 = c(t)^2,$$

then arguing as in section 7.2 yields polynomials $p(t), q(t)$ with

$$a(t) = p(t)^2 - q(t)^2, \quad b(t) = 2p(t)q(t), \quad c(t) = p(t)^2 + q(t)^2.$$

(Also, we need no longer distinguish between "even" and "odd," because polynomials that differ by a constant factor are essentially the same for purposes of factorization. Thus we can exchange the formulas for $a(t)$ and $b(t)$.)

With these basic results we can translate Fermat's proof from integers to polynomials and show that there is no rational function parameterization of the curve $y^2 = 1 - x^4$ (figure 7.2). It follows that there is no rational function substitution $x = a(t)$ that will rationalize the integral

$$\int \frac{dx}{\sqrt{1 - x^4}}.$$

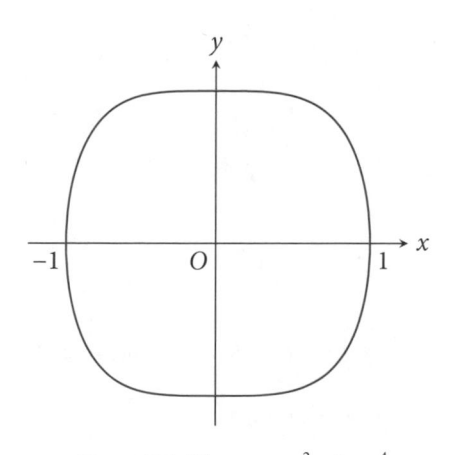

Figure 7.2 : The curve $y^2 = 1 - x^4$

7.4 GEOMETRY AND CALCULUS IN NUMBER THEORY

We have now seen how Pythagorean triples can be understood through rational points on the unit circle and how this leads in turn to parameterization of the circle by rational functions:

$$x = \frac{1 - t^2}{1 + t^2}, \quad y = \frac{2t}{1 + t^2}. \tag{*}$$

Moreover, we have seen that the equations provide a change of variable that rationalizes integrals of functions that are rational functions of x and $\sqrt{1 - x^2}$.

Readers who know the circle parameterization by circular functions,

$$x = \cos \theta, \quad y = \sin \theta,$$

may wonder where it fits into this story. The answer is that the parameters t and θ are linked by the equation $t = \tan \frac{\theta}{2}$. This can be seen by the diagram of the line and the circle used to find the equations (*), shown in figure 7.3.

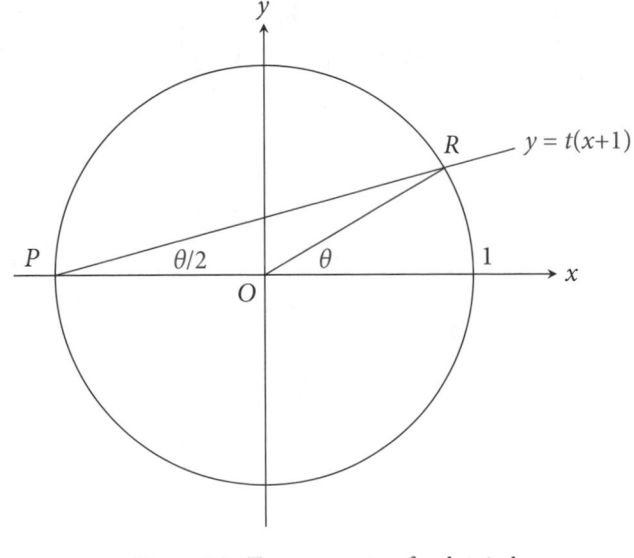

Figure 7.3 : Two parameters for the circle

On the one hand, we know that

$$R = \left(\frac{1 - t^2}{1 + t^2}, \frac{2t}{1 + t^2} \right).$$

On the other hand, by definition of sine and cosine,

$$R = \langle \cos \theta, \sin \theta \rangle.$$

Then by basic geometry (isosceles triangle, angle sum of triangle $= \pi$) we find that angle OPR equals $\theta/2$. Thus the slope t of the red line is $\tan \frac{\theta}{2}$.

While rational functions are often preferred to transcendental functions such as sine and cosine, the latter functions are a better guide to parameterization when we come to curves that cannot be parameterized by rational functions, such as $y^2 = 1 - x^4$. To see how to proceed with the latter curve, we review the role of the circle and circular functions in calculus.

Calculus of the Circle and Other Curves

If $\mathscr{C}$ is a curve of the form $y^2 = p(x)$, where $p(x)$ is a polynomial, there is a surprisingly simple way to find a pair of functions that parameterize $\mathscr{C}$. Moreover, if $x = f(u)$ then $y = f'(u)$ (the derivative of f). The idea is to consider the integral

$$\int_0^x \frac{dt}{\sqrt{p(t)}} \quad \text{and call it } f^{-1}(x) = u.$$

When this is done, obviously $x = f(u)$. But also

$$\frac{dx}{du} = 1 \bigg/ \frac{du}{dx}$$

$$= 1 \bigg/ \frac{1}{\sqrt{p(x)}} \qquad \text{(by the fundamental theorem of calculus)}$$

$$= \sqrt{p(x)} = y.$$

In the case of the unit circle $x^2 + y^2 = 1$ the polynomial $p(x)$ equals $1 - x^2$ and the integral defining the function $f(u) = x$ is

$$u = f^{-1}(x) = \int_0^x \frac{dt}{\sqrt{1 - t^2}}.$$

It is easy to see, by making the substitution $x = \sin \theta$, that this integral is $\sin^{-1} x$, so $u = f^{-1}(x) = \sin^{-1} x$ and hence $x = \sin u$. Finally, $y = dx/du = \cos u$, so we get the parameterization

$$x = \sin u, \quad y = \cos u,$$

which is the same as the usual parameterization by circular functions, but with x and y switched.

Elliptic Functions and Elliptic Curves

The above method of parameterizing a curve $y^2 = p(x)$ leads only to what we already know for the circle $y^2 = 1 - x^2$. The method gives us something new for the curve $y^2 = 1 - x^4$. We know that this curve does *not* have a rational parameterization, so the functions $f(u)$ and $f'(u)$ for x and y are probably interesting.

Indeed, the integral $\int_0^x \frac{dt}{\sqrt{1-t^4}} = f^{-1}(x) = u$ is an **elliptic integral**, as mentioned in section 6.9, and the functions $f(u)$ and $f'(u)$ are called **elliptic functions**. Looking at the functions, rather than the integrals inverse to them, seems like a good idea in hindsight, since it is clearly easier to study the sine function than the arcsine integral. However, the first mathematician to look at elliptic functions rather than elliptic integrals was Gauss (around 1800, unpublished), and this happened only after some properties of elliptic integrals had been laboriously worked out by Fagnano (1718) and by Euler, after he first saw Fagnano's work in 1751. The elliptic function idea was not published until the 1820s, when rediscovered by Abel and Jacobi.

Gauss discovered that the inverse function $\mathrm{sl}(u)$ of the integral

$$u = \mathrm{sl}^{-1}(x) = \int_0^x \frac{dt}{\sqrt{1 - t^4}}$$

resembles the sine function quite strongly—so much so that he called it the **lemniscatic sine**. (The lemniscate, you may recall from section 6.9, is a curve whose arc length was found by Jakob Bernoulli to be given by the integral $\int_0^x \frac{dt}{\sqrt{1-t^4}}$.)

In particular, the lemniscatic sine is *periodic*: $\mathrm{sl}(u + \varpi) = \mathrm{sl}(u)$ for a certain minimal number ϖ. Gauss chose the letter ϖ because it is a variant of the Greek letter π. But more than that, if we allow u to be complex then sl has a *second* period $i\varpi$. And we know it is natural to allow x and y to be complex for algebraic curves, for reasons already explained in section 5.5.

This **double periodicity** also holds for other elliptic functions (but not for the sine and cosine, which remain only singly periodic when we allow them complex arguments). It leads to a dramatically new interpretation of the curves they parameterize, called **elliptic curves**.[2]

For example, here is how to view the curve $\mathscr{C}$ with Cartesian equation $y^2 = 1 - x^4$. $\mathscr{C}$ has parametric equations

$$x = \mathrm{sl}(u), \quad y = \mathrm{sl}'(u),$$

so each point P on $\mathscr{C}$ is defined by a value of the parameter u. But, due to the periodicity of sl (and of sl′, which has the same periodicity), the parameter values $u + m\varpi + ni\varpi$ define the *same* point, for any integers m and n.

Thus each point P of $\mathscr{C}$ corresponds to a *class of points*

$$\{u + m\varpi + ni\varpi : m, n \in \mathbb{Z}\}$$

in the plane $\mathbb{C}$ of complex numbers. We can choose a representative of each class from the square with corners $0, \varpi, i\varpi, (1+i)\varpi$, in which case there is just one representative in the square for each point P with the exception of points on the boundary. Points $u, u + \varpi$ on the left and right sides represent the same point of $\mathscr{C}$, as do points $u, u + i\varpi$ on the top and bottom, and consequently, all the four corners $0, \varpi, i\varpi, (1+i)\varpi$ represent the same point. The square is drawn in gray on the left in figure 7.4, with the left and right sides blue, and top and bottom sides red.

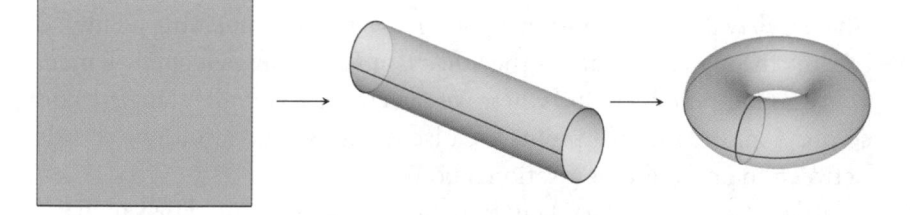

Figure 7.4 : From the square to the torus

Loosely speaking (or *topologically* speaking—see chapter 10), the complex curve $\mathscr{C}$ is the result of *pasting the like-colored sides of the square*,

---

2. The evolution of this terminology—from the arc length of the ellipse to "elliptic integrals" to "elliptic functions" to "elliptic curves"—has the unfortunate outcome that the ellipse is *not* an elliptic curve. Like its special case, the circle, any ellipse can be parameterized by rational functions or by circular functions.

which is the surface known as the **torus**. Thus the search for rational points on curves leads us not only to geometry and calculus but also to topology. In the next subsection we revisit the geometry of rational points on elliptic curves.

Rational Points on Elliptic Curves

The elliptic curves include many *cubic* curves, whose study goes back to Diophantus, so a lot was known about them before their connection with elliptic functions was discovered. In particular, it was known how to use a known rational point to find new ones.

Suppose, for example, that $p(x, y)$ is a cubic polynomial with rational coefficients and that $y = mx + c$ is a line through two rational points on the curve $p(x, y) = 0$. Then m and c are also rational and the line meets the curve where $p(x, mx + c) = 0$, a cubic equation with rational coefficients. It follows that $p(x, mx + c)$ splits into three linear factors, two of which correspond to the known rational points. The third factor must then also be rational and correspond to a third rational point. To sum up: *if $\mathscr{C}$ is a curve whose equation $p(x, y) = 0$ is cubic with rational coefficients, then a line through two rational points on $\mathscr{C}$ passes through a third rational point.*

This is all very well, but what if we know only *one* rational point on $\mathscr{C}$? We can escape this predicament by pretending that the one point is two! That is, view the known rational point P as a pair of coinciding points, so the line through this "pair" is the *tangent* at P. The tangent at P will meet $\mathscr{C}$ at another rational point, where we can construct another tangent, and so on. As more rational points are discovered we can also draw chords between them to find new rational points.

Remarkably, the tangent construction was actually used by Diophantus (see Heath 1910: 242) to find rational solutions of the equation

$$y^2 = x^3 - 3x^3 + 3x + 1.$$

He did this without mentioning curves, tangents, or the obvious rational point $\langle 0, 1 \rangle$, and with barely any calculation. In effect, though, he made the substitution $y = \frac{3}{2}x + 1$, presumably foreseeing that this leads to cancellation of $3x + 1$ from both sides, leaving a cubic equation with the double root $x = 0$:

$$\frac{9}{4}x^2 + 3x + 1 = x^3 - 3x^2 + 3x + 1, \quad \text{or} \quad 0 = x^3 - \frac{21}{4}x^2 = x^2\left(x - \frac{21}{4}\right).$$

The double root corresponds to the rational point $\langle 0, 1 \rangle$, counted twice, and the line $y = \frac{3}{2}x + 1$ is the tangent there. The other intersection of this tangent with the curve, at $x = \frac{21}{4}$, is the rational point $\langle \frac{21}{4}, \frac{71}{8} \rangle$ found by Diophantus. Figure 7.5 shows the curve $y^2 = x^3 - 3x^2 + 3x + 1$ and its tangent at $x = 0$.

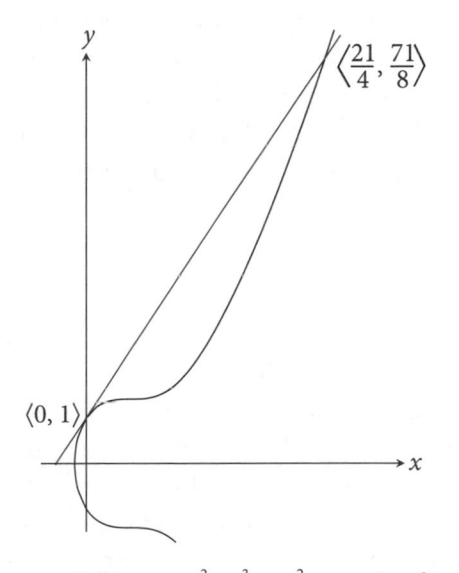

Figure 7.5 : Cubic curve $y^2 = x^3 - 3x^2 + 3x + 1$ and tangent

The work of Diophantus was rediscovered by Bombelli and Viète in the late sixteenth century, and it came to the attention of Fermat in the 1630s. Fermat recognized that the ideas of Diophantus were close to geometry, and Newton (1670) explained them as chord and tangent constructions. Much later, it was found that the geometry of lines and a cubic curve $\mathscr{C}$ is in perfect harmony with the elliptic functions that parameterize $\mathscr{C}$. Jacobi (1834) seems to have been the first to notice this, and Clebsch (1864) made it more explicit: *if P_1, P_2, P_3 are three collinear points on $\mathscr{C}$, and if $[u_1], [u_2], [u_3]$ are the corresponding classes of parameter values, then*

$$[u_1] + [u_2] + [u_3] = [u_1 + u_2 + u_3] = [0].$$

7.5 GAUSSIAN INTEGERS

Around 1770, Lagrange and Euler discovered a powerful new tool in number theory: the use of **algebraic numbers** as a new kind of "integer." We will see a spectacular example from Euler (1770) in the next section. But it is easiest to begin with the simpler example of the **Gaussian integers**, which are the complex numbers of the form $a + bi$, where a, b are ordinary integers and $i = \sqrt{-1}$.

Introduced by Gauss (1832), these were the first **algebraic integers** to be rigorously studied and compared with ordinary integers. In particular, Gauss was able to show that his integers enjoy *unique prime factorization*—a property as useful for algebraic integers as it is for ordinary integers. And, as we have already seen for ordinary integers (section 2.6) and polynomials (section 7.3), unique prime factorization is guaranteed by a *Euclidean algorithm*.

Division with Remainder

Section 1.4 introduced the Euclidean algorithm the way Euclid did, as a process of repeated subtraction. This is the simplest way to describe it and, indeed, the only way when the algorithm is applied to quantities in irrational ratio, such as lengths. But for integers the process can also be described as repeated **division with remainder**, and this is the only way that makes sense in certain other domains, such as polynomials and Gaussian integers.

In the ordinary integers, division with remainder has the following **division property** (often misleadingly called the "division algorithm"): *if $a > b$ are positive integers then there are integers q and r such that*

$$a = qb + r, \quad where \quad |r| < |b|.$$

The integer q is called the **quotient**, and r is called the **remainder**. Their existence follows by repeatedly subtracting b from a, producing a descending sequence of natural numbers. By induction, this sequence terminates at a natural number $r < b$.

The division property also holds for polynomials, if we interpret $|p|$ as the *degree* of the polynomial p. This is due to the way division works for polynomials. To divide a polynomial $a(x)$ by a polynomial $b(x)$ of lower degree one subtracts multiples of $b(x)$ by constant multiples of powers of x, until only a polynomial $r(x)$ with degree less than that of $b(x)$ remains.

Example. Divide $a(x) = x^3$ by $b(x) = x + 1$. We have

$$a(x) - x^2 b(x) = -x^2 \quad \text{(removing highest power of } x \text{ in } a(x)\text{)}$$
$$a(x) - x^2 b(x) + x b(x) = x \quad \text{(removing next highest power of } x\text{)}$$
$$a(x) - x^2 b(x) + x b(x) - b(x) = -1 \quad \text{(removing next highest power of } x\text{)}$$

and therefore $a(x) = (x^2 - x + 1)b(x) - 1$, so $q(x) = x^2 - x + 1$ and $r(x) = -1$. We see that $0 = |r(x)| < |b(x)| = 1$.

In the next subsection we will see that the division property

$$a = qb + r, \quad \text{where} \quad |r| < |b|,$$

also applies to Gaussian integers with $|a| > |b| > 0$, when $|z|$ denotes the **absolute value** of the complex number z.

Geometry of Complex Numbers

During the eighteenth century there was a gradual realization that the system $\mathbb{C}$ of complex numbers may be viewed as a plane, in such a way that addition and multiplication have natural geometric interpretations. We will say more about this in chapter 8. The idea came to full maturity in number theory with Gauss (1832), where the geometry was involved in the proof of unique prime factorization for Gaussian integers. Here is a more explicitly geometric version of Gauss's proof; the main application of the geometry is to prove the division property, which underlies the Euclidean algorithm.

Addition is easy to see. If $u = u_1 + iu_2$ and $v = v_1 + iv_2$, where u_1, u_2, v_1, v_2 are real numbers, then u and v may be viewed as the points $\langle u_1, u_2 \rangle$ and $\langle v_1, v_2 \rangle$, respectively, in the plane. Then $u + v$ is the fourth corner of a parallelogram whose other corners are O, u, and v (figure 7.6). Moreover, $|u| = \sqrt{u_1^2 + u_2^2}$ and $|v| = \sqrt{v_1^2 + v_2^2}$ are the *lengths* of the sides from O to u and v.

To see multiplication, we suppose first that $|u| = |v| = 1$, in which case u and v lie on the unit circle, at angles θ and φ, say. Then $u = \cos\theta + i\sin\theta$, as shown in figure 7.7, and similarly, $v = \cos\varphi + i\sin\varphi$. Their product is

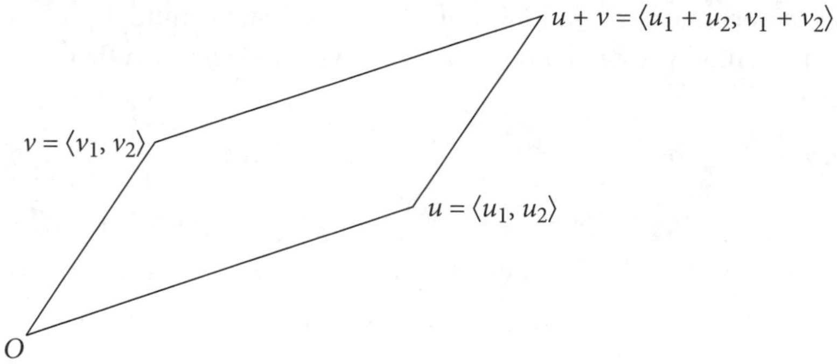

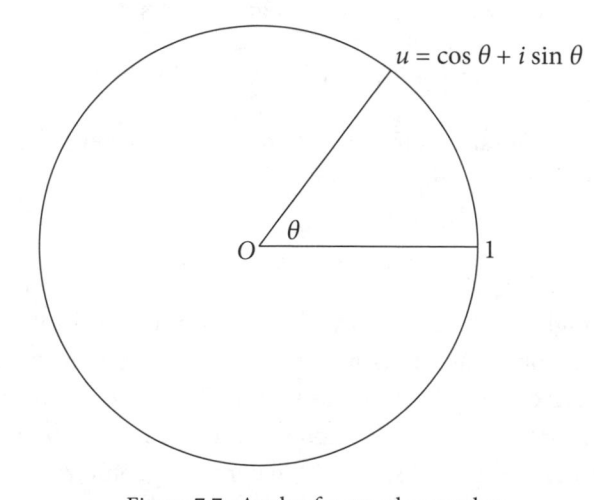

Figure 7.6 : Addition of complex numbers

$u = \cos\theta + i\sin\theta$

θ

1

O

Figure 7.7 : Angle of a complex number

found by ordinary algebra, using $i^2 = -1$, to be

$$uv = (\cos\theta + i\sin\theta)(\cos\varphi + i\sin\varphi)$$
$$= (\cos\theta\cos\varphi - \sin\theta\sin\varphi) + i(\sin\theta\cos\varphi + \cos\theta\sin\varphi)$$
$$= \cos(\theta + \varphi) + i\sin(\theta + \varphi).$$

The last line follows from the formulas

$$\cos(\theta + \varphi) = \cos\theta\cos\varphi - \sin\theta\sin\varphi,$$
$$\sin(\theta + \varphi) = \sin\theta\cos\varphi + \cos\theta\sin\varphi$$

that go back at least as far as Ptolemy's *Almagest* from around 150 CE (see Van Brummelen 2009). Thus, *when u, v have absolute value 1 and angles θ, φ, respectively, uv has absolute value 1 and angle θ + φ.*

More generally, when $|u| = r$ and $|v| = s$ we have

$$u = r(\cos\theta + i\sin\theta), \quad v = s(\cos\varphi + i\sin\varphi)$$

and $uv = rs(\cos(\theta + \varphi) + i\sin(\theta + \varphi))$. So, *to multiply complex numbers, multiply their lengths and add their angles.*

It follows in particular that *if we multiply all members of $\mathbb{C}$ by some $u \in \mathbb{C}$ then the plane $\mathbb{C}$ is magnified by $|u|$ and rotated about O by the angle of u.* We now apply this fact to visualize the set of multiples qb of some nonzero Gaussian integer b by all the Gaussian integers q. The result is shown in figure 7.8 in the case where $b = 3 + i$. The dots are the Gaussian integers, and the black dots among them are the multiples of b.

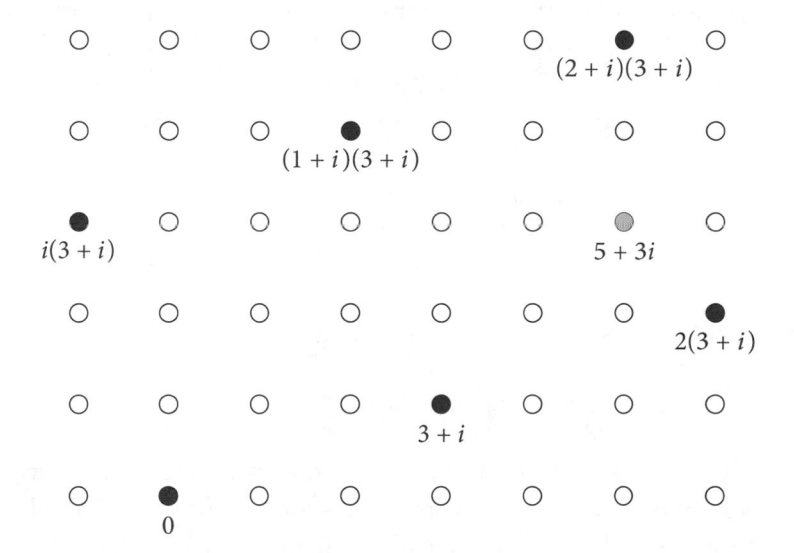

Figure 7.8 : Multiples of $3 + i$ near $5 + 3i$

The Gaussian integers form a grid of squares of side length 1. Their products with b therefore form a grid of squares of side length $|b|$, rotated about O by the angle of b. The crucial **division property**,

$$a = qb + r, \quad \text{for some } q \text{ and } r \text{ with } |r| < |b|,$$

follows for given integers a and nonzero b by choosing the black vertex qb nearest to a in the grid of multiples of b. Then if we put

$$r = a - qp$$

we have $|r| < |b|$ for the geometric reason that *the distance from a point in the square to the nearest vertex is less than the side length of the square.* In the example $a = 5 + 3i$, the nearest vertex is $2(3 + i)$, $r = -1 + i$, and $|r| = \sqrt{2} < \sqrt{10} = |b|$.

The division property, as we know, guarantees a Euclidean algorithm, which in turn gives a prime divisor property and unique prime factorization. We now need to clarify the concepts of "prime" and "unique factorization" in the Gaussian integers.

Gaussian Primes

We now have machinery suitable for proving unique prime factorization of Gaussian integers—except a definition of the Gaussian primes themselves. For this we start with the intuition that a prime is "larger" than 1 and not the product of "smaller" integers, so first we need a measure of size.

We could let the size of a Gaussian integer $a + ib$ be its absolute value,

$$|a + ib| = \sqrt{a^2 + b^2},$$

but it is better to use the squared absolute value, which is called the **norm**:

$$\text{norm}(a + ib) = |a + ib|^2 = a^2 + b^2.$$

Thus the norm is a natural number, which allows us to reduce certain questions about Gaussian integers to questions about natural numbers, where we can apply principles such as induction. The most important of these questions are about *divisibility*.

We say that an integer u (ordinary or Gaussian) **divides** integer w if $w = uv$ for some integer v. The key result about divisibility is:

Multiplicative property of the norm. *If $u = a + ib$ and $v = c + id$ are complex numbers then $|uv| = |u||v|$. So if u, v are Gaussian integers then*

$$\text{norm}(uv) = \text{norm}(u)\text{norm}(v).$$

Proof. The proof of this fact is an identity about sums of squares apparently known to Diophantus[3] and proved by Fibonacci (1225: 23–28):

$$(a^2 + b^2)(c^2 + d^2) = (ac - bd)^2 + (bc + ad)^2. \qquad (*)$$

This identity can be checked by expanding both sides and comparing the results, so it holds for all $a, b, c, d \in \mathbb{R}$. This shows that

$$|u|^2|v|^2 = |uv|^2, \text{ and hence } |u||v| = |uv|.$$

When $a, b, c, d \in \mathbb{Z}$, which is the case for Gaussian integers, we see that the left side of $(*)$ is $\mathrm{norm}(u)\mathrm{norm}(v)$ and the right side is $\mathrm{norm}(uv)$. $\quad\square$

Corollary. *A Gaussian integer u divides a Gaussian integer w only if* $\mathrm{norm}(u)$ *divides* $\mathrm{norm}(w)$ *in the ordinary integers.*

Proof. If u divides w then $w = uv$ for some Gaussian integer v. It follows that $\mathrm{norm}(w) = \mathrm{norm}(u)\mathrm{norm}(v)$ by the multiplicative property; hence $\mathrm{norm}(u)$ divides $\mathrm{norm}(v)$ in the ordinary integers. $\quad\square$

These results about the norm are the link between ordinary integers and **algebraic integers** of various kinds (not just the Gaussian integers but others we will see in the next section). In particular they imply:

Existence of prime factorization. *Any Gaussian integer has a factorization into Gaussian primes.*

Proof. If w is not itself a Gaussian prime, then $w = uv$ for some Gaussian integers u, v of smaller norm. Likewise, if u or v is not a Gaussian prime it splits into Gaussian integer factors of smaller norm. Since these norms are natural numbers, they cannot decrease forever. So the process of splitting into factors must terminate, necessarily with w factored into Gaussian primes. $\quad\square$

Now that we have *existence* of a prime factorization, we can bring to bear the division property of the previous subsection to prove

---

3. It is behind his remark in the *Arithmetica*, book 3, problem 19, that 65 is a sum of two squares because $65 = 5 \cdot 13$ and both 5 and 13 are sums of two squares. As mentioned in section 4.2, Fibonacci proved the general identity, taking several pages to do so because he described all his algebraic manipulations in words.

uniqueness. We follow the usual path, via the **prime divisor property**:

- The division property gives a Euclidean algorithm for the gcd of two Gaussian integers u, v. This divisor is "greatest" in the sense of having maximum norm.
- It follows from the nature of division with remainder that
  $$\gcd(u, v) = mu + nv, \quad \text{for some Gaussian integers } m \text{ and } n.$$
- If p is a Gaussian prime that divides a product uv of Gaussian integers, then p divides u or p divides v.
- Finally, it follows that the **Gaussian prime factorization is unique**, up to divisors of norm 1, which are called **units**.

The latter "not quite" uniqueness is unavoidable since we can always vary a factorization by the units -1, i, or $-i$, and the prime factors can vary accordingly. For example, $1 + i$ is a Gaussian prime (because its norm is 2, which is not the product of smaller norms), but $1 + i = i(1 - i)$. So, any factorization involving $1 + i$ can be rewritten with $i(1 - i)$ in its place. Primes that differ only by unit factors, such as $1 + i$ and $1 - i$, are called **associates**.

Circle Division

As an application of unique Gaussian prime factorization, we answer the following question: *can we divide the unit circle into equal parts by rational points?* If the number of parts is 1, 2, or 4, then obviously yes: use the rational points $\langle 1, 0 \rangle$, $\langle -1, 0 \rangle$, $\langle 0, 1 \rangle$, and $\langle 0, -1 \rangle$. But otherwise the answer is no, and we prove this by representing the rational point $\left\langle \frac{a}{c}, \frac{b}{d} \right\rangle$ by the "rational complex number" $\frac{a}{c} + i\frac{b}{d}$. I learned this proof from Jack Calcut.

Irrationality of circle division. *If $n \neq 1, 2, 4$ then there is no division of the unit circle into n equal parts by rational points.*

Proof. Suppose, for the sake of contradiction, that the circle can be divided into n equal parts by rational points, where $n \neq 1, 2, 4$. We can assume without loss of generality that one of the points is 1; if not, let u be any one of the division points and multiply them all by u^{-1}. This moves the point u to 1 and rotates all the other division points (through the angle of u) to new rational division points, since the quotients of rational points are rational.

Now that 1 is one of the division points and $n \neq 1, 2, 4$, any one of the other division points $v = \frac{a}{c} + i\frac{b}{d}$ is on neither the real nor the imaginary axis, so $a, b, c, d \neq 0$. By taking a common denominator we can assume $v = \frac{a}{c} + i\frac{b}{c}$. And since the points divide the unit circle into n equal parts we also have

$$v = \cos \frac{2\pi m}{n} + i \sin \frac{2\pi m}{n}, \quad \text{for some integer } m \text{ with } 1 \leq m \leq n - 1.$$

It follows, from the geometry of complex numbers, that

$$1 = v^n = \left(\frac{a}{c} + i\frac{b}{c} \right)^n$$

and therefore

$$(a + ib)^n = c^n, \quad \text{with } a, b, c \neq 0.$$

Now consider the Gaussian prime factorizations of $a + ib$ and c. These cannot be the same, even up to unit factors, because the angle of c is 0 and the angle of $a + ib$ is the same as the angle of v, hence unequal to any multiple of $\pi/2$. But then the Gaussian prime factorizations of $(a + ib)^n$ and c^n are not the same up to unit factors either, contrary to unique Gaussian prime factorization. □

7.6 ALGEBRAIC NUMBER THEORY

The idea of using algebraic numbers to answer questions about ordinary integers, and the idea that algebraic numbers could "behave like" ordinary integers, first arose with Euler (1770). Euler's most striking example was a proof that $x = 5$, $y = 3$ is the only positive integer solution of the equation

$$y^3 = x^2 + 2.$$

Diophantus mentioned this equation and this solution in his *Arithmetica*, book 6, problem 17; Fermat (1657) claimed it was the only solution.

Euler (1770: 400–402) gave a proof of Fermat's claim by splitting $x^2 + 2$ into the algebraic factors $x + \sqrt{-2}$ and $x - \sqrt{-2}$ and effectively treating them as "integers." He gave no justification for this procedure—but it *can* be justified, so we will look at Euler's amazing argument first. The justification can wait until the next subsection, but suffice to say that it is along the same lines as the treatment of Gaussian integers in the previous section.

Solution of $y^3 = x^2 + 2$. *The only positive integer solution of* $y^3 = x^2 + 2$ *is* $x = 5$, $y = 3$.

Proof. Suppose that x and y are positive integers such that

$$y^3 = x^2 + 2 = (x + \sqrt{-2})(x - \sqrt{-2}).$$

Suppose also that numbers of the form $a + b\sqrt{-2}$, for $a, b \in \mathbb{Z}$, behave like ordinary integers, in particular, that $\gcd(x + \sqrt{-2}, x - \sqrt{-2}) = 1$ and that unique prime factorization holds.

Then, since $(x + \sqrt{-2})(x - \sqrt{-2})$ is a cube, y^3, it follows that $x + \sqrt{-2}$ and $x - \sqrt{-2}$ are themselves cubes. Therefore

$$x + \sqrt{-2} = (a + b\sqrt{-2})^3, \quad \text{for some } a, b \in \mathbb{Z}$$

$$= a^3 + 3a^2 b\sqrt{-2} + 3ab^2(-2) + b^3(-2\sqrt{-2})$$

$$= a^3 - 6ab^2 + (3a^2 b - 2b^3)\sqrt{-2}.$$

Equating the coefficients of $\sqrt{-2}$ (the "imaginary parts") on each side we find

$$1 = b(3a^2 - 2b^2).$$

Since the only divisors of 1 in $\mathbb{Z}$ are ± 1, we must have

$$b = \pm 1, \quad 3a^2 - 2b^2 = \pm 1 \text{ and hence } a = \pm 1.$$

Now equating the real parts, we get $x = a^3 - 6ab^2$, which is positive only for $a = -1$. So $x = 5$ and therefore $y = 3$. $\qquad\square$

Justification for Euler's Proof

With the hindsight of experience with Gaussian integers, we can anticipate that questions about the divisibility of numbers $a + b\sqrt{-2}$, where $a, b \in \mathbb{Z}$, will be clarified with the help of a **norm** for these numbers. Indeed, a suitable norm is again the squared absolute value:

$$\mathrm{norm}(a + b\sqrt{-2}) = |a + b\sqrt{-2}|^2 = a^2 + 2b^2.$$

We observed, in proving the multiplicative property for the Gaussian norm, that we actually proved $|uv| = |u||v|$ for any complex numbers u, v. So it follows, exactly as before, that

$$\mathrm{norm}(uv) = \mathrm{norm}(u)\mathrm{norm}(v),$$

when u, v are numbers of the form $a + b\sqrt{-2}$. It likewise follows that u divides w only if norm(u) divides norm(w) in the ordinary integers.

So far, so good: the set of numbers of the form $a + b\sqrt{-2}$, which we call $\mathbb{Z}[\sqrt{-2}]$, is behaving like the set of Gaussian integers, which we call $\mathbb{Z}[i]$.

The first job for the norm in $\mathbb{Z}[\sqrt{-2}]$ is to prove that

$$\gcd(x + \sqrt{-2}, x - \sqrt{-2}) = 1 \quad \text{when } x, y \in \mathbb{Z} \text{ are such that } y^3 = x^2 + 2.$$

This really depends on having $y^3 = x^2 + 2$; it fails for other values of x, such as $x = 0$. But if $y^3 = x^2 + 2$ then x must be odd, since an even x makes y even and the left side $\equiv 0 \pmod 4$, whereas the right side $\equiv 2 \pmod 4$. With x odd, $x + \sqrt{-2}$ and $x - \sqrt{-2}$ have odd norms, so their common divisors have odd norms. On the other hand, any common divisor of $x + \sqrt{-2}$ and $x - \sqrt{-2}$ divides their difference, $2\sqrt{-2}$, whose norm has divisors 1, 2, 4, and 8. Thus any common divisor of $x + \sqrt{-2}$ and $x - \sqrt{-2}$ has norm 1, as required.

The other claim to justify is the big one:

Unique prime factorization in $\mathbb{Z}[\sqrt{-2}]$. *Each element of $\mathbb{Z}[\sqrt{-2}]$ has a prime factorization, which is unique up to factors of ± 1.*

Proof. This amounts to rewriting the proof for the Gaussian integers, $\mathbb{Z}[i]$, with appropriate changes. In $\mathbb{Z}[\sqrt{-2}]$ we define a **prime** to be an element of norm greater than 1 that is not the product of elements of smaller norm.

As in $\mathbb{Z}[i]$, everything we need follows from the **division property**: for any $a, b \in \mathbb{Z}[\sqrt{-2}]$ with $b \neq 0$ there are $q, r \in \mathbb{Z}[\sqrt{-2}]$ such that

$$a = qb + r, \quad \text{with } |r| < |b|.$$

Then there is a geometric setup that makes the existence of q and r visible. In brief:

- The elements of $\mathbb{Z}[\sqrt{-2}]$ form a grid of *rectangles*, each of width 1 and height $\sqrt{2}$.
- When this grid is multiplied by b, to form the set of multiples of b, the result is like the grid $\mathbb{Z}[\sqrt{-2}]$ but magnified by $|b|$ and rotated about O through the angle of b. So each rectangle in the magnified grid has short side of length $|b|$ and long side of length $\sqrt{2}|b|$.

- Any element $a \in \mathbb{Z}[\sqrt{-2}]$ falls in some rectangle in the grid of multiples of b. We let qb be a corner (possibly not unique) nearest to a in this rectangle and let $r = a - qb$.
- Then $|r| =$ distance between a and qb (figure 7.9).

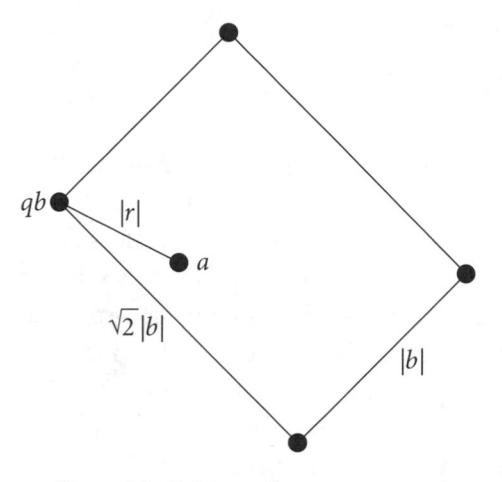

Figure 7.9 : Point a and nearest corner qb

If we view the segment from a to qb as the hypotenuse of a triangle, with one side of length $\leq |b|/2$, and the other of length $\leq \sqrt{2}|b|/2$, then the Pythagorean theorem gives $|r|^2 \leq \frac{3}{4}|b|^2$, hence $|r| < |b|$.

This completes the proof of the division property. As we know, it implies a Euclidean algorithm, prime divisor property, and unique prime factorization. In this case, prime factorization is unique up to factors ± 1, because the only elements of norm 1 in $\mathbb{Z}[\sqrt{-2}]$ are 1 and -1. $\qquad\square$

7.7 ALGEBRAIC NUMBER FIELDS

In section 7.1 we saw that the notion of congruence modulo a prime p divides $\mathbb{Z}$ into congruence classes mod p that form a *field* $\mathbb{Z}/p\mathbb{Z}$. The part of the argument that requires p to be prime is the existence of *inverses*, where the Euclidean algorithm gives, for each $a \not\equiv 0 \pmod{p}$, an m such that $ma \equiv 1 \pmod{p}$.

We have also seen, in sections 7.3 and 7.5, that polynomials enjoy many of the properties of $\mathbb{Z}$, such as the existence of "primes" (*irreducible* polynomials) and a Euclidean algorithm. This suggests we may push the analogy further: for each irreducible polynomial $p(x)$, and the notion of congruence modulo $p(x)$, we should find that the congruence classes form a field under the obvious sum and product operations.

Until now we have been vague about the coefficients in our polynomials. In examples they were mostly integers, but they could have been real or even complex numbers. For the rest of this section we will assume ordinary rational or integer coefficients.

Notation. The polynomials in x with rational coefficients form a ring called $\mathbb{Q}[x]$. Those with integer coefficients form a ring called $\mathbb{Z}[x]$.

Definitions. A number α is called **algebraic** if it satisfies a polynomial in $\mathbb{Z}[x]$. Equivalently, α satisfies a polynomial of the form

$$x^m + c_{m-1}x^{m-1} + \cdots + c_1 x + c_0 \quad \text{for some } c_0, c_1, \ldots, c_{m-1} \in \mathbb{Q},$$

called a **monic** polynomial in $\mathbb{Q}[x]$. The minimal-degree monic polynomial for α in $\mathbb{Q}[x]$ (which is unique—if not, consider the difference of two monic polynomials of the same degree) is called the **minimal polynomial** for α. Its degree is called the **degree of** α.

For example, $\sqrt{2}$ satisfies the polynomial $x^2 - 2$ and no polynomial of lower degree in $\mathbb{Q}[x]$ because $\sqrt{2}$ is irrational. This shows simultaneously that $\sqrt{2}$ has degree 2 and that $x^2 - 2$ is irreducible in $\mathbb{Q}[x]$, that is, not the product of linear polynomials with rational coefficients. More generally, any minimal polynomial $p(x)$ for a number α is irreducible; otherwise α would satisfy a factor of $p(x)$ of lower degree.

Now we are ready to extend the analogy of congruence modulo a prime p in $\mathbb{Z}$ to congruence modulo an irreducible $p(x)$ in $\mathbb{Q}[x]$.

Definitions. If $p(x) \in \mathbb{Q}[x]$ we say $a(x), b(x) \in \mathbb{Q}[x]$ are **congruent modulo** $p(x)$, written

$$a(x) \equiv b(x) \ (\mathrm{mod} \ p(x)),$$

if $p(x)$ divides $a(x) - b(x)$. The **congruence class** $[a(x)]$ of $a(x)$ is defined by

$$[a(x)] = \{b(x) \in \mathbb{Q}[x] : a(x) \equiv b(x) \ (\mathrm{mod} \ p(x))\}.$$

Thanks to the Euclidean algorithm for polynomials, we can now use the same arguments as for integers mod p in $\mathbb{Z}$ (section 7.1) to show that *the congruence classes mod $p(x)$ in $\mathbb{Q}[x]$ form a field when $p(x)$ is irreducible, and the sum and product of congruence classes are defined by*

$$[a(x)] + [b(x)] = [a(x) + b(x)], \quad [a(x)][b(x)] = [a(x)b(x)].$$

This field of congruence classes is denoted by $\mathbb{Q}[x]/p(x)\mathbb{Q}[x]$ and is called an **algebraic number field**. In it, the class $[x]$ plays the role of the number α whose minimal polynomial is $p(x)$. Thus, in a sense we have described the number α in terms of rational numbers, since the elements of $\mathbb{Q}[x]$ all have rational coefficients. However, we should keep in mind that $[x]$ equally well plays the role of *any other* root of the equation $p(x) = 0$.

For example, in $\mathbb{Q}[x]/(x^2 - 2)\mathbb{Q}[x]$ the class $[x]$ plays the role of $\sqrt{2}$, but also the role of $-\sqrt{2}$, since $-\sqrt{2}$ also satisfies $x^2 - 2 = 0$. Thus the description of an algebraic number α in terms of the rational numbers— as the class $[x]$ in the field $\mathbb{Q}[x]/p(x)\mathbb{Q}[x]$—while pleasingly concrete and simple, is somewhat ambiguous.

The upside of this ambiguity is that we can evaluate a rational expression in α, for example, $(1 - 2\alpha)/(7 + \alpha^3)$, for *any* root α of the equation $p(x) = 0$, by evaluating $(1 - 2[x])(7 + [x]^3)^{-1}$ in $\mathbb{Q}[x]/p(x)\mathbb{Q}[x]$. This is mainly a matter of finding the inverse of the polynomial $7 + x^3$ mod $p(x)$, which can be done using the Euclidean algorithm.

Notation. The algebraic number field $\mathbb{Q}[x]/p(x)\mathbb{Q}[x]$, for an irreducible polynomial $p(x)$, is also written $\mathbb{Q}(\alpha)$ for any root α of the equation $p(x) = 0$.

For example, the field $\mathbb{Q}[x]/(x^2 - 2)\mathbb{Q}[x]$ may be more concisely written as $\mathbb{Q}(\sqrt{2})$, but also as $\mathbb{Q}(-\sqrt{2})$. The elements $\sqrt{2}$ and $-\sqrt{2}$ each correspond to the element $[x]$ in $\mathbb{Q}[x]/(x^2 - 2)\mathbb{Q}[x]$. The map of $\mathbb{Q}(\sqrt{2})$ induced by sending $\sqrt{2}$ to $-\sqrt{2}$ is an **automorphism** (self-isomorphism): a one-to-one map of $\mathbb{Q}(\sqrt{2})$ onto itself that preserves sum and products.

The elements of the algebraic number field $\mathbb{Q}(\sqrt{2})$ are all of the form $c + d\sqrt{2}$, where $c, d \in \mathbb{Q}$. It is easy to see that the sum, difference, and product of numbers of this form are also of the same form, and the same is true for quotients because

$$\frac{1}{c+d\sqrt{2}} = \frac{1}{c+d\sqrt{2}}\frac{c-d\sqrt{2}}{c-d\sqrt{2}} = \frac{c-d\sqrt{2}}{c^2-2d^2} = \frac{c}{c^2-2d^2} - \frac{d}{c^2-2d^2}\sqrt{2}.$$

It follows that

$$\mathbb{Q}(\sqrt{2}) = \{c+d\sqrt{2}: c,d \in \mathbb{Q}\}$$

is a vector space of dimension 2 *over* $\mathbb{Q}$, *with basis elements* 1 *and* $\sqrt{2}$. Or, reverting to the congruence class description,

$$\mathbb{Q}[x]/(x^2-2)\mathbb{Q}[x] = \{c[1]+d[x]: c,d \in \mathbb{Q}\}$$

is a vector space of dimension 2 *over* $\mathbb{Q}$, *with basis elements* $[1]$ *and* $[x]$.

These results are instances of the general theorem about algebraic number fields, essentially due to Dedekind (1871):

Algebraic number field as a vector space. *If* $p(x) \in \mathbb{Q}[x]$ *is irreducible and of degree n, then* $\mathbb{Q}[x]/p(x)\mathbb{Q}[x]$ *is a vector space of dimension n over* $\mathbb{Q}$, *with basis elements* $[1],[x],\ldots,[x]^{n-1}$.

Proof. Since every element of $\mathbb{Q}[x]/p(x)\mathbb{Q}[x]$ is a polynomial in $[x]$, with rational cocfficients, it suffices to show that each positive power of $[x]$ is a linear combination of $[1],[x],\ldots,[x]^{n-1}$ with rational coefficients. First suppose that

$$p(x) = x^n + c_{n-1}x^{n-1} + \cdots + c_1x + c_0 = 0, \quad \text{for some } c_0, c_1, \ldots, c_{n-1} \in \mathbb{Q}.$$

Then, since $p([x]) = 0$ in $\mathbb{Q}[x]/p(x)\mathbb{Q}[x]$, we have

$$[x]^n = -c_{n-1}[x]^{n-1} - \cdots - c_1[x] - c_0[1],$$

showing that $[x]^n$ is a linear combination of $[1],[x],\ldots,[x]^{n-1}$ with rational coefficients. Multiplying both sides of this equation by $[x]$, we get

$$[x]^{n+1} = -c_{n-1}[x]^n - \cdots - c_1[x]^2 - c_0[x].$$

So, rewriting $[x]^n$ on the right as the linear combination of $[1],[x],\ldots,$ $[x]^{n-1}$ just found, we also get $[x]^{n+1}$ as a linear combination of $[1],[x],\ldots,[x]^{n-1}$ with rational coefficients. Repeating this process, we get each power of $[x]$, and hence each element of $\mathbb{Q}[x]/p(x)\mathbb{Q}[x]$, as a linear combination of $[1],[x],\ldots,[x]^{n-1}$ with rational coefficients. $\square$

Algebraic Integers

Dedekind was interested in algebraic number fields as a setting for the study of algebraic *integers*, because their finite dimensionality makes possible a concept of *primes* among these integers. The set of *all* algebraic integers lacks primes because, with the definition of algebraic integer we are about to give, when α is an algebraic integer so is $\sqrt{\alpha}$, and

$$\alpha = \sqrt{\alpha}\sqrt{\alpha}$$

is a factorization of α, so α is not "prime." When we restrict our integers to those in an algebraic number field $\mathbb{F}$ of dimension n, then $\sqrt{\alpha}$ does not necessarily belong to $\mathbb{F}$ when α does.

Definition. An **algebraic integer** is a number that satisfies an equation of the form

$$x^n + d_{n-1}x^{n-1} + \cdots + d_1 x + d_0 = 0, \quad \text{where } d_0, d_1, \ldots d_{n-1} \in \mathbb{Z}.$$

The algebraic integers in an algebraic number field $\mathbb{F}$ are called the integers *of* $\mathbb{F}$.

The algebraic integers include numbers that "look like" integers, such as the Gaussian integers, and indeed, these are the integers of the field $\mathbb{Q}(i)$. Likewise, the integers $a + b\sqrt{-2}$, for $a, b \in \mathbb{Z}$ are the integers of $\mathbb{Q}(\sqrt{-2})$. However, some fractions are also algebraic integers, for example,

$$\frac{-1 + \sqrt{-3}}{2}, \quad \text{which is a root of } x^3 - 1 = 0.$$

The above definition is the best way to get all the properties one would like algebraic integers to have, such as the ring properties and the existence of a norm that takes ordinary integer values. One can then prove that every algebraic integer can be split into "prime" or "irreducible" factors, by the argument about decreasing norms we used for the Gaussian integers and the integers of $\mathbb{Q}(\sqrt{-2})$. Unfortunately, that is as far as it goes: *prime factorization is not always unique*.

7.8 RINGS AND IDEALS

The failure of unique prime factorization for algebraic integers was discovered by Kummer in the 1840s, when he was investigating algebraic integers involving the imaginary roots of the equation $x^n - 1 = 0$. Specifically, he found that unique prime factorization fails when $n = 23$. Later, Dedekind (1877b) pointed out much simpler examples in the integers of the field $\mathbb{Q}(\sqrt{-5})$, which happen to form the ring $\mathbb{Z}[\sqrt{-5}]$ of numbers $a + b\sqrt{-5}$ for $a, b \in \mathbb{Z}$.

Among these integers we have two factorizations of the integer 6:

$$6 = 2 \cdot 3 = (1 + \sqrt{-5})(1 - \sqrt{-5}).$$

For these integers, $\text{norm}(a + b\sqrt{-5}) = a^2 + 5b^2$, so the norms of the factors $2, 3, 1 + \sqrt{-5}, 1 - \sqrt{-5}$, are respectively, $4, 9, 6, 6$. None of these norms is a product of smaller norms other than 1, so $2, 3, 1 + \sqrt{-5}, 1 - \sqrt{-5}$ are "prime" or "irreducible." Also, the only numbers of norm 1 are ± 1, so the above "prime" factorizations are definitely distinct.

Kummer refused to take the failure of unique prime factorization lying down. He speculated that apparently "prime" algebraic integers might be split further into what he called *ideal numbers*, among which unique prime factorization holds. In the case of the factorizations

$$6 = 2 \cdot 3 = (1 + \sqrt{-5})(1 - \sqrt{-5}),$$

we hope for a further splitting of $2, 3, 1 + \sqrt{-5}, 1 - \sqrt{-5}$ into their *common ideal divisors*, such as $\gcd(2, 1 + \sqrt{-5})$, whatever those are. Kummer understood ideal numbers and successfully worked with them, but Dedekind (1871) was the first to make them clear to the mathematical community by means of his concept of **ideals**.

Ideals

Before giving a general definition of ideal, we will look at ideals in $\mathbb{Z}$. In this ring each member c corresponds to the set of *multiples of c*: $c = \{nc : n \in \mathbb{Z}\}$. Any set $c \subseteq \mathbb{Z}$ of multiples has the properties:

- If $u, v \in c$ then $u + v \in c$.
- If $u \in c$ and $n \in \mathbb{Z}$ then $nu \in c$.

Conversely, a set $c \subseteq \mathbb{Z}$ with these two properties is indeed the set of multiples of some $c \in \mathbb{Z}$. If $c \neq \{0\}$ then c is the least nonzero member of c. This follows by division with remainder: if $b \in c$ is not a multiple of c, then its remainder $r = b - qc = b + (-1)qc$ is a nonzero member of c yet smaller than c, which is a contradiction.

Thus, in $\mathbb{Z}$, *there is a one-to-one correspondence between numbers c and sets* $\mathfrak{c}$ *with the above two properties.* An important case of this correspondence is when

$$\mathfrak{c} = \{ma + nb : m, n \in \mathbb{Z}\}, \quad \text{for some } a, b \in \mathbb{Z}.$$

In this case $\mathfrak{c}$ *is the set of multiples of* $\gcd(a, b)$. It is clear that $\mathfrak{c}$ has the two properties above; it is also clear that every member of $\mathfrak{c}$ is a multiple of $\gcd(a, b)$, since a and b are. Finally, we know that $\gcd(a, b) \in \mathfrak{c}$, because $\gcd(a, b) = ma + nb$ for some $m, n \in \mathbb{Z}$.

These results motivate Dedekind's definition of ideal, and they suggest how to bring $\gcd(2, 1 + \sqrt{-5})$ into existence.

Definitions. A subset $\mathfrak{c}$ of any ring R is called an **ideal** of R if we have the following:

- If $u, v \in \mathfrak{c}$ then $u + v \in \mathfrak{c}$.
- If $u \in \mathfrak{c}$ and $r \in R$ then $ru \in \mathfrak{c}$.

An ideal $\mathfrak{c} \subseteq R$ is called a **principal ideal** if $\mathfrak{c} = \{rc : r \in R\}$ for some $c \in R$. In this case we write $\mathfrak{c} = (c)$.

The two defining properties of an ideal may be called *closure* properties: $\mathfrak{c}$ is closed under sums, and under products with members of R.

In $\mathbb{Z}$, every ideal is a principal ideal. This just restates the correspondence between numbers $c \in \mathbb{Z}$ and sets $\mathfrak{c}$ we proved above. But in the ring

$$\mathbb{Z}[\sqrt{-5}] = \{a + b\sqrt{-5} : a, b \in \mathbb{Z}\}$$

there are *non*principal ideals; for example,

$$\mathfrak{d} = \{2m + (1 + \sqrt{-5})n : m, n \in \mathbb{Z}[\sqrt{-5}]\},$$

which happens to equal $\mathfrak{d} = \{2m + (1 + \sqrt{-5})n : m, n \in \mathbb{Z}\}$.

This set has the closure properties, and it should equal $\gcd(2, 1 + \sqrt{-5})$, by analogy with the principal ideal $(\gcd(a, b))$ in $\mathbb{Z}$ above. However, $\mathfrak{d}$ is *not* a principal ideal, as we can see from the picture of it in figure 7.10.

In this picture, the black and white dots are elements of $\mathbb{Z}[\sqrt{-5}]$, and the black dots are members of the ideal $\mathfrak{d}$. The elements of $\mathbb{Z}[\sqrt{-5}]$ form

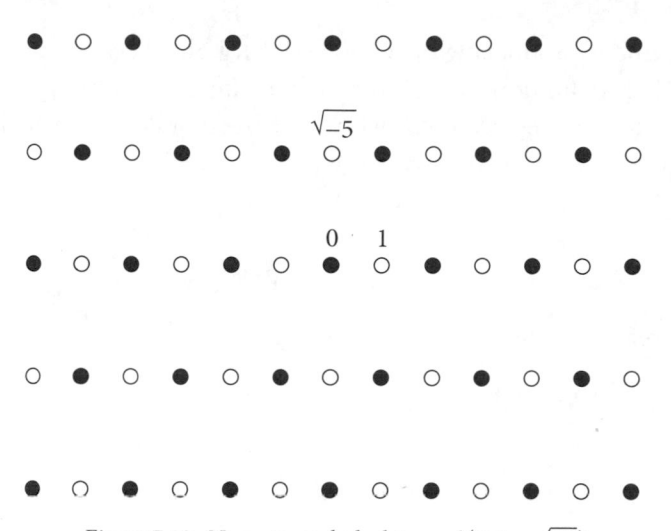

Figure 7.10 : Nonprincipal ideal $\mathfrak{d} = \gcd(2, 1 + \sqrt{-5})$

a grid of rectangles, of width 1 and height $\sqrt{5}$. Any principal ideal, being the set of multiples of some $c \in \mathbb{Z}[\sqrt{-5}]$, will form a grid of rectangles of width $|c|$ and height $|c|\sqrt{5}$. But it is clear that the black dots do not form rectangles at all; hence the ideal $\mathfrak{d}$ is nonprincipal.

This reflects the fact that the "right" gcd of 2 and $1 + \sqrt{-5}$ is not in $\mathbb{Z}[\sqrt{-5}]$. Their gcd is an "ideal number," which we cannot see, but we can see the set of its "multiples," which form the ideal $\mathfrak{d}$. This prompts a conceptual shift from *members* of $\mathbb{Z}[\sqrt{-5}]$ to certain *infinite subsets* of $\mathbb{Z}[\sqrt{-5}]$, the ideals. As part of this shift, we replace the members m by the corresponding principal ideals (m).

Products and Factors of Ideals

Since the purpose of ideals is to split numbers without proper factors, such as the number 2 in $\mathbb{Z}[\sqrt{-5}]$, into ideal factors, we need a concept of *product* for ideals. This was supplied by Dedekind (1871):

Definition. If $\mathfrak{a}$ and $\mathfrak{b}$ are ideals in a ring R, then

$$\mathfrak{a}\mathfrak{b} = \{a_1 b_1 + \cdots + a_n b_n : a_1, \ldots, a_n \in \mathfrak{a},\ b_1, \ldots, b_n \in \mathfrak{b}\}$$

is called the **product** of the ideals $\mathfrak{a}$ and $\mathfrak{b}$.

It is easy to check that $\mathfrak{a}\mathfrak{b}$ is an ideal in R. Closure under sums is immediate, thanks to the presence of arbitrary finite sums in the definition, and closure under products with members of R holds because $\mathfrak{a}$ and $\mathfrak{b}$ are ideals. This definition is also compatible with the product of elements $u, v \in R$ when we replace them with the corresponding principal ideals. Namely,

$$(u)(v) = (uv),$$

because

$$(u) = \{ru : r \in R\}, \quad (v) = \{sv : s \in R\},$$

and therefore, by the definition of ideal product,

$$
\begin{aligned}
(u)(v) &= \{r_1 u s_1 v + \cdots + r_n u s_n v : r_1, s_1, \ldots, r_n, s_n \in R\} \\
&= \{(r_1 s_1 + \cdots + r_n s_n) uv : r_1, s_1, \ldots, r_n, s_n \in R\} \\
&= \{ruv : r \in R\} \\
&= (uv).
\end{aligned}
$$

But of course we are more interested in products of *non*principal ideals, like the ideal from the previous subsection we expect to be $\gcd(2, 1 + \sqrt{-5})$:

$$\mathfrak{d} = \{2m + (1 + \sqrt{-5})n : m, n \in \mathbb{Z}\}.$$

Hopefully, $\mathfrak{d}$ is a factor of the principal ideal (2), and in fact $\mathfrak{d}^2 = (2)$, which can be checked as follows.

First, observe that each member of $\mathfrak{d}^2$ is a multiple of 2, since

$$[2m_1 + (1 + \sqrt{-5})n_1][2m_2 + (1 + \sqrt{-5})n_2]$$
$$= 4m_1 m_2 + 2(m_2 n_1 + m_1 n_2)(1 + \sqrt{-5}) + 2n_1 n_2(-2 + \sqrt{-5}).$$

Conversely, $\mathfrak{d}^2$ includes each multiple of 2. This is because

$\mathfrak{d}$ includes 2, $1 + \sqrt{-5}$, and hence also their difference,
$$2 - (1 + \sqrt{-5}) = 1 - \sqrt{-5}$$

so $\mathfrak{d}^2$ includes $4 = 2 \cdot 2$, $6 = (1 + \sqrt{-5})(1 - \sqrt{-5})$, and hence also $6 - 4 = 2$
and all multiples of 2, since $\mathfrak{d}$ is an ideal.

Thus $(2) = \mathfrak{d}^2$ is an ideal factorization of 2 in $\mathbb{Z}[\sqrt{-5}]$.

This accomplishes the first step in reconciling the two factorizations of 6 in $\mathbb{Z}[\sqrt{-5}]$: $6 = 2 \cdot 3 = (1 + \sqrt{-5})(1 - \sqrt{-5})$. We have split the factor 2 into $\mathfrak{d}^2$ by looking at the expected gcd of 2 and $1 + \sqrt{-5}$,

$$\mathfrak{d} = \{2m + (1 + \sqrt{-5})n : m, n \in \mathbb{Z}\}.$$

Now we also consider

$$\overline{\mathfrak{d}} = \{2m + (1 - \sqrt{-5})n : m, n \in \mathbb{Z}\} \quad \text{(expected gcd of 2 and } 1 - \sqrt{-5})$$

$$\mathfrak{e} = \{3m + (1 + \sqrt{-5})n : m, n \in \mathbb{Z}\} \quad \text{(expected gcd of 3 and } 1 + \sqrt{-5})$$

$$\overline{\mathfrak{e}} = \{3m + (1 - \sqrt{-5})n : m, n \in \mathbb{Z}\}. \quad \text{(expected gcd of 3 and } 1 - \sqrt{-5})$$

Actually $\overline{\mathfrak{d}} = \mathfrak{d}$, since we have already observed that $1 - \sqrt{-5} \in \mathfrak{d}$. For the rest, calculations like those we made for $\mathfrak{d}^2$ show that

$$\mathfrak{e}\overline{\mathfrak{e}} = (3), \quad \mathfrak{d}\mathfrak{e} = (1 + \sqrt{-5}), \quad \mathfrak{d}\overline{\mathfrak{e}} = (1 - \sqrt{-5}),$$

and therefore

$$\mathfrak{d}^2 \mathfrak{e}\overline{\mathfrak{e}} = (2)(3) = (1 + \sqrt{-5})(1 - \sqrt{-5}) = \mathfrak{d}\mathfrak{e}\mathfrak{d}\overline{\mathfrak{e}}.$$

In other words, *the factorizations* $2 \cdot 3$ *and* $(1 + \sqrt{-5})(1 - \sqrt{-5})$ *of 6 both split into the* same *ideal factorization* $\mathfrak{d}^2 \mathfrak{e}\overline{\mathfrak{e}}$. Moreover, the factors $\mathfrak{d}$, $\mathfrak{e}$, and $\overline{\mathfrak{e}}$ are *prime* ideals, as we will see in the next section.

7.9 DIVISIBILITY AND PRIME IDEALS

The ideal factors $\mathfrak{a}$ and $\mathfrak{b}$ of a product ideal $\mathfrak{ab}$ are naturally considered to *divide* $\mathfrak{ab}$, and they also *contain* it, because each term $a_i b_i$ belongs to both $\mathfrak{a}$ and $\mathfrak{b}$, and hence so does a sum of such terms. Dedekind (1871) in fact used containment as the definition of divisibility:

Definition. Ideal $\mathfrak{a}$ is said to **divide** ideal $\mathfrak{b}$ if $\mathfrak{a} \supseteq \mathfrak{b}$.

The idea that "to contain is to divide" gives a new way to look at some familiar facts about divisibility, for example:

- The whole ring R, which is also the principal ideal (1), contains any ideal $\mathfrak{a}$ in R and hence divides it (as it should, since 1 divides everything).
- The gcd of ideals $\mathfrak{a}$ and $\mathfrak{b}$ contains both $\mathfrak{a}$ and $\mathfrak{b}$, and it also contains any ideal $\mathfrak{c}$ that contains both $\mathfrak{a}$ and $\mathfrak{b}$. Thus from the containment viewpoint, $\gcd(\mathfrak{a}, \mathfrak{b})$ is the *smallest* ideal containing $\mathfrak{a}$ and $\mathfrak{b}$. It is easy to see that this smallest ideal is $\{ma + nb : a \in \mathfrak{a}, \, b \in \mathfrak{b}, \, m, n \in R\}$.

When containment is the fundamental concept, an aspect that comes to light is **maximality**:

Definition. An ideal $\mathfrak{a}$ in a ring R is called **maximal** if $\mathfrak{a} \neq R$ but the only ideals containing $\mathfrak{a}$ are $R = (1)$ and $\mathfrak{a}$ itself.

Since "containing" means "dividing," this definition sounds a lot like a definition of "prime." In practice, it is better to use a slightly broader definition of **prime ideal**, modeled on the prime divisor property:

Definition. An ideal $\mathfrak{p}$ is **prime** if, whenever $\mathfrak{p}$ divides a product $\mathfrak{a}\mathfrak{b}$ of ideals, $\mathfrak{p}$ divides $\mathfrak{a}$ or $\mathfrak{p}$ divides $\mathfrak{b}$.

We can then prove that maximality implies primality by much the same idea used to prove the prime divisor property for $\mathbb{Z}$ in section 2.6.

Maximality implies primality. *If $\mathfrak{p}$ is a maximal ideal then $\mathfrak{p}$ is prime.*

Proof. Let $\mathfrak{p}$ be a maximal ideal, and suppose that $\mathfrak{p}$ divides (contains) $\mathfrak{a}\mathfrak{b}$ but $\mathfrak{p}$ does not divide (contain) $\mathfrak{a}$. Then there is an $a \in \mathfrak{a}$ with $a \notin \mathfrak{p}$.

Since $\mathfrak{p}$ is maximal, any ideal containing a and $\mathfrak{p}$ is the whole ring, R. In particular, the ideal $\{ma + p : m \in R, \ p \in \mathfrak{p}\} = R$, so

$$1 = ma + p, \quad \text{for some } m \in R \text{ and } p \in \mathfrak{p}.$$

Then, multiplying both sides by any $b \in \mathfrak{b}$, we get

$$b = mab + pb.$$

But $ab \in \mathfrak{p}$, because $\mathfrak{p} \supseteq \mathfrak{a}\mathfrak{b}$, so $b = mab + p \in \mathfrak{p}$. Since b is any member of $\mathfrak{b}$, this shows $\mathfrak{p}$ contains (divides) $\mathfrak{b}$, so $\mathfrak{p}$ is prime. $\qquad\square$

With these definitions and facts established, we can now show that the ideal factors $\mathfrak{d}$, $\mathfrak{e}$, and $\bar{\mathfrak{e}}$ of 6 found in the previous subsection are prime ideals. In fact they are all maximal ideals. This is particularly clear for $\mathfrak{d}$, whose picture (figure 7.10) clearly shows that any element $r \in \mathbb{Z}[\sqrt{-5}]$ not in $\mathfrak{d}$ differs by 1 from a member of $\mathfrak{d}$. Thus any ideal containing r and $\mathfrak{d}$ is the whole ring. There are similar arguments for $\mathfrak{e}$ and $\bar{\mathfrak{e}}$.

Thus we have finally reconciled the two different factorizations in $\mathbb{Z}[\sqrt{-5}]$,
$$6 = 2 \cdot 3 = (1 + \sqrt{-5})(1 - \sqrt{-5}),$$

by a common *prime ideal* factorization $\mathfrak{d}^2 \mathfrak{e}\bar{\mathfrak{e}}$. Proving such prime ideal factorizations are *unique* is another problem, also solved by Dedekind (1871). It requires a deeper investigation that we do not have space for here.

Quotients of Rings

An elegant way to distinguish between maximal and prime ideals in a ring R is to consider the *quotient* of R by an ideal, which generalizes the quotient of $\mathbb{Z}$ by an ideal $n\mathbb{Z}$, via a generalization of congruence mod n.

Definitions. If $\mathfrak{c}$ is an ideal in a ring R, then elements $a, b \in R$ are called **congruent mod** $\mathfrak{c}$, written $a \equiv b \pmod{\mathfrak{c}}$, if $a - b \in \mathfrak{c}$, that is, if $a = b + c$ for some $c \in \mathfrak{c}$. The set of elements congruent to $a \pmod{\mathfrak{c}}$, $\{a + c : c \in \mathfrak{c}\}$, is written $[a]$ and called the **congruence class** of $a \pmod{\mathfrak{c}}$. Congruence classes are added and multiplied by the rules

$$[a] + [b] = [a+b], \quad [a][b] = [ab].$$

Finally, the ring of congruence classes, mod $\mathfrak{c}$, is denoted by $R/\mathfrak{c}$ and called the **quotient** of R by $\mathfrak{c}$.

It follows from the definition of ideal in the previous section that congruence mod $\mathfrak{c}$ is an equivalence relation and that sum and product of equivalence classes are well defined. For example, if $a = a' \pmod{\mathfrak{c}}$ then $a' = a + c'$ for some $c' \in \mathfrak{c}$ and then

$$
\begin{aligned}
[a'] + [b] = [a' + b] &= [a + c' + b] \\
&= \{a + b + c + c' : c \in \mathfrak{c}\} \\
&= \{a + b + c : c \in \mathfrak{c}\} \quad \text{(by closure of } \mathfrak{c} \text{ under sums)} \\
&= [a + b] = [a] + [b].
\end{aligned}
$$

This gives a well-defined quotient structure $R/\mathfrak{c}$, which is a ring because it inherits the ring properties from R (just as $\mathbb{Z}/n\mathbb{Z}$ inherits the ring properties from $\mathbb{Z}$).

Now the difference between maximal and prime ideals is given by the following theorems, which follow easily from the definition of $R/\mathfrak{c}$.

Quotient of ring by ideal. *if $\mathfrak{c}$ is an ideal of a ring R then*

- *if $\mathfrak{c}$ is maximal then $R/\mathfrak{c}$ is a field;*
- *if $\mathfrak{c}$ is prime then $R/\mathfrak{c}$ is an **integral domain**, that is, a ring in which the product of nonzero elements is nonzero.*

Obviously, there are integral domains that are not fields, such as $\mathbb{Z}$. However, it is easy to show that *any finite integral domain is a field,* and the prime ideals we have seen above do indeed give finite quotients, which explains why they are maximal. For example, the ideal $\mathfrak{d}$ in $\mathbb{Z}[\sqrt{-5}]$ has two congruence classes (black dots and white dots in figure 7.10), and the ideal $\mathfrak{e}$ has three.

7.10 REMARKS

Hopefully, this chapter has shown that even by the end of the nineteenth century number theory had reached, influenced, and tied together several fields of mathematics: algebra, geometry, and analysis. By 1900 its scope was already so great that different subfields of number theory began to develop somewhat independently. We revisit some of these subfields briefly in later chapters, but it is not possible to keep up with all twentieth-century developments in a book of this size. Instead, here are a few brief remarks on some of them.

A Brief History of Induction

The principle of induction, often called "complete" or "mathematical" induction to distinguish it from the looser concept of induction that draws a general conclusion from a few examples, has been present in mathematics since ancient times. It appears in several logically equivalent forms:

- *There is no infinite descending sequence of natural numbers.* This form, also called "infinite descent," appears in Euclid, as explained in section 2.6. It was revived by Fermat, using more sophisticated means of descent, to prove theorems such as the special case of Fermat's last theorem we saw in section 7.3.
- *Base step and induction step.* This form proves a property $P(n)$ of natural numbers by proving $P(0)$ (the "base step") and $P(n) \Rightarrow P(n+1)$ (the "induction step"). It appeared in rudimentary form in Levi ben Gershon (1321) and in mature form in Pascal (1654), where it was used to prove properties of the binomial coefficients. The realization that induction underlies even simple properties of the natural numbers, such as $m + n = n + m$, is due to Grassmann (1861), and it was enshrined as an **induction axiom** by Peano (1888) (see section 13.1).

- *Well-ordering of* $\mathbb{N}$. A sophisticated variation of infinite descent states that $\mathbb{N}$ is **well-ordered** by the $<$ relation, which means that the linear ordering by $<$ has the additional property that every subset of $\mathbb{N}$ has a least member. Cantor (1883) speculated that every set has a well-ordering. This conjecture implies that induction can be generalized to what is called **transfinite induction**, allowing arbitrary infinite sets to be handled to some extent like $\mathbb{N}$ (see chapter 14).

In the twentieth century, algebraists took up transfinite induction to prove existence of bases for arbitrary vector spaces (Hamel 1905) and maximal ideals in rings (Krull 1929) and to lay new foundations of modern algebra and algebraic geometry (Grothendieck 1957).

Modern Algebraic Geometry

Algebraic geometry was revolutionized by Dedekind and Weber (1882), when they reimagined the theory of algebraic curves as a theory of **algebraic functions**. Their idea was to study algebraic function fields of finite degree over the field of rational functions, modeled on algebraic number fields of finite degree over the rational numbers, and to carry over the ideas of algebraic integers and ideals.

The analogy begins with polynomial functions as the analogue of the integers, and rational functions as the analogue of rational numbers. Thus an algebraic function y of x satisfies an equation of the form

$$y^n + a_{n-1}(x)y^{n-1} + \cdots + a_1(x)y + a_0(x) = 0, \tag{$*$}$$

where $a_0(x), a_1(x), \ldots, a_{n-1}(x)$ are rational functions (quotients of polynomials) of x. The function y is of degree n if $(*)$ is the equation of minimal degree that y satisfies, and in this case y generates an **algebraic function field** of dimension n over the field of rational functions.

From such a field $\mathscr{F}$, Dedekind and Weber were able to extract a curve $\mathscr{C}$, and $\mathscr{F}$ also enabled them to define various properties of $\mathscr{C}$ such as its **genus**, a number originally defined in integral calculus by Abel (1826) and reinterpreted topologically by Riemann (1851, 1857). We will explore the intuition behind Riemann's view of genus in chapter 10.

The arguments of Riemann were astoundingly insightful, even though they rested on intuitions from geometry and physics. It is almost equally astounding that Dedekind and Weber were able to give rigorous proofs of the key theorems of Abel and Riemann by algebraic methods, originating in number theory. This approach to algebraic geometry bloomed in the twentieth century, with number theory, algebra, and algebraic geometry developing in close cooperation. The algebraic geometry of the twentieth century is too sophisticated for this book, but note that it was crucial to the proof of Fermat's last theorem, finally achieved by Wiles (1995).

Analytic Number Theory

In section 7.4 we saw how "rationalization" of algebraic functions in integral calculus was found to be related to the problem of finding rational points on algebraic curves, enabling number theory to guide analysis. There was also traffic in the opposite direction, with analysis guiding number theory. The elliptic functions uncovered by elliptic integrals gave rise to an extraordinary profusion of formulas in the *Fundamenta nova theoriae functionum ellipticarum* of Jacobi (1829), many of which had consequences in number theory. On the very last page of the book Jacobi triumphantly displays a formula whose corollary, as he points out in his last sentence, is a famous conjecture of Fermat that every number is a sum of four squares.

Later in the nineteenth century many other applications of analysis to number theory were found. We will focus on some that generalize Euclid's theorem that there are infinitely many primes.

The application of analysis to prime number theory began when Euler (1748) discovered the **product formula**:

$$\frac{1}{1-1/2^s} \cdot \frac{1}{1-1/3^s} \cdot \frac{1}{1-1/5^s} \cdot \frac{1}{1-1/7^s} \cdot \frac{1}{1-1/11^s} \cdots$$
$$= 1 + \frac{1}{2^s} + \frac{1}{3^s} + \frac{1}{4^s} + \frac{1}{5^s} + \cdots,$$

where the left side is the product of all terms $\frac{1}{1-1/p^s}$, where p is a prime, and the right side is the sum of all terms $1/n^s$, where s is an integer. We need $s > 1$ for the series on the right side to converge, in which case the sum of the series is called the **zeta function**, $\zeta(s)$.

The proof of the product formula is a simple application of geometric series and unique prime factorization. Since $|1/p^s| < 1$, each term in the product has the expansion

$$\frac{1}{1-1/p^s} = 1 + \frac{1}{p^s} + \frac{1}{p^{2s}} + \frac{1}{p^{3s}} + \cdots,$$

so the product of all these geometric series is 1 plus a sum of terms

$$\frac{1}{\left(p_1^{m_1} p_2^{m_2} \cdots p_k^{m_k}\right)^s},$$

where $p_1, p_2, \ldots, p_k$ are primes. Moreover, *each distinct product of primes $p_1^{m_1} p_2^{m_2} \cdots p_k^{m_k}$ occurs exactly once.* It follows, by unique prime factorization, that the terms $1/\left(p_1^{m_1} p_2^{m_2} \cdots p_k^{m_k}\right)^s$ are precisely the terms $1/n^s$ on the right side, since each $n > 1$ has a unique prime factorization

$$n = p_1^{m_1} p_2^{m_2} \cdots p_k^{m_k}.$$

The product formula gives a new proof, by contradiction, that there are infinitely many primes. Suppose for the sake of contradiction that there are only finitely many primes, $2, 3, 5, \ldots, p$. Then the left side of the product formula is a finite product and it is meaningful for $s = 1$. But the right side is *not* meaningful for $s = 1$, by the divergence of the harmonic series we saw in section 6.2. Thus we have a contradiction and therefore there are infinitely many primes.

While this proof is novel, it does not give a result stronger than Euclid's. Euler (1748) proved the infinitude of primes in a stronger sense when he showed that the sum of their reciprocals diverges. This shows, for example, that the primes are "denser" than the squares, because the sum of reciprocals of squares is

$$1 + \frac{1}{2^2} + \frac{1}{3^2} + \frac{1}{4^2} + \frac{1}{5^2} + \cdots = \frac{\pi^2}{6},$$

as Euler himself had found in 1734 (see section 6.4).

Around 1800, Legendre and Gauss made a more precise statement about the density of the primes, using numerical evidence to conjecture that the number of primes less than n is "of the order of $n/\ln n$." This means that, if the actual number of primes less than n is $\pi(n)$, then the ratio of $\pi(n)$ to $n/\ln n$ tends to 1 as n tends to infinity. The first big advance toward the proof of this conjecture, which is now known as the **prime number theorem**, occurred when Riemann (1859) related it to properties of the zeta function $\zeta(s)$, which he extended to *complex* values of s.

The prime number theorem was eventually proved by Hadamard (1896) and de la Vallée Poussin (1896). Both of their proofs used complex analysis to show that $\zeta(s) \neq 0$ when $s = 1 + it$ with $t > 0$. This property is enough to prove the prime number theorem but is weaker than Riemann's conjecture that $\zeta(s) = 0$ only when the real part of s equals 1/2 (apart from some so-called trivial zeros). This **Riemann hypothesis** implies even stronger results about the primes, and it is probably the most sought-after result in mathematics today.

CHAPTER 8

The Fundamental Theorem of Algebra

The continuum intruded into algebra when proofs of the fundamental theorem, such as that of Gauss (1816), were found to rely on general properties of continuous functions. Bolzano (1817) identified the hidden assumption of Gauss's proof—**the intermediate value theorem**—and defined the key concept of continuous function. He also showed that the intermediate value theorem is a consequence of the **least upper bound** property of $\mathbb{R}$, but he could go no further since no definition of $\mathbb{R}$ was available—only the intuition of the line that had not been sharpened since the time of Euclid.

A definition of $\mathbb{R}$ that made the least upper bound property provable was found by Dedekind in 1858 but not published until 1872. His definition of $\mathbb{R}$ made possible not only a proof of the least upper bound property but also a proof of the basic properties of continuous functions. Among these were the intermediate and extreme value theorems and many other basic theorems of analysis that were systematically worked out by Weierstrass in the 1870s.

However, a fundamental theorem of algebra dependent on concepts of analysis was not welcomed by some algebraists, particularly Kronecker. He produced his own version of the fundamental theorem, in which the roots of each polynomial equation $p(x) = 0$ are sought not in $\mathbb{C}$ but in an algebraic number field defined directly from the polynomial $p(x)$ itself.

8.1 THE THEOREM BEFORE ITS PROOF

The **fundamental theorem of algebra** states that any equation of the form

$$p(x) = x^n + a_{n-1}x^{n-1} + \cdots + a_1 x + a_0 = 0, \quad \text{where } a_0, a_1, \ldots, a_{n-1} \in \mathbb{R},$$

has a solution in the complex numbers, $\mathbb{C}$. As we know from section 4.5, Descartes proved that *if $p(x) = 0$ has a solution $x = a$ then the polynomial $p(x)$ has a factor $x - a$.* Repeating this argument with the quotient polynomial $p(x)/(x-a)$, we find that any polynomial $p(x)$ of degree d has a factorization into d linear factors, assuming the fundamental theorem holds. Moreover, the factorization is unique, up to constant factors, as we saw in section 7.3.

Although $p(x)$ is assumed to have real coefficients, the factors may not be real, since the roots of $p(x) = 0$ may not be real. For example,

$$x^2 + 1 = (x + i)(x - i),$$

because the roots of $x^2 + 1 = 0$ are $x = \pm i$. In fact, *imaginary roots always occur in conjugate pairs $u, \bar{u}$*, where the conjugate of $u = a + bi$ is

$$\overline{a + ib} = a - bi.$$

This was observed by d'Alembert (1746), and it holds because conjugation preserves sums, differences, and products:

$$\overline{u + v} = \bar{u} + \bar{v}, \quad \overline{u - v} = \bar{u} - \bar{v}, \quad \overline{u \cdot v} = \bar{u} \cdot \bar{v}, \quad \text{as is easily checked.}$$

Therefore,

$$\overline{x^n + a_{n-1} x^{n-1} + \cdots + a_1 x + a_0}$$

$$= \bar{x}^n + \overline{a_{n-1}} \, \bar{x}^{n-1} + \cdots + \overline{a_1} \, \bar{x} + \overline{a_0}$$

$$= \bar{x}^n + a_{n-1} \, \bar{x}^{n-1} + \cdots + a_1 \, \bar{x} + a_0, \quad \text{since } \overline{a_i} = a_i \text{ for real } a_i,$$

$$= p(\bar{x}).$$

Thus if $x = u$ is a root of $p(x) = 0$, so is $x = \bar{u}$, because

$$0 = \bar{0} = \overline{p(u)} = p(\bar{u}).$$

It follows that if $p(x)$ has imaginary factor $x - u$, where $u = a + bi$, then $p(x)$ also has the imaginary factor $x - \bar{u}$, and therefore $p(x)$ *has the real quadratic factor* $(x - a - bi)(x - a + bi) = x^2 - 2ax + a^2 + b^2$.

This gives an equivalent statement of the fundamental theorem.

Real fundamental theorem of algebra. *Any polynomial $p(x)$ with real coefficients can be factorized into real linear and quadratic factors.*

In the eighteenth century, this theorem was also the "real" one in the sense of being what mathematicians really wanted to prove. They particularly wanted it in calculus, in order to complete the theory of integration of rational functions. This was due to the method of *partial fractions*, which splits each polynomial quotient $p(x)/q(x)$ into a sum of terms whose denominators are factors of $q(x)$. If these factors are linear or quadratic, the integral of $p(x)/q(x)$ reduces to the known indefinite integrals

$$\int \frac{dx}{x+1} = \ln x, \qquad \int \frac{dx}{x^2+1} = \tan^{-1} x.$$

Indeed, the first widely accepted proof of the theorem, by Gauss (1799), stated the theorem in precisely these terms. Gauss's dissertation, written in Latin, was titled *Demonstratio nova theorematis omnem functionem algebraicam rationalem integram unius variabilis in factores reales primi vel secundi gradus resolvi posse (New demonstration of the theorem that every integral rational algebraic function of one variable may be resolved into real factors of first or second degree).*

But for a long time the search for a proof of the fundamental theorem was sidetracked by the search for solutions by radicals. As mentioned in section 4.5, this search succeeded only for the general equations of degrees 2, 3, and 4. We now know that it was doomed to fail for equations of degree 5 (quintic equations). Nevertheless, two eighteenth-century results on quintic equations are worth a mention.

Lambert (1758) found infinite series solutions for equations of the form $x^m + x = a$ by a method later known as **Lagrange inversion**, after Lagrange (1770) proved a general result about inversion of power series. In the special case $m = 5$, Lambert's result is

$$x = a - a^5 + 10\frac{a^9}{2!} - 15 \cdot 14\frac{a^{13}}{3!} + 20 \cdot 19 \cdot 18\frac{a^{17}}{4!} - \cdots.$$

Then Bring (1786) showed that the general equation of degree 5 can be reduced (by radicals) to the form $x^5 + x = a$ solved by Lambert—so Lambert had in fact solved the general quintic equation! However, Lambert and Bring were ships that passed in the night. Bring published in an obscure journal, and no one noticed the happy combination of his result with Lambert's until it was rediscovered by Eisenstein (1844).

8.2 EARLY "PROOFS" OF FTA AND THEIR GAPS

From now on, let us abbreviate the fundamental theorem of algebra by FTA. The best early attempt to prove FTA was by d'Alembert (1746), in a paper ostensibly about integral calculus. This attempt by d'Alembert has some gaps, but they are easily fixed by methods that undergraduate mathematicians know today. The first gap was in d'Alembert's understanding of complex numbers. The interpretation of $\mathbb{C}$ as a plane in which the distance between points u and v is given by $|u - v|$ did not become current until the early nineteenth century, when Argand (1806) was able to give a simple geometric proof of d'Alembert's basic lemma.

d'Alembert's lemma. *If $p(z)$ is a polynomial function and $|p(z_0)| > 0$ for some $z_0 \in \mathbb{C}$, then there is a Δz such that $|p(z_0 + \Delta z)| < |p(z_0)|$.*

Proof. We can suppose without loss of generality that

$$p(z) = z^n + a_{n-1}z^{n-1} + \cdots + a_1 z + a_0.$$

Then

$$p(z_0 + \Delta z) = (z_0 + \Delta z)^n + a_{n-1}(z_0 + \Delta z)^{n-1} + \cdots + a_1(z_0 + \Delta z) + a_0$$
$$= z_0^n + a_{n-1}z_0^{n-1} + \cdots + a_1 z_0 + a_0 + A\Delta z + \varepsilon,$$
$$= p(z_0) + \varepsilon + A\Delta z,$$

where, by the binomial theorem,

$$A = nz_0^{n-1} + (n-1)a_{n-1}z_0^{n-2} + \cdots + a_1$$

and ε is a sum of terms in $(\Delta z)^2, (\Delta z)^3, \ldots$.

The assumption that $|p(z_0)| > 0$ means that the complex number $p(z_0)$ is not at the origin. We can now choose Δz so that $p(z_0 + \Delta z)$ is *closer* to the origin than $p(z_0)$, so that $|p(z_0 + \Delta z)| < |p(z_0)|$. The idea is shown in figure 8.1.

First, choose the absolute value of Δz so that $|A\Delta z|$ is small compared with $|p(z_0)|$ but $|\varepsilon|$ is *very* small compared with both $|p(z_0)|$ and $|A\Delta z|$. This is possible because, when Δz is sufficiently small, $(\Delta z)^2, (\Delta z)^3, \ldots$ will be much smaller. Then $p(z_0) + \varepsilon$ is almost as near to O as $p(z_0)$ is. Second, choose the direction of Δz so that $A\Delta z$ has direction opposite to that of $p(z_0)$. Then $p(z_0 + \Delta z)$ is closer to O than $p(z_0)$ is. □

With his geometric interpretation of complex numbers, Argand cleared up any difficulties in d'Alembert's lemma. The next stage of

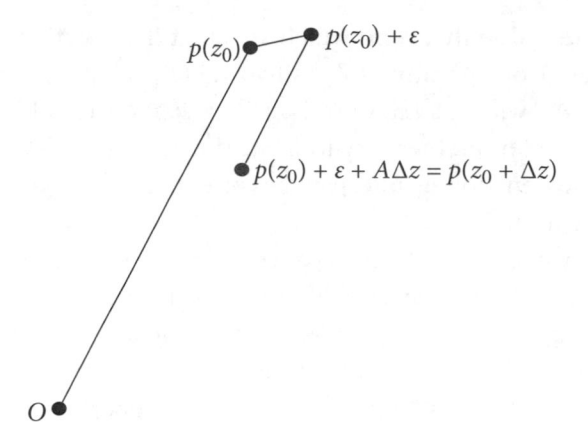

Figure 8.1 : Construction for d'Alembert's lemma

d'Alembert's proof, while plausible, is more subtle. We want to claim that $|p(z)|$ has a minimum value, which can only be zero by d'Alembert's lemma. Certainly, $p(z)$ has large values when $|z|$ is large, because in that case $p(z)$ is dominated by the term z^n. So a minimum value, if it exists, must lie within some radius R of the origin. But why must $|p(z)|$ have a minimum value on the disk where $|z| \le R$? The claim is it's *because $|p(z)|$ is a continuous function of z.*

This was the second gap in the understanding of both d'Alembert and Argand. It is true that $|p(z)|$ is a continuous function of z, but what does continuity mean? And under what conditions does a continuous function attain a minimum value? We will see the answer in section 11.7.

Gauss's 1799 proof also depends on properties of continuous functions, in fact properties harder to justify than the minimum value property assumed in the d'Alembert/Argand proof. It seems that Gauss himself became dissatisfied with his 1799 proof, because he gave another proof in 1816 depending on a simpler property of continuous functions. We look at the property in question, and its influence on the subsequent development of mathematics, in the next section.

8.3 CONTINUITY AND THE REAL NUMBERS

The proof of Gauss (1816) has two parts. What seems to be the hard part is an algebraic reduction of any polynomial equation to one of *odd* degree. This is followed by the seemingly easy part, an appeal to the

intermediate value theorem, which says that if a function $f(x)$ passes continuously from negative to positive values, then $f(c) = 0$ for some value of c. We will not examine Gauss's reduction to odd degree, but see section 8.5 for another way to make this reduction. The intermediate value theorem comes into play because it applies to an odd-degree polynomial $p(x)$.

Evidently inspired by Gauss's paper, Bolzano (1817) noticed the crucial role of the intermediate value theorem in the proof and realized that two things were lacking: a precise definition of continuous function and a proof of the intermediate value property. To our intuition, continuity of a function f seems to be a "global" property—that the graph $y = f(x)$ is somehow unbroken, or capable of being drawn without lifting the pen from the paper. Eventually, a very general "global" definition of continuity was given by Hausdorff (1914), but Bolzano found a "local" definition that was adequate for his purpose and, indeed, is still used today in calculus courses.

Definitions.   A function $f : \mathbb{R} \to \mathbb{R}$ is **continuous at a point** a if $f(x) \to f(a)$ as $x \to a$; that is, for any $\varepsilon > 0$ there is a $\delta > 0$ such that

$$|x - a| < \delta \quad \text{implies} \quad |f(x) - f(a)| < \varepsilon.$$

We say f is **continuous** on a set A if f is continuous at each point $a \in A$.

It is clear that the identity function $f(x) = x$, and all constant functions, is continuous on $\mathbb{R}$ according to this definition. And it is not hard to prove that sum, difference, and product of continuous functions are continuous (and also the quotient, if the denominator is nonzero), so *any polynomial function is continuous*. It is also clear that any polynomial $p(x) = x^{2m+1} + \cdots$ of odd degree takes positive values for large positive x and negative values for large negative x. Thus we are all set to prove that each polynomial of odd degree takes the value 0, provided the intermediate value theorem holds.

Bolzano (1817) attempted to prove the intermediate value theorem but could not do so from the available knowledge of $\mathbb{R}$, which had not changed since the time of Euclid. However, he was able to identify a property of $\mathbb{R}$ that, if proved, would enable a proof of the intermediate value theorem:

Least upper bound property. If A is a bounded set of real numbers, then A has a **least upper bound**, that is, a number $l \geq a$ for each $a \in A$, and such that $l \leq b$ for each b such that $b \geq a$ for each $a \in A$.

8.4 DEDEKIND'S DEFINITION OF REAL NUMBERS

Bolzano (1817) was not widely read, so his idea did not catch on. However, it was the right idea, because Dedekind in 1858 independently came to the same conclusion: that calculus demands an understanding of upper bounds in $\mathbb{R}$. As he wrote in Dedekind (1872: 1–2):

> My attention was first directed toward the considerations that form the subject of this pamphlet in the autumn of 1858. As professor in the Polytechnic School in Zürich I found myself for the first time obliged to lecture upon the elements of the differential calculus and felt more keenly than ever the lack of a really scientific foundation for arithmetic. In discussing the notion of the approach of a variable magnitude to a fixed limiting value, and especially in proving the theorem that every magnitude which grows continually, but not beyond all limits, must certainly approach a limiting value, I had recourse to geometric evidences. Even now such resort to geometric intuition in a first presentation of the differential calculus, I regard as exceedingly useful, from the didactic standpoint, and indeed indispensable, if one does not wish to lose too much time. But that this form of introduction into the differential calculus can make no claim to being scientific, no one will deny. For myself this feeling of dissatisfaction was so overpowering that I made a fixed resolve to keep meditating on the question till I should find a purely arithmetic and perfectly rigorous foundation for the principles of infinitesimal analysis. The statement is so frequently made that the differential calculus deals with continuous magnitude, and yet an explanation of this continuity is nowhere given; even the most rigorous expositions ... either appeal to geometric notions or those suggested by geometry, or depend upon theorems which are never established in a purely arithmetic manner. Among these, for example, belongs the above mentioned theorem [on the limit of a bounded increasing quantity], and a more careful investigation convinced me that this theorem, or any one equivalent to it, can be regarded in some way as a sufficient basis for infinitesimal analysis.

Dedekind goes on to say that he secured "a real definition of the essence of continuity" on November 24, 1858. But he held off publishing until 1872, when others began to put forward solutions to the same problem. What Dedekind called "continuity'" of $\mathbb{R}$ we now call **completeness**, but it is closely related to continuity of functions and the intermediate value property.

Dedekind's "purely arithmetic" viewpoint defines a real number in terms of *sets* of rational numbers. It is inspired by the idea of Eudoxus (see section 1.6) that a point on the line is determined by the rational points to its left and right, but Dedekind goes further by taking infinite sets to be legitimate mathematical objects. As we saw in section 7.8, Dedekind (1871) had already crossed this line when taking an "ideal number" to be an *ideal*, which is an infinite set of actual numbers.

Definition. A **Dedekind cut** is a partition of the set $\mathbb{Q}$ into two nonempty sets, L and U (the "lower" and "upper" set)—so each member of $\mathbb{Q}$ belongs to either L or U—such that each member of L is less than every member of U.

Figure 8.2 tries to show what the Dedekind cut for $\sqrt{2}$ looks like. The nonnegative rationals with square less than 2 are shown as black dots—integers as large dots, multiples of 1/10 as smaller dots, multiples of 1/100 smaller still—and those with square greater than 2 in red. Thus the rationals are separated into a cloud L of black points and a cloud U of red points.

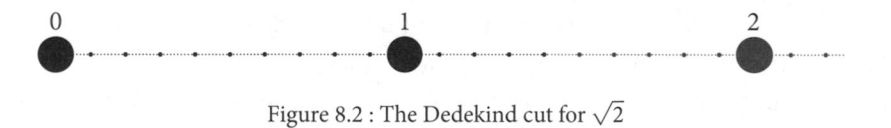

Figure 8.2 : The Dedekind cut for $\sqrt{2}$

Since each real number is determined by a pair $\langle L, U \rangle$, Dedekind had an almost immediate proof of the least upper bound property.

Least upper bound property. *If A is any bounded set of real numbers, then A has a least upper bound.*

Proof. For each $a_i \in A$, let $\langle L_i, U_i \rangle$ be the corresponding Dedekind cut. Since A is bounded, each L_i is bounded above by some rational number b. Then the union, L, of all the L_i is also bounded above by b, so the set U of rationals *not* in L is nonempty.

Thus $\langle L, U \rangle$ is a Dedekind cut, and the real number it determines is the least upper bound of A. ☐

We can now complete Bolzano's train of thought, and Gauss's 1816 proof of the FTA, by proving the intermediate value theorem.

Intermediate value theorem. *If* $f : [a, b] \to \mathbb{R}$ *is a continuous function on the interval* $[a, b] = \{x \in \mathbb{R} : a \leq x \leq b\}$: *with* $f(a) < 0$ *and* $f(b) > 0$, *then there is a* $c \in [a, b]$ *with* $f(c) = 0$.

Proof. Consider the set $A = \{x \in [a, b] : f(x) < 0 \text{ and } f(x') < 0 \text{ for any } x' < x\}$. The set A is nonempty because it includes a, and it is bounded above by b. Hence A has a least upper bound c by the least upper bound property. (Intuitively speaking, we expect c to be the first point where f is not negative.)

We prove that $f(c) = 0$ by using the continuity of f to rule out the other possibilities, that $f(c) < 0$ and $f(c) > 0$.

First, if $f(c) < 0$ then the continuity of f gives a δ such that $f(x) < 0$ for all $x < c + \delta$. All such x therefore belong to A, which contradicts the definition of c as the least upper bound of A. Second, if $f(c) > 0$ there is similarly a δ such that $f(x) > 0$ for all x between $c - \delta$ and c. These x therefore do not belong to A, which again contradicts the definition of c. ☐

8.5 THE ALGEBRAIST'S FUNDAMENTAL THEOREM

To some, it seems like overkill to base a theorem about polynomials on the entirety of $\mathbb{R}$ and the full strength of a theorem about continuous functions, the intermediate value theorem. An early opponent of this heavy machinery was the algebraist Leopold Kronecker, in the 1870s. Kronecker objected to several aspects of the proof: not just to the use of the full real number system but to infinite objects per se, and to a proof that claims existence of an object (root of the polynomial) without giving the means to construct it.

Kronecker (1887) sought to replace FTA with his **fundamental theorem of general arithmetic**, which *constructs*, for each polynomial $p(x)$, a field in which the equation $p(x) = 0$ has a root. We have already proved this theorem, at least for irreducible $p(x)$ with rational coefficients, by defining $\mathbb{Q}[x]/p(x)\mathbb{Q}[x]$ in section 7.6. In this field, the congruence class $[x]$ satisfies $p([x]) = [0]$. Also, $\mathbb{Q}[x]/p(x)\mathbb{Q}[x]$ is "constructible" in a reasonable sense. Of course this field is infinite, but one can give an algorithm for systematically listing its members,[1] and also an algorithm for testing congruence modulo $p(x)$, since this amounts to deciding whether a given polynomial $q(x)$ is divisible by $p(x)$, which in turn boils down to finding $\gcd(p(x), q(x))$ by the Euclidean algorithm.

We are tempted to call Kronecker's theorem the *algebraist's FTA*, since it is a purely algebraic theorem about the roots of polynomial equations. It looks at first very different from the classical FTA, but there is an interesting connection: *the algebraist's FTA gives a reduction of an arbitrary polynomial equation to one of odd degree.* This reduction simplifies the Gauss (1816) proof and also fills a gap in an interesting proof sketched by Laplace (1795). We follow the outline of Laplace's proof given by Ebbinghaus et al. (1990).

Laplace's proof of FTA. As we have seen, it suffices to show we can solve a polynomial equation of degree $n = 2^k q$, where q is an odd number, by solving an equation of odd degree. We prove this by induction on k, with the case $k = 0$ being true by the intermediate value theorem.

Let $p(x)$ be an equation of degree $n = 2^k q$, with rational coefficients. Without loss of generality we can assume that $p(x)$ has leading term x^n. Then, by the algebraist's FTA, there is a field $\mathbb{F} \supseteq \mathbb{Q}$ in which $p(x)$ splits into linear factors:

$$p(x) = (x - x_1)(x - x_2)\cdots(x - x_n), \quad \text{for some} \quad x_1, x_2, \ldots, x_n \in \mathbb{F}.$$

Following Laplace, we now consider the polynomials L_t, for rational t, with the $n(n-1)/2$ factors

$$x - x_i - x_j - tx_ix_j, \quad \text{where} \quad 1 \le i < j \le n.$$

Since $n - 1$ is odd, the degree $n(n-1)/2$ of L_t equals $2^{k-1}r$, for some odd integer r. Also, the coefficients of L_t are symmetric functions of $x_1, x_2, \ldots, x_n$, since any permutation of $x_1, x_2, \ldots, x_n$ produces the same set of factors.

It then follows, by the fundamental theorem of symmetric polynomials (see section 4.4), that the coefficients of L_t are polynomials in the

---

1. This makes the field only "potentially" infinite, rather than "actually" infinite, which is the kind of infinity Kronecker objected to. We say more about the distinction between potential and actual infinity in chapter 12.

elementary symmetric functions of $x_1, x_2, \ldots, x_n$, with rational coefficients, and hence they are rational combinations of the coefficients of $p(x)$, which are rational. Thus L_t has rational coefficients, and since its degree contains a lower factor of 2 than $p(x)$, we can assume that some root of $L_t = 0$ is in $\mathbb{C}$, that is, that

$$x_k + x_l + t x_k x_l \in \mathbb{C}, \quad \text{for some } k, l \text{ depending on } t.$$

Since there are infinitely many values of t but only finitely many pairs $\langle i, j \rangle$, for at least two values of t we will get the same pair $\langle k, l \rangle$. That is, we can find distinct rational s, t so that both $x_k + x_l + s x_k x_l$ and $x_k + x_l + t x_k x_l$ are in $\mathbb{C}$. Then their difference is also in $\mathbb{C}$, so

$$u = x_k x_l \in \mathbb{C} \quad \text{and hence} \quad v = x_k + x_l \in \mathbb{C}.$$

This means that x_k and x_l are the roots of the quadratic polynomial equation $z^2 - vz + u = 0$ with coefficients in $\mathbb{C}$. The usual quadratic formula then gives $x_k, x_l \in \mathbb{C}$, as required. $\square$

8.6 REMARKS

The problems raised by the fundamental theorem of algebra drew sustained attention to the foundations of calculus—the nature of the real numbers, continuity, and limits—that are studied in what we now call **analysis**. The main contributors to the foundations were Bolzano, Cauchy in his Paris lectures of the 1820s, and Weierstrass in his Berlin lectures of the 1870s. Cauchy's results were published in his textbooks (Cauchy 1821, 1823), but those of Weierstrass were spread mainly by his students. They became part of a movement to "arithmetize" analysis that is the subject of chapter 11.

In particular, in section 11.7 we will prove the **extreme value theorem** of Weierstrass, which is the key to the d'Alembert/Argand proof of FTA.

CHAPTER 9

Non-Euclidean Geometry

The first "non-Euclidean" geometry to be developed in detail was *projective* geometry. Though it is not what we mean today by "non-Euclidean," it took a new view of parallel lines by declaring that they meet—*at infinity*. This seemingly radical idea is natural in perspective drawing, where projective geometry originated, but it raised questions about the meaning and existence of objects called points and lines at infinity. These questions are easily answered by **models** of projective geometry, as we saw in section 3.8.

A second "non-Euclidean" geometry is *spherical* geometry. It has been studied since ancient times in connection with astronomy, and later in navigating the earth's surface. Its "lines," the *great circles* of the sphere, behave like Euclidean lines in some ways but in other ways are quite different. In particular, they are finite, and there are no "parallels."

What is now called non-Euclidean geometry grew from attempts to show that Euclid's parallel axiom is not in fact an *axiom* but a consequence of his other axioms. Saccheri (1733) tried to do this by showing that the *negation* of the parallel axiom leads to a contradiction. He succeeded in the case of no parallels but failed in the case where there is more than one parallel to a given line through a given point. Later, Bolyai and Lobachevsky systematically generated the consequences of the axiom that there is more than one parallel, finding many interesting theorems and no contradiction.

Then in 1868 Beltrami showed that no contradiction could ever arise, by finding **models** of Bolyai's and Lobachevsky's geometry. Beltrami's models were much less obvious than the standard model of projective geometry, though they were part of classical mathematics. Indeed, one of them was part of projective geometry.

9.1 THE PARALLEL AXIOM

It is thought that Euclid himself was dissatisfied with his parallel axiom. In book 1 of his *Elements* he went as far as possible without using it, bringing it out only when it was needed to prove the Pythagorean theorem. But neither Euclid nor anyone else was able to dispense with the parallel axiom, because all attempts to prove it from Euclid's other axioms led to failure. Over the millennia it was noticed that many consequences of the parallel axiom also imply it and hence could be used in its place. Among them:

1. The existence of rectangles (al-Haytham, 10th century)
2. Angle sum of a triangle equals π (Legendre, 1820s)
3. An equidistant curve of a line is a line (Thā̄ bit ibn Qurra, 9th century)
4. Any three points not in a line lie on a circle (Farkas Bolyai, 1830s)
5. Similar triangles of arbitrary sizes exist (Wallis, 1690s).[1]

Some of these are arguably more plausible than the parallel axiom itself, so possibly more easily provable. However, all attempts to prove them from Euclid's other axioms failed. A thorough account of these early investigations is in Rosenfeld (1988: 40–97).

Saccheri (1733) made the most determined attempt to prove the parallel axiom, assuming its negation and pursuing consequences until he found one he considered to be false. There are two ways to negate the parallel axiom: assuming *no* parallels, and assuming *more than one* parallel (to a given line through a given point not on the line). Saccheri found that the first assumption leads to the conclusion that lines are finite, directly contradicting one of Euclid's axioms, so that assumption could be ruled out.

The second assumption—more than one parallel—did not, however, lead to a contradiction, only to a conclusion that Saccheri considered "repugnant to the nature of the straight line." He found that it implies the existence of **asymptotic lines**—lines that approach arbitrarily closely to

---

1. This gives the characteristic of Euclidean geometry referred to in section 1.6, that length is a *relative* concept, because there is no distinguished unit of length.

each other but never meet. What Saccheri found repugnant is that such lines "have a common perpendicular at infinity." This is perhaps strange, but it is *not* a contradiction, and later events showed that this is natural behavior in non-Euclidean geometry (see sections 9.6 and 9.7).

9.2 SPHERICAL GEOMETRY

We will see in this chapter that both Euclidean and non-Euclidean geometry can first be viewed as geometries of surfaces: Euclidean geometry as the geometry of the plane and non-Euclidean geometry as the geometry of a curved surface known as the **hyperbolic plane**. To prepare for the study of curved surfaces, we begin with the simplest case: the sphere.

Since the earth is nearly spherical, and the heavens were assumed to be spherical, the geometry of the sphere has been studied since ancient times, for both navigation and astronomy. In fact, *trigonometry* was for a long time studied more on the sphere than on the plane. As Van Brummelen (2013: vii) wrote:

> Born of the need to locate stars and planets in the heavens, it [spherical trigonometry] was the big brother to the plane trigonometry that high school students slog through today.

The formulas of spherical trigonometry played a role in the discovery of non-Euclidean geometry, as we will see below, but we will not go into their derivation. Instead we will look at a result even simpler than trigonometry: the formula for the area of a spherical triangle.

Definitions. Lines on a sphere $\mathscr{S}$ are taken to be great circles, that is, the intersections of $\mathscr{S}$ with planes through its center. The **angle** between two lines is the angle between the corresponding planes (which is also the angle between the tangents to the corresponding great circles at the point where they meet). A **spherical triangle** is a region bounded by three distinct lines with no common point.

A remarkable discovery of Thomas Harriot[2] in 1603 shows that the area of a spherical triangle $\Delta_{\alpha\beta\gamma}$ depends only on the sum of its angles,

---

2. Harriot was mathematical consultant to the explorer Sir Walter Raleigh, so he too was interested in the geometry of the sphere because of its applications to navigation. His theorem on the equiangular spiral in section 6.1 was a spinoff from his study of *rhumb lines* on the sphere. These are the spirals that cross the parallels of latitude at a constant angle, as a ship does when it sails in a fixed compass direction.

α, β, and γ. A triangle that suggests some relation between angle sum and area for spherical triangles is the one with three right angles. Eight such triangles fill the spherical surface (figure 9.1), so each has area $S/8$, where S is the surface area of the sphere. Area in fact depends on the *excess* of the angle sum over π.

Figure 9.1 : Filling the sphere with eight triangles

Area of a spherical triangle. *If $\Delta_{\alpha\beta\gamma}$ is a spherical triangle on a sphere of area S, then the area of $\Delta_{\alpha\beta\gamma}$ equals $S(\alpha + \beta + \gamma - \pi)/4\pi$.*

Proof. If we extend the three sides of $\Delta_{\alpha\beta\gamma}$, the resulting great circles divide the sphere into the eight spherical triangles shown in figure 9.2.

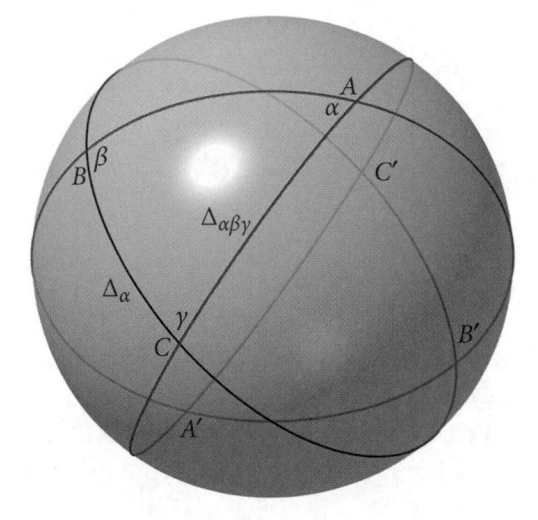

Figure 9.2 : Division of sphere by three great circles

We notice that $\Delta_{\alpha\beta\gamma}$, combined with the neighboring triangle Δ_α, forms a wedge of the sphere bounded by two great semicircles meeting at angle α (figure 9.3). The area of the wedge is obviously proportional to α, so it equals $S\alpha/2\pi$. Similarly, the other two triangles neighboring $\Delta_{\alpha\beta\gamma}$ form wedges of angles β and γ. These observations give us three equations (slightly abusing notation by using names of triangles for their areas):

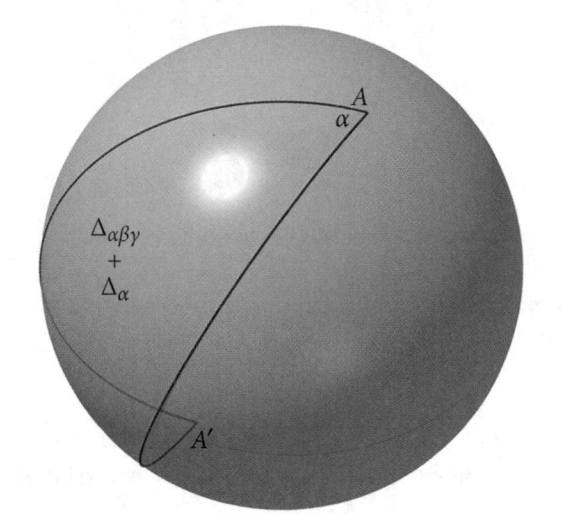

Figure 9.3 : Wedge of sphere between two great circles

$$\Delta_{\alpha\beta\gamma} + \Delta_\alpha = S\alpha/2\pi,$$
$$\Delta_{\alpha\beta\gamma} + \Delta_\beta = S\beta/2\pi,$$
$$\Delta_{\alpha\beta\gamma} + \Delta_\gamma = S\gamma/2\pi,$$

which sum to

$$3\Delta_{\alpha\beta\gamma} + \Delta_\alpha + \Delta_\beta + \Delta_\gamma = S(\alpha + \beta + \gamma)/2\pi. \qquad (*)$$

On the other hand, the eight spherical triangles in figure 9.2 consist of two copies of each of $\Delta_{\alpha\beta\gamma}$, Δ_α, Δ_β, and Δ_γ. For example, $\Delta_{\alpha\beta\gamma}$ is triangle ABC, which is congruent to triangle $A'B'C'$ because the pairs A, A', B, B', and C, C' are pairs of antipodal points. So we also get the equation

$$2\Delta_{\alpha\beta\gamma} + 2\Delta_\alpha + 2\Delta_\beta + 2\Delta_\gamma = S = 2\pi S/2\pi.$$

Subtracting half the latter equation from (*) gives

$$2\Delta_{\alpha\beta\gamma} = S(\alpha + \beta + \gamma - \pi)/2\pi,$$

so $\Delta_{\alpha\beta\gamma} = S(\alpha + \beta + \gamma - \pi)/4\pi$, as required. $\square$

We can now see that the tiling of the sphere shown in figure 9.1 agrees with Harriot's theorem. In this case, $\alpha + \beta + \gamma - \pi = 3\pi/2 - \pi = \pi/2$, so the area of $\Delta_{\alpha\beta\gamma}$ is $\frac{S\pi/2}{4\pi} = S/8$, as expected.

Another interesting example is shown in figure 9.4.

Figure 9.4 : Filling the sphere with 120 triangles

Here the triangles come from dividing each face of an icosahedron into six equal parts and projecting the result onto a sphere. Notice that each triangle has an angle $\alpha = \pi/2$, where four equal angles meet; an angle $\beta = \pi/3$, where six equal angles meet; and an angle $\gamma = \pi/5$, where 10 equal angles meet. Therefore,

$$\alpha + \beta + \gamma - \pi = \frac{\pi}{2} + \frac{\pi}{3} + \frac{\pi}{5} - \pi = \frac{31\pi}{30} - \pi = \frac{\pi}{30},$$

which gives area $S\frac{\pi/30}{4\pi} = S/120$ and confirms that there are 120 triangles.

9.3 A PLANAR MODEL OF SPHERICAL GEOMETRY

Although the sphere is quite readily understood as an object in three-dimensional space, it is also useful to project it onto the plane, as we know from studying maps of the earth in geography. It might even be useful to have a planar projection of figure 9.4, to check that there really are 120 triangles. For mathematical purposes, the most useful projection is **stereographic projection**, shown in figure 9.5. This kind of projection first appears in Claudius Ptolemy's *Planisphere* from around 150 CE.

Each point P of the sphere, except the north pole, is projected to a point P' of the plane by a ray from the north pole N. Thus the plane is not quite a complete map of the sphere unless we add a "point at infinity" to the plane, which is generally done. This is like adding a point to the ordinary line to create a projective line, as we did in section 3.8. In fact, when the plane is viewed as the plane $\mathbb{C}$ of complex numbers, $\mathbb{C}$ with the added point is called the **complex projective line**, as mentioned in section 5.5.

The plane with its point at infinity is a model of the geometry of the sphere in the sense that geometric features of the sphere—points, great circles, and angles—correspond to geometric features of the plane. Obviously, points correspond to points. More remarkable, angles correspond to angles, and circles correspond to circles (if we include straight lines through the south pole among the "circles"—circles with center at infinity, if you like).

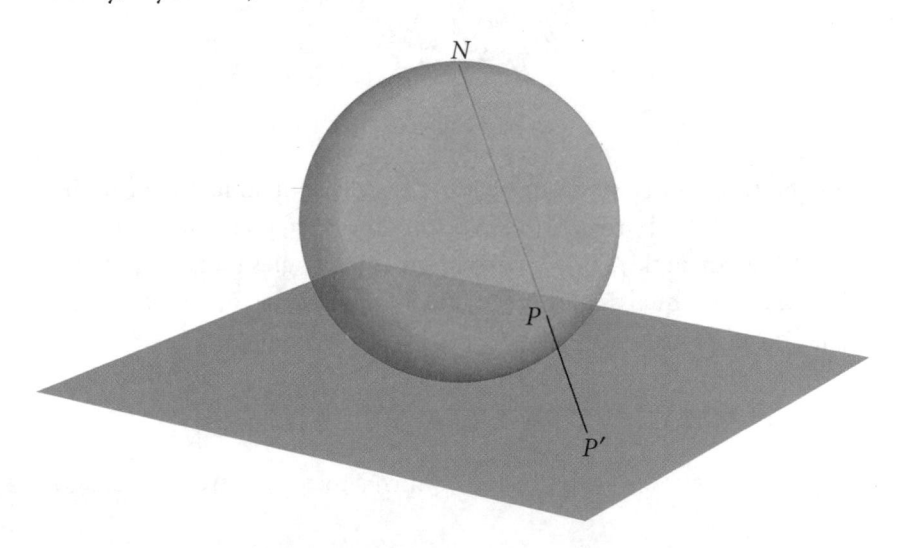

Figure 9.5 : Stereographic projection of the sphere

The circle-to-circle property was used in Ptolemy's *Planisphere*, though not proved there. According to Rosenfeld (1988: 122), it follows from a proposition in the *Conics* of Apollonius. The angle-preserving property was apparently discovered by Harriot in the 1590s (see Lohne 1979). Preservation of circles and preservation of angles can both be seen in the figures below, where the sphere tiled by 120 triangles is projected onto the plane.

First we cut out every other triangle from the sphere (figure 9.6); then we project its shadow onto the plane by a light from the north pole and inspect the image (figure 9.7).

Figure 9.6 : Sphere with triangles cut out

Figure 9.7 : Stereographic projection of the cutout sphere

Another elegant feature of stereographic projection is that rotations of the sphere induce maps of the plane $\mathbb{C}$ that are very simple functions of a complex variable. Gauss (1819) discovered that these are functions of the form

$$f(z) = \frac{az + b}{-\bar{b}z + \bar{a}} \, .$$

9.4 DIFFERENTIAL GEOMETRY

After the invention of calculus it was possible to investigate curved surfaces more complicated than those studied by the ancients: the sphere, cylinder, and cone. In particular, it became possible to define and investigate the concept of **curvature** itself, first for curves and then (considerably later) for surfaces and higher-dimensional spaces.

Calculus also gave the first tools for systematically investigating *transcendental curves*, such as the equiangular spiral already mentioned in sections 6.1 and 5.3, and a host of others defined by properties calculus was able to compute, such as arc length and tangents. Space does not permit us to discuss many of them here, but we examine two that later played a role in non-Euclidean geometry: the **tractrix** and the **catenary**.

Tractrix and Catenary

The tractrix (left side of figure 9.8) was introduced by Newton (1676b) as the curve for which the length of the tangent between its point of contact with the curve and the *x*-axis is constant, *a*. Its equation is

$$x = a \ln \frac{a + \sqrt{a^2 - y^2}}{y} - \sqrt{a^2 - y^2}.$$

Essentially this equation was given by Huygens (1693a), who also pointed out a physical interpretation of the curve: the path of a stone at the end of a string of length *a* pulled (hence the name "tractrix") by someone walking along the *x*-axis. Huygens (1693b) also studied the surface obtained by revolving the tractrix about the *x*-axis (right side of figure 9.8), a surface now known as the **pseudosphere**. The reason for the name is a certain analogy with the sphere we will meet in the next subsection.

The catenary, whose name comes from the Latin word for "chain," is the shape of a hanging chain (blue curve in figure 9.9). Its equation is

$$y = \cosh x = \frac{e^x + e^{-x}}{2}.$$

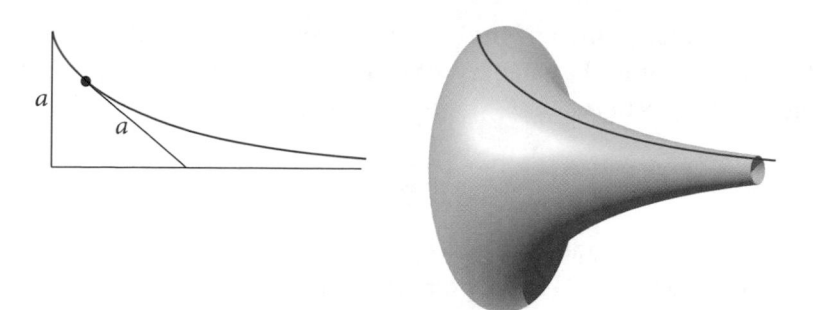

Figure 9.8 : The tractrix and the pseudosphere

The problem of finding this shape was posed by Jakob Bernoulli in 1690 and independently solved by Johann Bernoulli, Huygens, and Leibniz in 1691. Neither the hyperbolic cosine nor the exponential function was a recognized function at the time, so the solutions came in the form of integrals of algebraic functions, such as $\int \frac{dx}{\sqrt{x^2-1}}$.

Figure 9.9 also shows the tractrix, in red, and its relation to the catenary. The tractrix is the **involute** of the catenary—intuitively speaking, the path C of the end of a thread being unwound from the catenary C'.

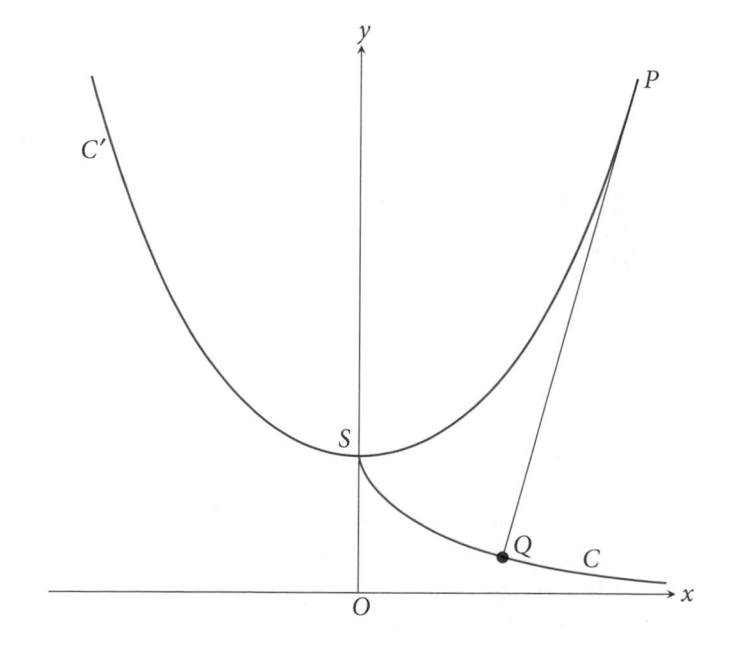

Figure 9.9 : Tractrix and catenary

At each instant of the unwinding, the unwound portion of the thread, PQ, is a tangent to the catenary, of length equal to the arc length of the catenary from S to P.

Curvature

Figure 9.9 also illustrates the concept of curvature of a plane curve, introduced by Newton (1671). The point P is called the **center of curvature** of the tractrix at point Q. It is the limiting position of the intersection of the normal at Q with the normal at nearby point. The length of PQ is called the **radius of curvature**, and its reciprocal is called the **curvature** at Q. It is clear from this definition that a circle of radius r has constant curvature $1/r$ and a straight line has constant curvature 0. It can be shown that these are the only two plane curves of constant curvature.

Euler (1760) extended the concept of curvature from curves to surfaces by cutting a surface orthogonally by planes at a given point and considering the curvatures of the curves of intersection. He found that the maximum and minimum curvature values (allowing curvatures to have opposite signs if the corresponding centers of curvature lie on opposite sides of the surface) occur for cutting planes that are orthogonal to each other. These curvatures are called the **principal curvatures**. Figure 9.10 shows them in the case of a cylinder and a saddle-shaped surface.

Figure 9.10 : Principal curvature sections of cylinder and saddle

Gauss (1827) made the wonderful discovery that the *product* of the principal curvatures is an intrinsic geometric property of the surface. That is, it can be defined by measurements within the surface and is therefore unchanged by bending. The product of the principal curvatures is called the **Gaussian curvature**. With the cylinder, one of the principal curvature sections is a straight line, hence with curvature zero, so the Gaussian curvature is zero. This is what the intrinsic curvature of the cylinder should be, since the cylinder can be unbent to form a plane, which certainly has curvature zero. The principal curvatures of the saddle, on the other hand, have opposite signs, so its Gaussian curvature is negative.

This raises the question, What are the surfaces of constant curvature? The sphere of radius r obviously has both principal curvatures equal to $1/r$; hence its Gaussian curvature is the constant $1/r^2$. The same is true of any surface that is locally like a sphere, the most interesting example of which is the **real projective plane**. Recall from section 3.8 that the real projective plane has "points" that correspond to lines through the origin in $\mathbb{R}^3$. Equivalently, we can consider the intersections of these lines with the unit sphere, which are *pairs* of antipodal points (figure 9.11). In this interpretation, the "points" of the real projective plane are pairs of antipodal points on the sphere.

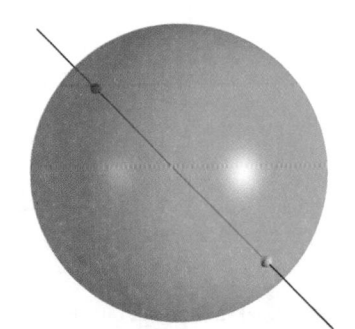

Figure 9.11 : Antipodal points on the sphere

It follows that any portion of the real projective plane small enough to include at most one representative of each pair is identical with a portion of the sphere, and hence has the same curvature.

Surfaces of constant zero curvature, similarly, are locally like the plane. They include cylinders and cones.

Surfaces of constant negative curvature are not so easy to find. The best-known example is the pseudosphere. It was among a few examples studied by Minding (1839). To see why its Gaussian curvature is constant, first observe that its principal radii of curvature are the lines PQ and QR

shown in figure 9.12. Since the equation of the blue catenary is

$$y = \cosh x,$$

we can suppose that the point P is $\langle u, \cosh u \rangle$ for some value u. We also have

$$\frac{dy}{dx} = \frac{e^x - e^{-x}}{2} = \sinh x \qquad \cosh^2 x - \sinh^2 x = 1.$$

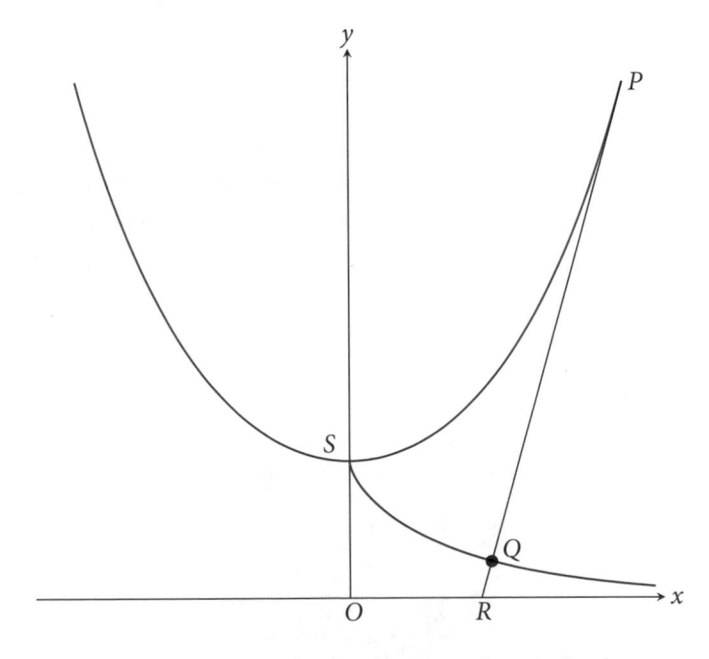

Figure 9.12 : Principal radii of curvature for pseudosphere

Then, since the red tractrix is the involute of the catenary,

$$\text{length } PQ = \text{arc length } PS = \int_0^u \sqrt{1 + \left(\frac{dy}{dx}\right)^2}\, dx$$

$$= \int_0^u \sqrt{1 + \sinh^2 x}\, dx$$

$$= \int_0^u \cosh x\, dx = \sinh u.$$

Also, by calculating the equation of the tangent PR we find that

$$R = \left\langle u - \frac{\cosh u}{\sinh u}, 0 \right\rangle.$$

This in turn gives us the length PR, and hence QR. Miraculously, it turns out that the length of $QR = 1/\sinh u$. So the product of the radii of curvature is 1 and hence so is the (unsigned) product of the curvatures themselves.[3]

9.5 GEOMETRY OF CONSTANT CURVATURE

From now on we will refer to Gaussian curvature simply as "curvature."

To talk about geometry on arbitrary surfaces we need to interpret terms such as "point," "line," "distance," "area," and "angle." In the case of constant curvature we are lucky to be able to refer to specific surfaces for a *standard* interpretation of these concepts: the sphere for surfaces of constant positive curvature, the plane for surfaces of zero curvature, and the pseudosphere for surfaces of constant negative curvature. We often use such a bijective, distance-preserving correspondence—called an **isometry**—between portions of a surface $\mathscr{S}$ and portions of a standard surface to *define* distance on $\mathscr{S}$ by "transporting" distance from the standard surface.

For example, when we mapped the sphere onto the plane by stereographic projection we called the plane (completed by a point at infinity) a "model" of spherical geometry. To be precise, we are taking **distance** between any two (sufficiently close) points in the completed plane to be the great circle distance between the corresponding points on the sphere. Distance on the sphere is a little subtle because two points P, Q on the sphere are generally connected by two great circle arcs: a short one and a long one. We take the length of the short one as the distance between P and Q.

Distance on the pseudosphere is subtle for a different reason. First, we must define arc length of a curve on the pseudosphere, which is done by the same limit process used for curves in the plane (section 6.7). Next we need to know there is a *shortest* curve between any two points, called a **geodesic**. One approach, due to Gauss (1825), defines the **geodesic curvature** of a curve at a point P on a surface $\mathscr{S}$ as the ordinary curvature of the orthogonal projection of the curve onto the tangent plane of $\mathscr{S}$ at P. Geodesics are then defined as curves of zero geodesic curvature. Finally, one proves that there is a unique geodesic between any two points P, Q

---

3. Another very simple, and more geometric, proof that the curvature of the pseudosphere is constant may be found in Needham (1997: 295–296).

on the pseudosphere and defines the distance between P and Q as the arc length of this geodesic.

Geodesic Triangles

Harriot's theorem on the area of a spherical triangle may now be seen more generally as a theorem about geodesic triangles on surfaces of constant positive curvature, because the portion of a surface of constant positive curvature containing a geodesic triangle is isometric to a portion of the sphere and a corresponding spherical triangle.

There is a similar theorem about geodesic triangles on the pseudosphere and hence on any surface of constant negative curvature: *a geodesic triangle with angles α, β, γ has area $c(\pi - \alpha - \beta - \gamma)$ for some positive constant c.* This means that the angle sum of any geodesic triangle is less than π and, more surprising, that there is an upper bound $c\pi$ to the area of all triangles. Minding (1839) also found trigonometric formulas for surfaces of constant negative curvature. Remarkably, they are like the formulas of spherical trigonometry, except that the circular functions $\cos$ and $\sin$ are replaced by the hyperbolic functions $\cosh$ and $\sinh$.

A simple example of this replacement, already known to Gauss (1831), is in the formula for the circumference of a circle of radius r. On the sphere of radius 1, a great circle arc of length r subtends an *angle* r. So a circle of radius r *on the sphere* bounds a planar disk of radius $\sin r$, and hence its circumference is $2\pi \sin r$ (see figure 9.13). Gauss found that on a surface of curvature -1 the circumference of a circle of radius r is $2\pi \sinh r$.

However, "hyperbolic trigonometry" did not immediately lead to a "hyperbolic geometry," probably because the known surfaces of constant negative curvature are not good analogues of the plane or the sphere. They are all **incomplete** in the sense that geodesics cannot be extended indefinitely in each direction—they stop at an "edge." On the pseudosphere (figure 9.8), for example, geodesics extend infinitely far to the right, but on the left they stop at the edge that corresponds to the end of the tractrix. One would hope for a surface of constant negative curvature in which geodesics extend infinitely far in each direction, but no such surface was found.

Eventually, Hilbert (1901) proved that there is no complete smooth surface in $\mathbb{R}^3$ with constant negative curvature. Thus if "hyperbolic

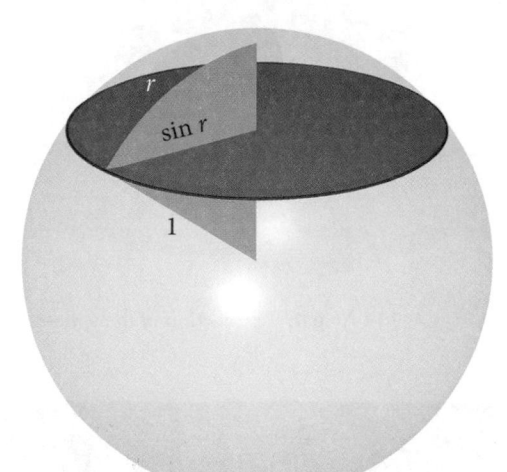

Figure 9.13 : Circle of radius r on the sphere of radius 1

geometry" has a model $\mathcal{M}$ it cannot be one whose "distance" is the standard distance in $\mathbb{R}^3$. Rather, its distance must be transported to pieces of $\mathcal{M}$ from pieces of the pseudosphere. We will see how such a model was discovered in the next subsection.

Mapping Surfaces of Constant Curvature to the Plane

Beltrami (1865) looked for surfaces that can be mapped to the plane in such a way that their geodesics go to straight lines and discovered that they are precisely the surfaces of constant curvature. In the case of the sphere the map is **central projection**, in which each point in the lower hemisphere is sent to the plane by projection from the center of the sphere (figure 9.14). This map obviously sends great (semi)circles to straight lines.

Figure 9.15 shows the lines that result from central projection of the sphere tiled with 48 equal spherical triangles. The 24 triangles in the lower hemisphere go to 24 triangles with straight edges, some with vertices at infinity. In fact it makes sense to add a **line at infinity** to the image plane; it thereby becomes another model of the real projective plane, related to the sphere model of section 9.4 by projection of the hemisphere, which includes a member of each pair of antipodal points.

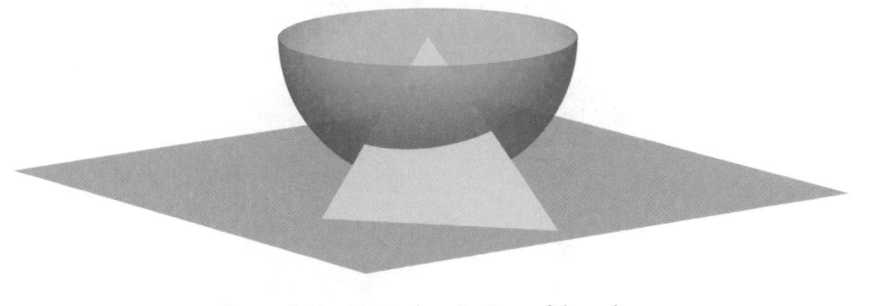

Figure 9.14 : Central projection of the sphere

Figure 9.15 : Tiling of the projective plane by triangles

The equator of the hemisphere is mapped two to one onto the line at infinity of the plane.

Beltrami's result is most interesting in the case of surfaces of constant negative curvature, where the surfaces map into parts of the disk. Figure 9.16 shows three examples; the middle one is the pseudosphere. This figure is taken from Klein (1928: 286), slightly modified for clarity. In particular, the surfaces have been rotated so that their geodesics better align with their images in the disk, which are straight line segments.

When the distance function is transported from the pseudosphere to the disk, it turns out to be meaningful over the whole disk, not just the wedge-shaped piece corresponding to the pseudosphere. This

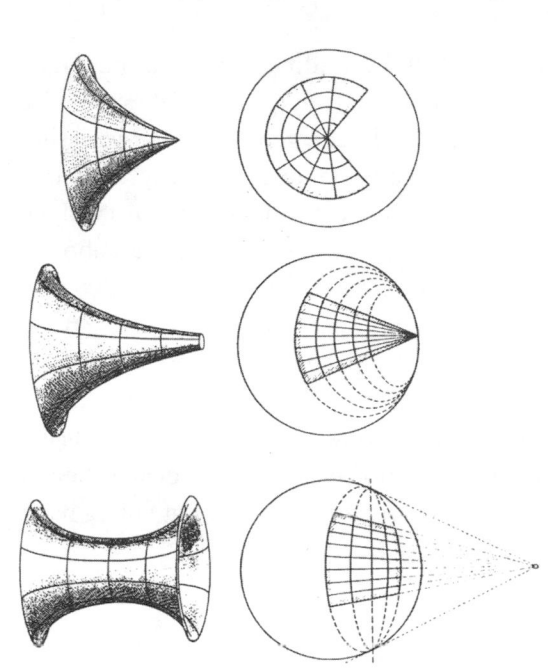

Figure 9.16 : Maps of surfaces with constant negative curvature

suggests that the disk can serve as a **complete** surface of constant negative curvature—one in which "lines" are modeled by line segments within the disk. In the next section we will see what happened when Beltrami took up this idea in 1868.

9.6 BELTRAMI'S MODELS OF HYPERBOLIC GEOMETRY

Before differential geometry hinted that non-Euclidean geometry might live in the surfaces of constant negative curvature, there had been extensive exploration of the consequences of the assumption that there is more than one parallel to any line through any point not on the line. Some was conducted privately by Gauss and his friends, and independently of them, more was conducted and published by Lobachevsky (1829) and Janos Bolyai (1832b). Both Lobachevsky and Bolyai were convinced that non-Euclidean geometry was valid, and they have been rightly hailed for their courage in publishing their discoveries.

However, they did not *prove* the validity of non-Euclidean geometry, or perhaps even see how this might be possible. The first to do so was Beltrami in 1868, by finding **models** of non-Euclidean geometry in ordinary mathematics. Today, we would say that Beltrami proved the consistency of non-Euclidean geometry *relative to* the consistency of ordinary geometry. The consistency of ordinary mathematics was not an issue at that time, though it became one later, as we will see in chapter 13.

In 1868 Beltrami actually published two papers about models of non-Euclidean geometry. The first (Beltrami, 1868a) described a particular model for plane geometry, based on an idea for extending the pseudosphere to a planar surface. The second (Beltrami, 1868b), described a more comprehensive approach in any number of dimensions, but including several natural models of plane non-Euclidean geometry. The comprehensive approach was based on the idea of Riemann (1854b) for describing curved spaces *intrinsically* without falling back on the idea of distance in Euclidean space.

The Model Based on the Pseudosphere

The idea of Beltrami (1868a) can be visualized with the help of the map of the pseudosphere shown in the middle of figure 9.16. The pseudosphere itself maps to the wedge in the disk, bounded by two line segments that meet ("at infinity") on the boundary of the disk, which we can take to be the unit circle. This wedge-shaped portion of the disk can be naturally extended in two directions:

- By "rotating" the wedge about its endpoint on the disk boundary, which corresponds to "unwrapping" a surface wrapped infinitely many times around the pseudosphere, like unwrapping a plane wrapped around a cylinder. The dashed lines in figure 9.16 show where the circles on the pseudosphere unwind as the unwrapping proceeds.
- After unwrapping the pseudosphere on the disk, extend each of its geodesics backward to the unit circle.

Then the whole interior of the unit circle is filled, and we have a model of a complete surface of constant negative curvature. It can be checked that it is also a model of the non-Euclidean plane: its "points" are points inside the unit circle, its "lines" are open line segments bounded by points on the unit circle, and its "distance" function (extended in a natural way to the whole open disk) is the distance between preimage points on the pseudosphere.[4]

It can also be checked that this model satisfies all of Euclid's axioms *except* the parallel axiom. Since "lines" are line segments within the disk, there is obviously more than one "parallel" to a given line $\mathscr{L}$ through a given point P outside $\mathscr{L}$ (figure 9.17). Thus the model proves, for the first time, that *the parallel axiom is not a consequence of Euclid's other axioms.*

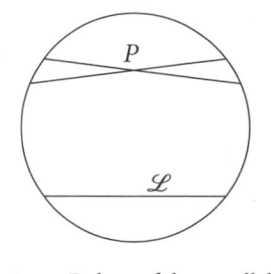

Figure 9.17 : Failure of the parallel axiom

Beltrami's Conformal Models

Beltrami (1868a) made his model of non-Euclidean geometry as concrete as possible by basing it on an intuitive understanding of the pseudosphere in $\mathbb{R}^3$. In his second paper on the subject, Beltrami (1868b) gave a more general and abstract approach, based on the Riemann (1854b) concept of what is now called a **Riemannian manifold**. Riemann freed geometry from the limitations of $\mathbb{R}^3$, defining "points" in spaces of arbitrary dimension by coordinates and defining "distance" as a function of coordinates. From the distance function one can compute the curvature, and Riemann actually gave an example with constant negative curvature. Beltrami saw, as Riemann apparently did not, that this example was a model of non-Euclidean geometry, and from it he constructed three concrete models of the non-Euclidean plane: the hemisphere model, the conformal disk model, and the half-plane model. All of them are quite

---

4. It follows, incidentally, that the **isometries of the disk**—the maps of the disk that preserve the distance function—are those maps of the plane that send straight lines to straight lines and map the unit circle into itself. These maps had already been studied by Cayley (1859), who noticed that they define a concept of "distance." But he did not notice that this was distance in the sense of non-Euclidean geometry. Since the isometries map straight lines to straight lines, they are *projective* maps. As Klein (1871) observed, this means that non-Euclidean geometry can be considered part of projective geometry.

simply related to his model described in the previous subsection, which may be called the **projective disk model**. The new models are called **conformal models** because they faithfully represent angles.

- **The Hemisphere Model.** This model is a hemisphere above the projective disk, and its "lines" are the semicircles directly above the lines of the projective disk model (figure 9.18).

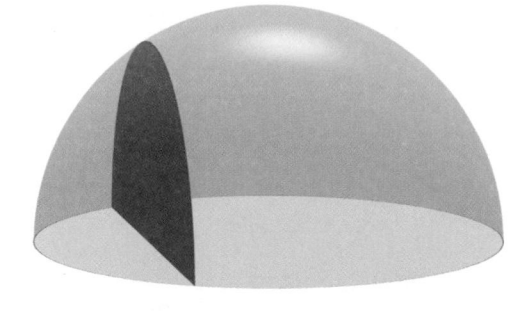

Figure 9.18 : The hemisphere model

- **The Conformal Disk Model.** This model is obtained by stereographically projecting the hemisphere model, inverted and from the center of the opposite hemisphere, onto the plane (figure 9.19). The "lines" of this model are circular arcs orthogonal to the boundary circle of the image (plus the diameters of this circle).

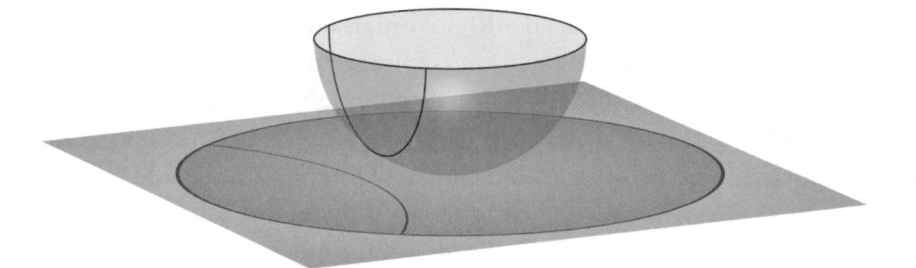

Figure 9.19 : The conformal disk model

- **The Half-Plane Model.** This model is obtained by stereographically projecting the hemisphere model, from a point on its boundary circle, to a plane tangent to the hemisphere at the opposite point. The "lines" of this model (see figure 9.20) are semicircles orthogonal to the boundary line of the image (plus the vertical half lines).

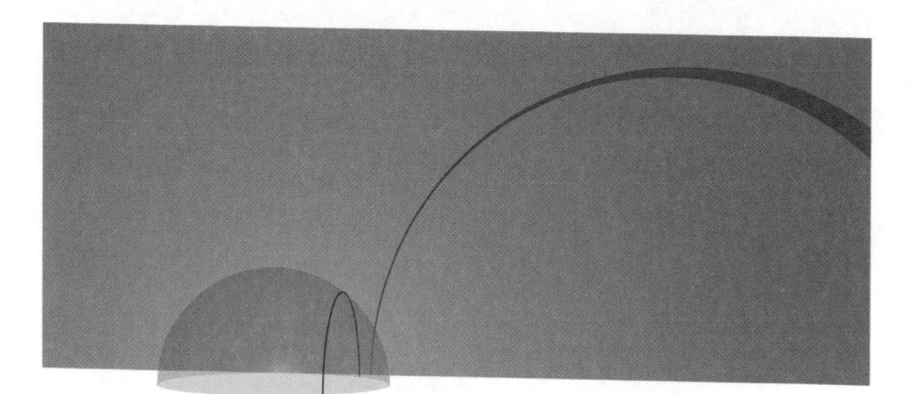

Figure 9.20 : The half-plane model

From now on we will consider all these models, and any structure isometric to them, to represent an abstract structure called the **hyperbolic plane**. Hilbert (1899) gave axioms for the hyperbolic plane (the same as his axioms for the Euclidean plane discussed in chapter 3, but with the parallel axiom replaced by the "more than one parallel" axiom) and showed that all its models are isometric.

9.7 GEOMETRY OF COMPLEX NUMBERS

By mapping the geometries of constant curvature to the plane, which we know can be interpreted as $\mathbb{C}$, we raise the possibility of using complex numbers to study these geometries. Indeed, in section 9.3 we saw the discovery of Gauss (1819) that rotations of the sphere can be described by functions of the form

$$f(z) = \frac{az + b}{-\overline{b}z + \overline{a}},$$

when the sphere is interpreted as $\mathbb{C}$ plus a point at infinity via stereographic projection. To complete this story we point out that the conjugation function $c(z) = \overline{z}$ similarly represents **reflection** of the sphere in a vertical plane through its center, and the rotations and reflection between them generate all isometries of the sphere.

It is a remarkably similar story for the hyperbolic plane, though we now have two models that are naturally interpreted as parts of $\mathbb{C}$: the conformal disk model and the half-plane model. What makes them natural for a complex interpretation (unlike the projective disk model, which is more at home in projective geometry) is the fact that they faithfully represent angles. Differentiable functions of a complex variable are precisely those that *preserve* angles, so isometries of the complex plane ought to be represented by differentiable complex functions.

Throw in some constraints that a function must satisfy in order to be an isometry, such as being bijective and orientation-preserving, and one finds exactly what these functions are:

- For the conformal disk, the functions $f(z) = \frac{az+b}{\bar{b}z+\bar{a}}$ with $|a|^2 - |b|^2 \neq 0$.
- For the half plane, the functions $f(z) = \frac{az+b}{cz+d}$, for $a, b, c, d \in \mathbb{R}$ such that $ad - bc \neq 0$.

These simple functions, called **linear fractional functions**, were known long before Beltrami's models, but it was Poincaré (1882) who first realized they can be interpreted as isometries of non-Euclidean geometry. After Poincaré's discovery, it was noticed that certain pictures of patterns in the disk could be viewed as *tilings of the hyperbolic plane by congruent triangles*. Probably the most spectacular example is shown in figure 9.21, which appeared in Schwarz (1872). This is a tiling of the conformal disk model by congruent triangles, each of which has angles $\pi/2$, $\pi/3$, and $\pi/5$.

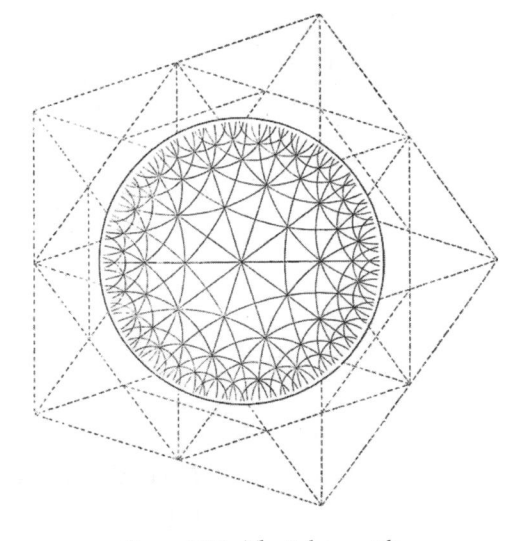

Figure 9.21 : The Schwarz tiling

Outside the circle is a Euclidean world of straight lines. Schwarz is using them to construct, unwittingly, a non-Euclidean world inside the circle. Tilings of the disk like this one arise from particular linear fractional transformations. One of the first and most important examples involves the functions $f(z) = z + 1$ and $g(z) = -1/z$ applied to the half plane.

These functions express a kind of "periodicity" of the **modular function**, a function $j(z)$ from the theory of elliptic functions that also comes up in number theory and other places. The function j is defined on the upper half plane of $\mathbb{C}$ and has "period 1" in the sense that $j(z + 1) = j(z)$ for each z in the half plane. But j also has a second kind of periodicity, because $j(-1/z) = j(z)$ as well. This creates a complicated pattern of places where j repeats its value, illustrated by the pattern of non-Euclidean triangles in figure 9.22, which is taken from Klein and Fricke (1890: 113).

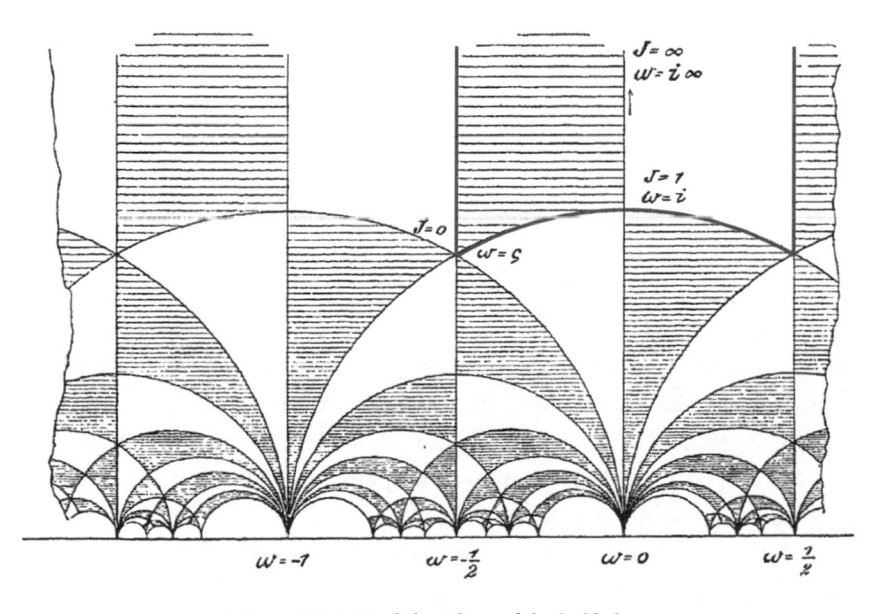

Figure 9.22 : Modular tiling of the half plane

Before this picture appeared, Dedekind (1877a) had noticed that the pattern arises from the region outlined in red, called a **fundamental region**, by repeatedly applying the functions f and g and their inverses. We now view the region as a hyperbolic triangle with one vertex at infinity, and f and g as hyperbolic isometries, so the pattern is in fact a tiling of the hyperbolic plane by congruent triangles.

The realization of Poincaré (1882) that non-Euclidean geometry applies to functions such as the modular function led to great advances not only in the theory of complex functions but also in group theory and even topology, as we will see in the next chapter.

9.8 REMARKS

Non-Euclidean geometry put an end to the idea that axioms are universal truths from which other truths may be derived by logic. Rather, axioms are truths *about certain domains* called *models* of the axioms, and their logical consequences are true in those models. Later, algebra made this idea its modus operandi: take some axioms known to hold in certain domains, such as the group axioms, and study their consequences. In effect, axioms are used to *define* a certain kind of structure, and their consequences are theorems *about* that kind of structure. However, algebra differs from non-Euclidean geometry in having obvious models. There are simple, finite models of the axioms for groups, rings, and fields, so the consistency of these axioms was never in doubt. For non-Euclidean geometry, the existence of models seemed to defy intuition.

In fact, one lesson of non-Euclidean geometry is that one may write down axioms without knowing one has a model. Then, by exploring their consequences and not finding a contradiction, one may become convinced that models exist, and hopefully a model will turn up. Something like this also happened with the axiom of choice in set theory, which we will study in chapter 14. In theory, one may even be able to prove that axioms do not lead to contradiction *without* finding a model first. This was the aim of the **Hilbert program**, launched by Hilbert (1900) as a way around the difficulty of finding models of certain axioms, such as axioms for infinite sets. However, the Hilbert program has difficulties of its own, as we will see in chapter 16.

In the case of Hilbert's axioms for geometry, the heart of the difficulty is that Hilbert wanted his axioms to support more than geometry; he wanted them to support the theory of real numbers. We saw in chapter 3 how his axioms for Euclidean geometry give the line the structure of a complete ordered field, which means the line is essentially $\mathbb{R}$. If he had wanted merely to model the Euclidean plane he could have skipped the Archimedean and completeness axioms: it would have been enough to build a "minimal model" based on the **constructible numbers**—which are those obtainable by straightedge and compass constructions. But Hilbert seems to have been determined to obtain $\mathbb{R}$ from his geometric axioms, because he also gave a separate and entirely different construction of $\mathbb{R}$ from his axioms for *non*-Euclidean (hyperbolic) geometry.

For deep reasons, which we will partially explore in chapter 16, Hilbert's program for proving consistency without appealing to infinite models is problematic for any theory that includes the arithmetic of natural numbers, and even more problematic for theories that include the real numbers. Because of this, the real numbers are technically a less reliable foundation for geometry than the constructible numbers.

Riemannian Geometry

Riemann (1854b) freed geometry from dependence on Euclidean space and its usual distance function, by allowing a much more general definition of "distance" and arbitrary dimension. Yet a stunning theorem of Nash (1956) showed that Riemann's generalization is, in a sense, not more general at all. Nash showed that *any Riemannian manifold* $\mathbb{M}$ *can be smoothly embedded in some Euclidean space* $\mathbb{R}^n$. That is, $\mathbb{M}$ is isometric to a smooth object $\mathbb{M}' \subseteq \mathbb{R}^n$, where the distance between two points P and Q on $\mathbb{M}'$ equals the Euclidean length of a geodesic curve on $\mathbb{M}'$ between P and Q. Thus $\mathbb{M}$ "inherits" its distance function from the surrounding $\mathbb{R}^n$ via arc length, in the same way that the sphere gets its distance function from $\mathbb{R}^3$.

A particular instance of Nash's theorem is that the hyperbolic plane can be modeled by a smooth surface in $\mathbb{R}^5$. Unfortunately, Nash's theorem is very difficult—mathematicians consider it to be much more profound than the work that won Nash a Nobel prize for economics—so it does not give a practical alternative to the models of the hyperbolic plane described above.

CHAPTER 10

■ ■ ■ ■ ■

Topology

Topology began as a *discrete* form of geometry, with highly visual methods. In this chapter we will see how topology revitalized visual methods of proof, not only in geometry but also in algebra. As Poincaré (1895: 1) wrote:

> We know how useful geometric figures are in the theory of imaginary functions ... and how much we desire their assistance when we want to study, for example, functions of two complex variables.
>
> If we try to account for the nature of this assistance, figures first of all make up for the infirmity of our intellect by calling on the aid of our senses, but not only this. It is worth repeating that geometry is the art of reasoning well from badly drawn figures; however, these figures, if they are not to deceive us, must satisfy certain conditions; the proportions may be grossly altered, but the relative positions of the different parts must not be upset.

Topology became mainstream mathematics when Riemann (1851) found that algebraic curves are best viewed as *surfaces*, now called **Riemann surfaces**. This view of curves comes to light when coordinates are taken to be complex, rather than real, numbers. Among other things, Riemann surfaces give a simple visual interpretation of a quantity discovered by Abel, the **genus**.

Genus is an example of a **topological invariant**: a characteristic of geometric objects that remains the same under transformations that do not destroy "the relative positions of the different parts." The search for such invariants led to new developments in the concept of proof.

10.1 GRAPHS

What we now call **graph theory** began with the paper of Euler (1736) on the Königsberg bridges problem, a puzzle that Euler solved by converting it to a problem about abstract "points" and "lines." His solution is considered to be the origin of graph theory—a field of mathematics whose history is expounded in Biggs et al. (1976)—and the origin of topology. Graph theory may be considered as one-dimensional topology, and while one-dimensional geometry is quite trivial, one-dimensional topology is not.

The Seven Bridges of Königsberg

The city of Königsberg (now Kaliningrad in Russia) was in Euler's time a city in East Prussia. Figure 10.1 shows a panoramic view of the city in 1652, courtesy of Wikimedia. The view shows the Pregel River that runs through the city, two islands in the river, and seven bridges connecting them to each other and the rest of the city. Walks covering all the bridges were a popular recreation in Königsberg, which prompted the question, Is it possible for a walk to cover each of the seven bridges exactly once?

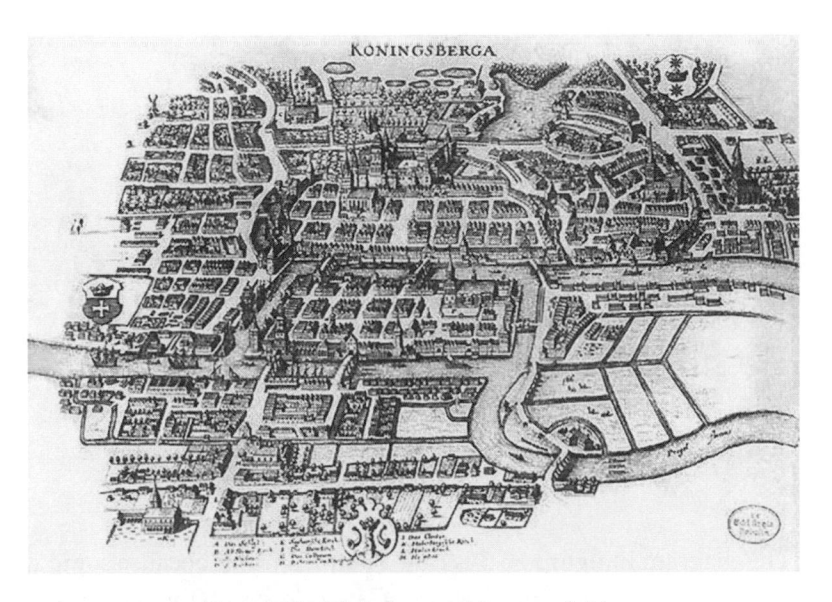

Figure 10.1 : Königsberg and its seven bridges

When this question was posed to Euler he first simplified the situation to one involving only bridges and regions of land (figure 10.2), and then to its mathematical essence: a list A, B, C, D of "points" (now called **vertices**) and a list a, b, c, d, e, f, g of "lines" (now called **edges**) joining them. For example, a and b both join A and B. Figure 10.3 shows how we represent this information today. The vertices represent land regions, and the edges represent the bridges.

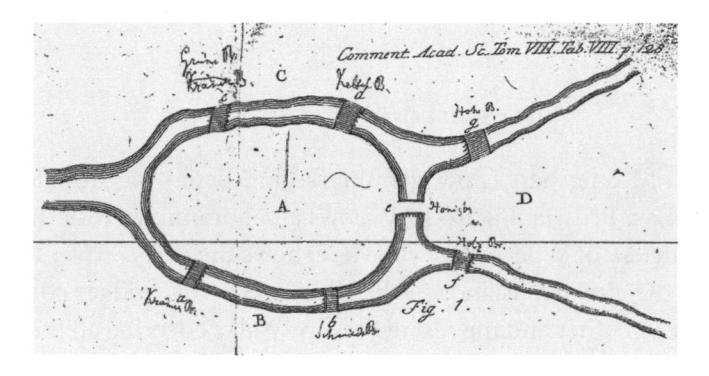

Figure 10.2 : Euler's diagram of the bridges

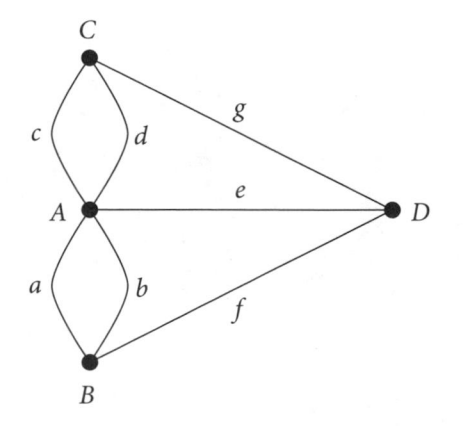

Figure 10.3 : The seven bridges multigraph

The diagram in figure 10.3 is called a **multigraph** because some vertices are connected by multiple edges. Generally graph theory deals with **simple graphs**, in which at most one edge connects any pair of vertices, but in either case each vertex X has a **valency**,[1] which is the number of ends of edges at X. In our case, A has valency 5 and B, C, D each have valency 3.

Now if there is a walk that uses each edge (bridge) exactly once, each time the walk passes through a vertex X it uses up two from the valency of X—one going in and one coming out. Consequently, only an even valency can be used up in a walk that uses each edge, with the exception of the first and last vertex of the walk, supposing they are different. In any case, a walk using all edges is possible only when at most *two* vertices have odd valency. Since all four vertices of the seven bridges multigraph have odd valency, *a walk using each edge exactly once is impossible*.

Although the problem is easily solved, the advantage of reducing it to its mathematical essence is obvious. The graph abstraction solves not just the problem of the Königsberg bridges but any problem of a similar kind.

Trees

Graph theory was until recently a small field at the recreational end of the mathematical spectrum. The first book on the subject was König (1936), published 200 years after Euler's paper on the seven bridges. But in recent decades the subject has expanded greatly in both breadth and depth. In this chapter we can offer only a small fragment of the subject, chosen because of its connection with two vastly different fields: **topology**, which occupies the rest of this chapter, and the **axiom of choice**, which occupies chapter 14. First, some definitions, which formalize the somewhat intuitive concept of graph we have used so far, and some associated concepts:

Definitions. A (simple) **graph** $\mathscr{G}$ consists of a nonempty set of objects called **vertices** and a set of unordered pairs of distinct vertices called **edges**. A **path** in $\mathscr{G}$ is a finite sequence of *ordered* pairs of vertices of the form

$$\langle v_1, v_2 \rangle, \langle v_2, v_3 \rangle, \ldots, \langle v_{n-2}, v_{n-1} \rangle, \langle v_{n-1}, v_n \rangle,$$

where each unordered pair $\{v_i, v_{i+1}\}$ is an edge—in other words, a sequence of **directed edges** in which the final vertex of each edge is the initial vertex of the next. The path is called **closed** if $v_1 = v_n$, and **simple**

---

1. Also called the vertex **degree** by many, but the word "degree" is already overused in mathematics, and chemistry gives us exactly the right word.

if it contains no repeated vertex (except the initial and final vertex of a closed path).

A graph $\mathcal{G}$ is called **connected** if any two vertices in $\mathcal{G}$ are the initial and final vertex of a path. (Thus a graph with just one vertex is connected.) A connected graph with no simple closed path is called a **tree**. The **valency** of a vertex v in a graph $\mathcal{G}$ is the number of edges of $\mathcal{G}$ that include v.

While this rather long-winded definition of a tree is needed for precision, it is intended to capture the idea of objects that really do look (more or less) like trees. Figure 10.4 shows three examples, in which vertices are represented by black dots, and edges by line segments.

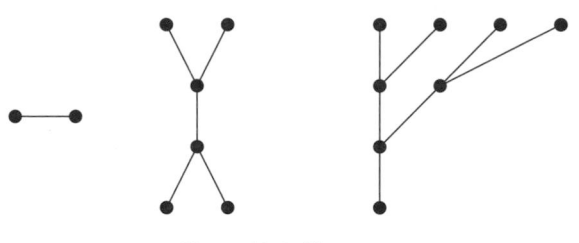

Figure 10.4 : Three trees

In many cases we are interested in graphs with a finite set of vertices (and hence a finite set of edges), in which case there are some interesting relations between numbers of vertices and edges. Here are two examples.

End vertices of a finite tree. *Each finite tree $\mathcal{T}$ with at least one edge has a vertex of valency* 1.

Proof. Given a finite tree $\mathcal{T}$ with at least one edge, choose any vertex u in $\mathcal{T}$. Since $\mathcal{T}$ is connected there is a path from u to each other vertex in $\mathcal{T}$, in fact a *simple* path when we omit portions between repeated vertices. Since $\mathcal{T}$ is finite, any such path has a finite number of edges, and there is a maximum number of edges among all the simple paths out of u.

A vertex v at the end of such a maximum path has valency 1, because the final edge of the path from u to v is necessarily the *only* edge containing v; otherwise the simple path from u to v could be extended further, creating either a longer simple path or a simple closed path. □

Vertex and edge numbers of a finite tree. *If $\mathcal{T}$ is a tree with V vertices and E edges, then $V - E = 1$.*

Proof. The theorem is trivially true when $\mathcal{T}$ has one vertex and hence no edges. Now assume inductively that the theorem is true of any tree $\mathcal{T}_n$ with n edges, and suppose we have a tree $\mathcal{T}_{n+1}$ with $n+1$ edges. Also suppose $\mathcal{T}_{n+1}$ has V vertices and E edges.

Find a vertex v of $\mathcal{T}_{n+1}$ with valency 1 by the theorem above, and remove v and the edge e containing it from $\mathcal{T}_{n+1}$. The resulting graph $\mathcal{T}_n'$ is connected, because any two of its vertices are connected by a path in $\mathcal{T}_{n+1}$ that, when simplified, cannot include v or e, so the path is in $\mathcal{T}_n'$. Thus $\mathcal{T}_n'$ is also a tree, with $V-1$ vertices and $E-1$ edges. But $\mathcal{T}_n'$ has n edges so, by our induction hypothesis,

$$1 = (V-1) - (E-1) = V - E,$$

which is the required result. $\qquad\qquad\qquad\qquad\qquad\qquad\qquad\square$

The proofs of these two theorems implicitly use induction: the first hides it in the assumption that a finite graph contains a finite number of simple paths; the second uses induction on the number n of edges. With infinite graphs, induction is not always available, and we may have to use strong assumptions about infinite sets, which is the case in the following theorem.

Existence of a spanning tree. *If $\mathcal{G}$ is any connected graph then $\mathcal{G}$ contains a **spanning tree**, that is, a tree containing all vertices of $\mathcal{G}$.*

Proof. Since $\mathcal{G}$ is connected, there is a path from any particular vertex u of $\mathcal{G}$ to any other vertex v. Since paths are finite, if we now fix u we may group the other vertices $v \in \mathcal{G}$ into sets according to the minimum length of a path from u to v:

$$\mathcal{V}_1 = \{v : \text{minimum length of path from } u \text{ to } v \text{ equals } 1\}$$
$$\mathcal{V}_2 = \{v : \text{minimum length of path from } u \text{ to } v \text{ equals } 2\}$$
$$\mathcal{V}_3 = \{v : \text{minimum length of path from } u \text{ to } v \text{ equals } 3\}$$
$$\vdots$$

We now make a spanning tree $\mathcal{T}$ of $\mathcal{G}$ in stages, connecting the vertices in $\mathcal{V}_n$ to u at stage n. To ensure that $\mathcal{T}$ is a tree, we keep only enough edges of $\mathcal{G}$ to allow a unique simple path from u to each v in $\mathcal{G}$. This ensures $\mathcal{T}$ is connected, and it also guarantees no simple closed path in $\mathcal{T}$, because any v in a simple closed path can be reached by at least two simple paths from u, as figure 10.5 suggests.

Figure 10.5 : Nonunique simple paths from a simple closed path

Each $v \in \mathscr{V}_1$ already is connected to u by a unique edge, since $\mathscr{G}$ is a simple graph. We now assume inductively that, in the edges of $\mathscr{T}$ chosen so far, u is connected by a unique path to each $v_n \in \mathscr{V}_n$, passing through $\mathscr{V}_1, \mathscr{V}_2, \ldots, \mathscr{V}_n$ in succession. Then it remains to connect each $v_{n+1} \in \mathscr{V}_{n+1}$ to a unique $v_n \in \mathscr{V}_n$.

Certainly each $v_{n+1} \in \mathscr{V}_{n+1}$ is connected by an edge to *some* $v_n \in \mathscr{V}_n$, since there is known to be a path of length $n+1$ from u to v_{n+1}, which necessarily includes some $v_n \in \mathscr{V}_n$. So it is only necessary to *choose*, for each $v_{n+1} \in \mathscr{V}_{n+1}$, a single edge connecting it to some $v_n \in \mathscr{V}_n$ and to put that edge into $\mathscr{T}$. □

When the graph $\mathscr{G}$ is finite, we can describe exactly how to choose v_n for a given v_{n+1}. Namely, let the finitely many vertices of $\mathscr{G}$ be $u_1, u_2, \ldots, u_m$. Then, whenever we have to choose a u_i, we take the one *with least subscript* that satisfies our requirements. When the graph $\mathscr{G}$ is infinite, there may be no such rule available, and we may need the **axiom of choice**. We say more about the axiom of choice in chapter 14, where it will be shown that the existence of a spanning tree for a connected graph is in fact *equivalent* to the axiom of choice, so the axiom really is unavoidable in general.

10.2 THE EULER POLYHEDRON FORMULA

Euler (1752) made a second important contribution to topology, though at the time it was thought to be a theorem about polyhedra: *if $\mathscr{P}$ is a polyhedron with V vertices, E edges, and F faces, then $V - E + F = 2$.* What makes this really a theorem of topology is that the "edges" do not have to be straight and the "faces" do not have to be flat: only the *numbers* of them are important. To be precise, we do not distinguish between objects that are **homeomorphic**, that is, in a bijective correspondence that is continuous in both directions (called a **homeomorphism**).

What is also important is that the surface of $\mathscr{P}$ be convex or, more generally, homeomorphic to the plane (completed by a point at infinity). When this is done—with no point of the graph going to the point at infinity, for convenience—the theorem becomes one about **plane graphs**.

Plane Graphs

While graph theory is a very visual subject, rigorous proofs require a rather large stock of definitions. Here are some more, which capture the idea of drawing a graph in the plane without edges crossing.

Definitions. A **plane graph** is a model $\mathscr{G}'$ of an abstract graph $\mathscr{G}$ in which each vertex v of $\mathscr{G}$ is represented by a point v' of the plane, and each edge $\{u, v\}$ of $\mathscr{G}$ by a curve in the plane from u' to v'. In addition, the edges of $\mathscr{G}'$ meet only where they have an endpoint in common. We also say that $\mathscr{G}'$ is an **embedding** of $\mathscr{G}$ in the plane.

Examples of plane graphs that model the vertices and edges of the regular polyhedra by lines in the plane are shown in figure 10.6.

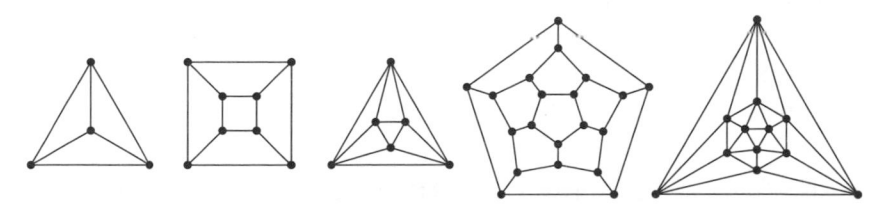

Figure 10.6 : Plane graphs of the regular polyhedra

These graphs also model the **faces** of the corresponding polyhedra, as the regions into which they divide the plane. These regions include the "infinite region," the one that includes the point at infinity. We can formalize the concept of face of a plane graph with the following definitions.

Definitions. A subset $\mathscr{R}$ of the plane is called **path-connected** if, for any two points $P, Q \in \mathscr{R}$ there is a polygonal path in $\mathscr{R}$ from P to Q. A **face** of a plane graph $\mathscr{G}$ is a maximal path-connected region in the subset of the plane left after removal of $\mathscr{G}$.

If $\mathscr{G}$ has polygonal edges, which we will assume to be the case, then the faces of $\mathscr{G}$ are regions of the plane with polygonal boundaries. The plane graphs include not just the graphs of the regular polyhedra but in fact any graphs that can be drawn on a sphere without edges crossing, because any such graph can be transported to the plane by stereographically projecting it from a point not on any of its edges.

Another important class of plane graphs comprises trees.

Finite trees in the plane. *Each finite tree can be embedded in the plane, and the resulting plane graph has one face.*

Proof. We can embed any finite tree in the plane inductively by building it one edge at a time (the reverse of the process of finding an end vertex and removing it, from the previous subsection). We can certainly map a tree with *one* edge into the plane, and if we can map any tree with n edges we can also map a tree with $n + 1$ edges: the last edge e_{n+1} can always be fitted in by making it sufficiently small.

The same inductive process shows that *a plane graph that is a tree has only one face*. This is certainly true for a tree with one edge, and if it is true for a tree T_n with n edges then it is also true for a tree T_{n+1} with $n + 1$ edges: any two points outside T_{n+1} are connected by a polygonal path outside T_n, so they can also be connected by a polygonal path outside T_{n+1}, if necessary by replacing edges that cross the newly added edge e_{n+1} by paths that go around it, but close enough to e_{n+1} to avoid hitting other parts of T_n. □

The "one face" property of trees is the base step of an inductive proof of the Euler polyhedron formula.

Euler polyhedron formula. *If $\mathscr{G}$ is a connected plane graph with V vertices, E edges, and F faces, then $V - E + F = 2$.*

Proof. If $\mathscr{G}$ has one face then $\mathscr{G}$ is a tree because it is connected and has no simple closed paths, since a simple closed path encloses a face. In this case we know that $V - E = 1$ from the previous section, so $V - E + F = 2$ because $F = 1$.

Now suppose that the theorem is true for connected plane graphs with n faces, and suppose that $\mathscr{G}$ is a connected plane graph with $n + 1$ faces. Let e be an edge in the boundary of two different faces. If we remove e those two faces become one, decreasing F by one. And of course E is also decreased by one, but V does not change, since we remove e but not the vertices at its ends. Thus $V - E + F$ remains the same.

But $V - E + F = 2$ for the new graph with e removed and n faces, by induction, so $V - E + F$ also equals 2 for the graph $\mathscr{G}$. ☐

The general structure of the above proof is sound, but it assumes something at the beginning that should be further examined: *that a simple closed polygon separates the plane into at least two regions.* This is a nontrivial result known as the **polygonal Jordan curve theorem**. It is a special case of the corresponding theorem for simple closed curves, known as the **Jordan curve theorem**, which is even less trivial. Indeed, it was the proof of the latter theorem by Jordan (1887) that first drew attention to seemingly obvious but hard-to-prove results in topology.

Part of the difficulty lies in the transition from the world of discrete, combinatorial objects, such as graphs and polygons, to the world of continuous objects such as curves. We already know there are difficulties in moving from the discrete world of integers to the continuous world of real numbers, and these difficulties are magnified in two or more dimensions. At present, however, we will be content to prove just the polygonal Jordan curve theorem, so as to complete the proof of the Euler polyhedron formula.

The Polygonal Jordan Curve Theorem

A proof of the polygonal Jordan curve theorem by the simplest possible means was found by Dehn around 1899. The proof used only Hilbert's axioms of incidence and order, discussed in chapter 3. Dehn's proof was unpublished and was brought to light much later by Guggenheimer (1977). Here we give a more standard proof, which uses the distance measure in the plane and the kind of reasoning typical of analysis and topology, where objects are chosen to be "sufficiently small."

Polygonal Jordan curve theorem. *If $\mathscr{P}$ is a simple closed polygon then the graph $\mathscr{P}$ has two faces.*

Proof. Take any edge e of $\mathscr{P}$ and points A, B on opposite sides of e but close enough that the line segment AB does not meet $\mathscr{P}$ at any point except where AB crosses e (figure 10.7).

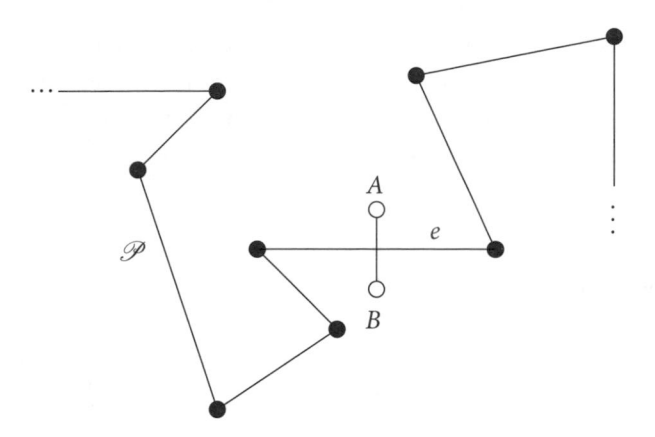

Figure 10.7 : Representatives *A* and *B* of the two faces of $\mathscr{P}$

We show that *A* and *B* represent the two faces of $\mathscr{P}$ by showing the following:

1. Any point not in $\mathscr{P}$ can be connected to either *A* or *B* by a polygonal path that does not meet $\mathscr{P}$.
2. Any polygonal path from *A* to *B* meets $\mathscr{P}$.

We now introduce the notation $\mathscr{P} - \{e\}$ to denote the graph that results from $\mathscr{P}$ by removal of the edge *e*. Clearly $\mathscr{P} - \{e\}$ is a tree, so any two points not in it are connected by a polygonal path that does not meet $\mathscr{P} - \{e\}$, by the theorem in the previous subsection that a plane tree graph has one face.

Such a path, *p*, may of course cross the edge *e*, but we can *deform p*, in small steps and without touching $\mathscr{P} - \{e\}$, so as to remove intersections two at a time, as in figure 10.8. Thus any polygonal path between two points in the plane not in $\mathscr{P}$ may be deformed into one that meets $\mathscr{P}$ only on the edge *e*, which it crosses either once or not at all.

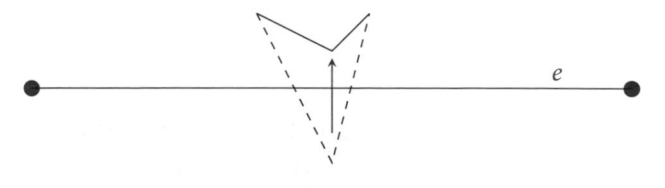

Figure 10.8 : Removing crossings two at a time

This proves the first claim above, that any point X not in $\mathcal{P}$ may be connected to either A or B by a path that does not meet $\mathcal{P}$. If X is connected to A, say, by a path p that crosses e once, then extending p by the edge AB gives a path from X to B crossing e twice. The latter path may therefore be deformed into a polygonal path from X to B that does not meet $\mathcal{P}$ at all.

Conversely, any deformation of a path from A to B, such as the line segment AB, may be accomplished in steps so small that only two crossings change at a time, as in figure 10.8. Thus any polygonal path from A to B crosses $\mathcal{P}$ an odd (and hence nonzero) number of times, so A and B are in different faces of $\mathcal{P}$. □

10.3 EULER CHARACTERISTIC AND GENUS

The previous section was about graphs in the plane, with results that also transfer to the sphere by stereographic projection. It is also of interest to study graphs on surfaces *not* homeomorphic to the completed plane, such as those in figure 10.9. These surfaces appear to be nonhomeomorphic because of their different numbers of "holes." The number of holes is called the **genus** of the surface, and it may be detected intrinsically by the quantity $V - E + F$, where V, E, and F are the numbers of vertices, edges, and faces of a graph drawn on the surface. Vertices and edges are of course the same as in the plane, but on other surfaces a **face** must be homeomorphic to a finite region bounded by a polygon in the plane.

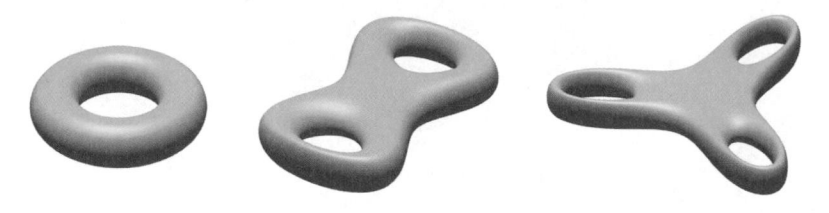

Figure 10.9 : Surfaces of genus 1, 2, 3

This idea is due to Riemann (1851), who gave the following informal argument that $V - E + F$ has the same value for all graphs on the same surface. Suppose that $\mathcal{G}_1$ and $\mathcal{G}_2$ are two finite graphs on the same surface $\mathcal{S}$. By deforming the edges slightly, if necessary, we can assume that $\mathcal{G}_1$

and $\mathcal{G}_2$ meet at only finitely many points. This means the union $\mathcal{G}$ of the two graphs is a *common subdivision* of both $\mathcal{G}_1$ and $\mathcal{G}_2$.

That is, $\mathcal{G}$ may be obtained from either $\mathcal{G}_1$ or $\mathcal{G}_2$ by operations of the following kinds:

Subdivision of an edge. Insert a new vertex v in an edge e, creating two edges. This increases both V and E by 1, therefore leaving $V - E + F$ the same.

Subdivision of a face. Insert a new edge e in a face, connecting two of its boundary vertices to create two faces. This increases both E and F by 1, therefore leaving $V - E + F$ the same.

It follows that $\mathcal{G}_1$ and $\mathcal{G}_2$ have the same value of $V - E + F$ as $\mathcal{G}$, and hence of each other. This invariant quantity $V - E + F$ is called the **Euler characteristic** $\chi(\mathcal{S})$ of the surface $\mathcal{S}$. By calculating the value of χ for specific graphs on surfaces with g "holes," one finds the following relation between Euler characteristic and genus:

$$\chi = 2 - 2g.$$

For example, the **torus**, which has genus 1, has Euler characteristic 0, as can be seen from the graph in figure 10.10, because it has one vertex (where the red and blue edges meet), two edges, and one face.

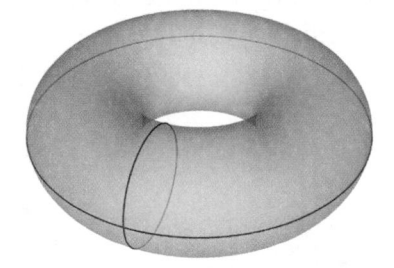

Figure 10.10 : A graph on the torus

Riemann became interested in surfaces when he discovered that they model complex algebraic curves, as we will explain in more detail in the next section. He discovered that the possible parametric equations $x = f(t), y = g(t)$ of the curve are governed by its genus. For example, curves of genus 0 can be parameterized by rational functions, curves of genus 1 by elliptic functions. This amazing discovery brought topology to the attention of mathematicians before they were ready, so to speak. At first they could see only the discrete, combinatorial aspect of the subject—points, polygons, polyhedra, and so on—with corresponding methods of proof. But it gradually became clear that topology is really about **continuity** and that invariants such as the Euler characteristic remain invariant when arbitrary continuous bijections (homeomorphisms) are allowed.

This realization forced an upgrade in methods of proof. For example, instead of proving the Jordan curve theorem for polygons, one had to prove it for continuous curves. And, as mentioned in the previous section, the upgraded theorems are often much more difficult to prove. They involve not only the subtlety of the continuous in comparison with the discrete but also complexities due to having more than one dimension. Since we later discuss a few of these complexities in connection with analysis, we will postpone them until chapter 11 and continue for the remainder of this chapter to use combinatorial methods. These were the methods generally in use until the early twentieth century, and they are still useful in topology today.

10.4 ALGEBRAIC CURVES AS SURFACES

To explain why complex algebraic curves should be viewed as surfaces, we begin with the complex numbers themselves. As we know, $\mathbb{C}$ is naturally viewed as a plane, and stereographic projection makes it natural to complete the plane by a point at infinity, giving a **complex projective line** that is best viewed as a sphere. It is also useful to complete complex curves by a point at infinity—for example, to obtain Bézout's theorem, as we saw in section 5.5—and this leads to viewing each complex curve as a surface covering the sphere, called a **Riemann surface**.

The simplest example that illustrates this idea is the curve $y^2 = x$. The sphere of possible x-values, including ∞, is covered by a surface of y-values, generally two for each x: $\sqrt{x}$ and $-\sqrt{x}$. Thus, as a first approximation, the y-values cover the sphere by two "sheets," like layers of an onion.

However, over $x = 0$ we have only the value $y = 0$, and over the value $x = \infty$ we have only the value $y = \infty$. Thus in the neighborhood of $x = 0$ the two "sheets" of the covering come together, creating what is called a **branch point**. The map from the surface to the sphere, which is two to one except at branch points, is called a **branched covering**. Figure 10.11 gives an idea of what the branch point looks like: above $x = 0$ is the single point $y = 0$, but elsewhere there are two y-values, $\sqrt{x}$ and $-\sqrt{x}$, above x.

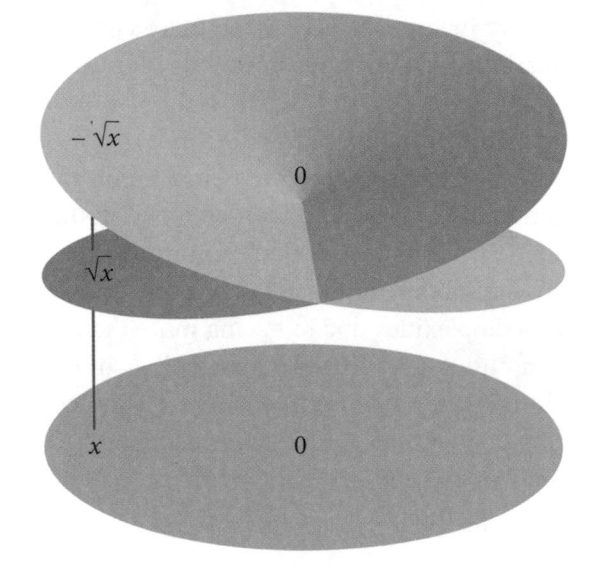

Figure 10.11 : Branch point for $y^2 = x$

The apparent line of intersection, where the "upper" and "lower" sheets change places, is a side effect of drawing the branch point in three dimensions—in the four-dimensional space $\mathbb{C} \times \mathbb{C}$ the sheets meet *only* at 0 and ∞. To represent the covering surface more faithfully we cut the surface open at the branch point, as shown in figure 10.12, painting the edges blue and green to indicate the edges that are supposed to be joined.

Figure 10.12 : Branch point cut open

These cuts are extended from one branch point to the other as shown in figure 10.13.

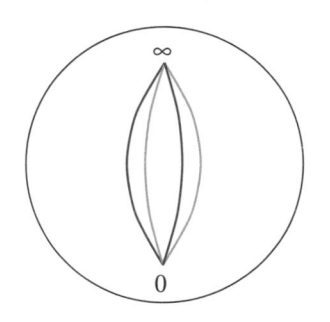

Figure 10.13 : The sphere covering, cut open

We can now separate the two cut sheets and rejoin them as shown in figure 10.14, thereby avoiding the surface crossing itself. At this stage we can see that the covering surface is homeomorphic to a sphere.

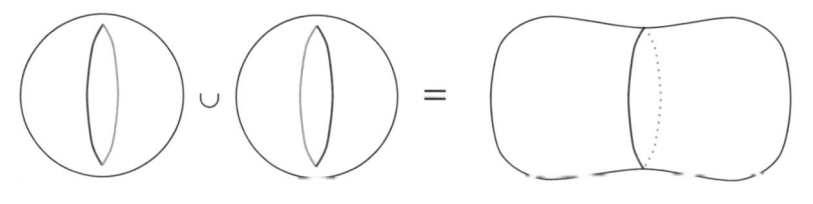

Figure 10.14 : Joining the separated sheets

The situation is more interesting in the case of the cubic curve

$$y^2 = x(x-1)(x+1),$$

which again has two sheets but *four* branch points: over $x = 0, 1, -1, \infty$. If we similarly cut the branch points open, extend the cuts between pairs of branch points, separate the sheets, and rejoin the cuts, then the resulting surface is topologically a **torus**. The process is shown in figure 10.15.

This is how Riemann first observed the presence of topology in algebraic geometry and complex analysis. Topology makes its presence felt in many ways. In particular, **genus**, the most blindingly obvious topological characteristic of surface, turns out to agree with a number discovered by Abel (1826) in a very deep and obscure study of integrals. Abel's theorem describes the complexity of relations between integrals $\int_a^b y \, dx$, where y

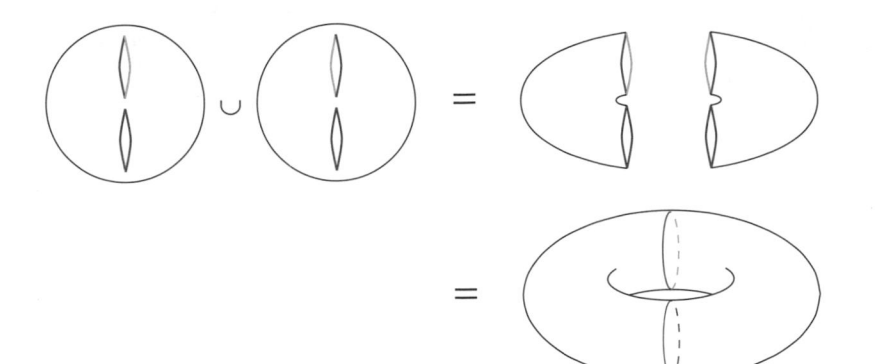

Figure 10.15 : Joining the sheets of a cubic curve

is an algebraic function of x, in terms of an integer p, which Riemann showed to be precisely the genus of the Riemann surface for y.

10.5 TOPOLOGY OF SURFACES

The kind of surfaces that arise as Riemann surfaces are called **orientable** or **two-sided** because, informally speaking, they have two sides. An example of a one-sided surface is the real projective plane, $\mathbb{RP}^2$, whose "points" are the pairs of antipodal points on the sphere. This is because $\mathbb{RP}^2$ contains a **Möbius band** (figure 10.16), which arises from a band around the equator on the sphere.

Figure 10.16 : A Möbius band

Bearing in mind that a band *half* way around the equator includes a representative of each point of $\mathbb{RP}^2$, and that a point above the equator on one side corresponds to a point below the equator on the opposite side, we see that the band is homeomorphic to the result of joining opposite ends of a rectangle with a twist. This is the one-sided surface shown.

Although Möbius is best known for the one-sided surface that bears his name, his main contribution to topology was a classification of the *two*-sided surfaces. Confining his attention to surfaces of finite extent in $\mathbb{R}^3$ and without boundary (unlike the Möbius band, which has a boundary curve, and unlike $\mathbb{RP}^3$, which cannot be embedded in $\mathbb{R}^3$), Möbius (1863) showed that any such surface is homeomorphic to one of those listed in figure 10.17, which are characterized by their genus.

Figure 10.17 : The orientable surfaces

Later proofs of this theorem dropped some of the assumptions made by Möbius, such as the embedding in $\mathbb{R}^3$. These later proofs typically assume an abstract description of a surface as a polygon with edges identified in pairs. This idea comes to mind when one sees the opposite process of slicing a surface along certain curves so as to produce a polygon. For example, a genus 1 surface gives a polygon with four sides, and a genus 2 surface gives a polygon with eight sides, as can be seen from figure 10.18 (the eight corners of the octagon can be seen close together near the center of the genus 2 surface).

Figure 10.18 : Cutting surfaces to form polygons

The later proofs also include a classification of the nonorientable surfaces, which are related to the orientable surfaces in much the same way that $\mathbb{RP}^2$ is related to the sphere. We say that the sphere **covers** $\mathbb{RP}^2$ by the 2-to-1 map that sends a point P and its antipodal point P' on the sphere to the single point $\{P, P'\}$ of $\mathbb{RP}^2$. This covering is called

unbranched because it is a homeomorphism "locally," that is, on all sufficiently small patches of the sphere. There is a similar "double covering" of each nonorientable surface by an orientable surface.

Coverings of Orientable Surfaces

Unbranched coverings are common in topology and particularly in the topology of orientable surfaces. The most important examples are coverings of the surfaces of genus $1, 2, 3, \ldots$ by the plane, and we illustrate the idea in the case of genus 1, the torus.

As we saw in the previous subsection, cutting the torus along suitable curves gives a four-sided polygon homeomorphic to a rectangle. We color the edges of the rectangle red and blue like the corresponding curves on the torus, as in figure 10.19, and place copies of the rectangle side by side so as to fill the plane.

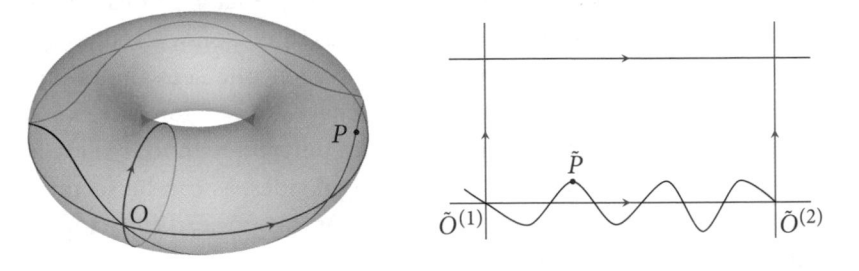

Figure 10.19 : Plotting on the covering surface

This creates a clear, unobstructed view of curves on the torus, such as the wavy curve shown. This curve begins and ends at O, but it also winds once around the torus in the process, rather like the red curve. To see more clearly what is happening, we let a point P move along the curve while a corresponding point $\tilde{P}$ moves on the plane. Each time P crosses a red or blue curve on the torus, $\tilde{P}$ crosses the corresponding red or blue edge of a rectangle. We call this process **lifting** a curve from the torus to the plane. In this case the lifted curve runs from $\tilde{O}^{(1)}$ to $\tilde{O}^{(2)}$ in the plane, as does a red edge of the rectangle, which is the result of similarly lifting the red curve.

When two curves $\mathscr{C}_1$ and $\mathscr{C}_2$ on the torus lift to curves $\tilde{\mathscr{C}}_1$ and $\tilde{\mathscr{C}}_2$ on the plane with the same initial and final point, then obviously $\tilde{\mathscr{C}}_1$ may be deformed to $\tilde{\mathscr{C}}_2$ while keeping their endpoints fixed. If we perform the deformation in stages, each within a small region of the plane, then it may be "projected" to a deformation of $\mathscr{C}_1$ to $\mathscr{C}_2$ on the torus, with their initial and final point fixed. Such curves $\mathscr{C}_1$ and $\mathscr{C}_2$ are called **homotopic**.

Conversely, if $\mathscr{C}_1$ and $\mathscr{C}_2$ are two homotopic curves on the torus, we can "lift the deformation" of $\mathscr{C}_1$ to $\mathscr{C}_2$ to the plane. In particular, lifted curves $\tilde{\mathscr{C}}_1$ and $\tilde{\mathscr{C}}_2$ with the same initial point $\tilde{O}^{(1)}$ have the same final point. Thus *points on the covering plane correspond to classes of homotopic curves with initial point O and fixed final point.* For example, if $\tilde{O}^{(1)}$ corresponds to O on the torus, then $\tilde{O}^{(2)}$ corresponds to all curves that begin and end at O and are homotopic to the red curve.

The torus is a rather simple case, because we are working with familiar polygons (rectangles) in the familiar Euclidean plane. When we come to genus 2, *non*-Euclidean geometry comes to the rescue.

Looking again at figure 10.18, we see that not only does cutting the genus 2 surface give an octagon, but also *on the covering, eight octagons meet at each vertex*, because eight corners meet at the same point on the genus 2 surface. Thus, we need octagons whose eight angles sum to 2π (for example, with angles of 45°). Euclidean octagons are certainly not like this, but there are such octagons in the hyperbolic plane. In fact we can tile the hyperbolic plane with octagons that have equal sides and equal angles of 45°. Figure 10.20 shows the tiling of the hyperbolic plane by these octagons.

Figure 10.20 : Tiling of the hyperbolic plane by octagons

Poincaré (1904) was the first to use hyperbolic geometry to study curves on surfaces of genus 2 or more. In particular, he showed how to tell whether a closed curve on a surface is homotopic to a simple curve by deforming its lift on the covering surface into a hyperbolic line.

The Fundamental Group

In the previous subsection we observed that the points on the plane covering the torus correspond to homotopy classes of curves on the torus that begin at O and have a fixed final point. Particularly interesting points on the cover are the vertices of the rectangles, which all lie above the point O. These points correspond to the homotopy classes of *closed* curves that begin and end at O. Each such homotopy class can be "lifted" to a pair of vertices $\tilde{O}^{(i)}$ and $\tilde{O}^{(j)}$ on the covering, such as the pair $\tilde{O}^{(1)}$ and $\tilde{O}^{(2)}$ we found for the homotopy class of the red curve in the previous subsection.

Instead of making an arbitrary choice of initial vertex, we could associate the red curve with a *translation* of the whole tiling of the covering by rectangles. In this case it is the translation that moves the whole tiling one tile to the right. The translations corresponding to all possible (homotopy classes of) closed curves on the torus with initial point O form a group called the **covering motion group**. It is also called the **fundamental group** of the torus. The group operation on the covering is simply composition of the functions that represent translations of the tiling; the corresponding group operation on the torus is composition of (homotopy classes of) closed paths, which amounts to joining closed paths end to end.

Groups of motions were studied by Klein and Poincaré in the 1880s, and the general idea of the fundamental group, as the group of homotopy classes of closed paths in a space, was introduced by Poincaré (1895).

The tiling of the covering plane by rectangles, with red and blue directed edges, serves as a diagram of the fundamental group of the torus. If we label one vertex arbitrarily as the identity element, **1**, of the group, and read each directed red edge as a and each directed blue edge as b (or a^{-1} and b^{-1} when the edge is traversed in the direction opposite to the arrow), then each path of edges out of **1** spells a "word" in the letters a, b, a^{-1}, b^{-1}. This word names the end vertex of the path. Figure 10.21 shows some of the names.

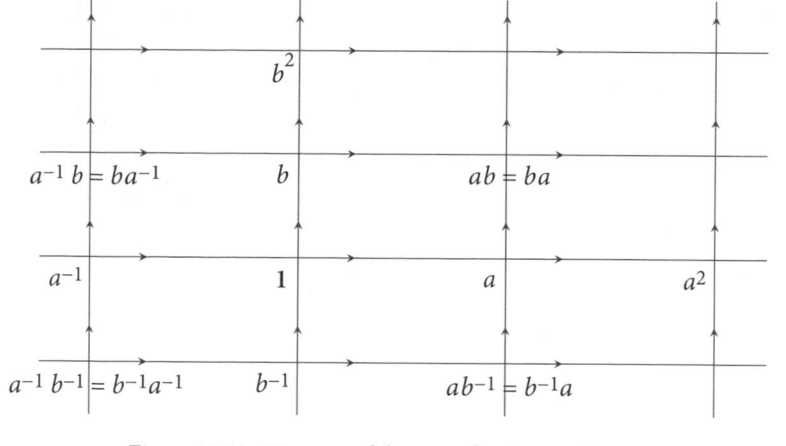

Figure 10.21 : Diagram of the torus fundamental group

Obviously, each vertex has many names, so we have the problem of deciding when two names represent the same group element. Dehn (1912) called this the **word problem**. For the fundamental group of the torus the word problem is easy to solve. It is clear that $ab = ba$ and, consequently, that any word is equal to one of the form $a^m b^n$ for some integers m and n. Conversely, since $a^m b^n$ represents the vertex that is m edges to the right of 1 and n edges up, different pairs m, n represent different elements of the group. Thus we can solve the word problem for this group by using the relation $ab = ba$ (or, equivalently, $aba^{-1}b^{-1} = 1$) to reduce the words u and v to the form $a^m b^n$, and then see whether they each give the same ordered pair $\langle m, n \rangle$.

Deciding whether words u and v are equal is equivalent to deciding whether $uv^{-1} = 1$. So one can say, more simply, that the word problem is the problem of deciding, given a word w, whether $w = 1$. A geometric equivalent of this problem is to decide whether w spells out a closed path in the group diagram. And by solving this problem we decide whether a given curve on the torus, written as a product of the red and blue curves represented by a and b, can be deformed to a point. Thus the algorithm just described solves a topological problem about curves on the torus.

Dehn (1912) solved the same problem for curves on surfaces of higher genus. He did so by solving the word problem for each of their fundamental groups, using similar steps. The main difference, as mentioned above, is that the tiling of the covering surface is in the *hyperbolic* plane. But the tiling is again a diagram of the fundamental group, whose word problem is solved by reducing words to their shortest form. The shortening algorithm, called **Dehn's algorithm**, is very simple. It exploits a property of the tilings in question: that any simple closed path around edges of the tiles includes more than half the boundary of some polygon. The path can then be moved to the other side of the polygon, thereby shortening the word.

In solving a problem in group theory by purely geometric means, Dehn created a new approach to algebra, today called **geometric group theory**. For a modern introduction to the subject, see Clay and Margalit (2017).

10.6 CURVE SINGULARITIES AND KNOTS

In the 1890s, Wilhelm Wirtinger at the University of Vienna made a remarkable discovery: the singularities of certain algebraic curves can be described by *knots*. When other mathematicians learned about them a topological theory of knots began to develop. It began with the work of Tietze (1908) and Dehn (1910) and came into full bloom in the 1920s when the German mathematicians Emil Artin and Kurt Reidemeister visited Vienna.

The connection between curve singularities and knots can be illustrated in the case of the curve $y^2 = x^3$, which has a **cusp** singularity at the origin. Figure 10.22 shows what the cusp looks like in the plane of real values of x and y.

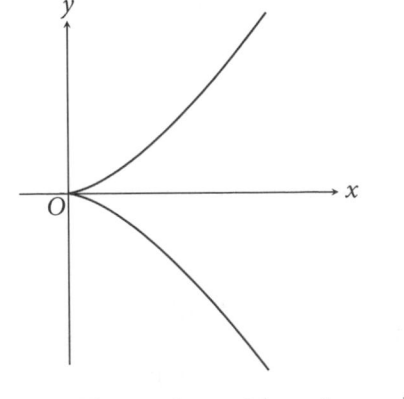

Figure 10.22 : The singularity of the real curve $y^2 = x^3$

Now, as we have seen in section 10.4, the picture is more interesting when x and y are taken to be complex. In that case we have a plane $\mathbb{C}$ of x values and a plane $\mathbb{C}$ of y values, and $\mathbb{C} \times \mathbb{C}$ contains the curve, which is the image of the plane $\mathbb{C}$ of t values under the map

$$t \mapsto \langle t^2, t^3 \rangle$$

corresponding to the parametric equations $x = t^2$, $y = t^3$ of the curve $y^2 = x^3$.

The points $\langle t^2, t^3 \rangle$ of the curve form a surface in the four-dimensional space $\mathbb{C} \times \mathbb{C}$ of ordered pairs $\langle x, y \rangle$ for $x, y \in \mathbb{C}$. The map $t \mapsto \langle t^2, t^3 \rangle$ is a homeomorphism, so the surface is topologically a plane, but its "knottedness" comes to light through the following observations:

1. The plane of complex t values is the union of disjoint circles C_r with center $\langle 0, 0 \rangle$ and radius r. C_r consists of the numbers $r(\cos \theta + i \sin \theta)$ for $0 \le \theta < 2\pi$.

2. The map $t \mapsto \langle t^2, t^3 \rangle$ is injective because if $t_1^2 = t_2^2$ for $t_1 \ne t_2$ then $t_2 = -t_1$, in which case $t_1^3 \ne t_2^3$. Thus the map sends the disjoint circles C_r in the plane of t values to disjoint simple curves K_r in the Riemann surface $y^2 = x^3$.

3. Since

$$x = t^2 = r^2(\cos 2\theta + i \sin 2\theta), \quad y = t^3 = r^3(\cos 3\theta + i \sin 3\theta),$$

the curve K_r lies in the surface T_r defined by

$$|x| = r^2, \quad |y| = r^3.$$

T_r consists of the points

$$x = r^2(\cos \alpha + i \sin \alpha), \quad y = r^3(\cos \beta + i \sin \beta) \quad \text{for} \quad 0 \le \alpha, \beta < 2\pi.$$

Each of x, y runs round a circle as α, β, respectively, run from 0 to 2π. Thus T_r is a *torus* surface, at least for values of r small enough to make a minor radius r^3 less than half a major radius r^2. On such a surface T_r, each point P is given by the ordered pair $\langle \alpha, \beta \rangle$, as indicated in figure 10.23. The red circle is the x-axis (of radius r^2), and the blue circle is the y-axis (of radius r^3).

4. The curve K_r, as we've said, consists of the points

$$\langle t^2, t^3 \rangle = \langle r^2(\cos 2\theta + i \sin 2\theta), r^3(\cos 3\theta + i \sin 3\theta) \rangle \quad \text{for}$$
$$0 \le \theta < 2\pi;$$

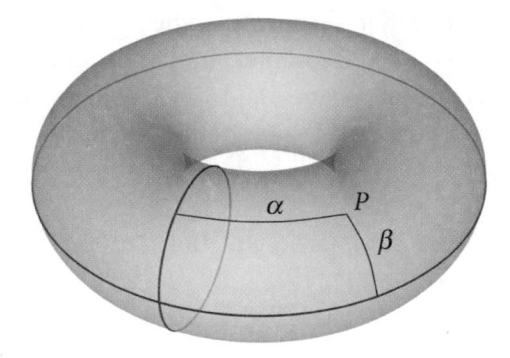

Figure 10.23 : Torus coordinates

hence K_r runs twice around the torus T_r in the α direction and three times around T_r in the β direction. This makes K_r a **trefoil knot** on T_r (figure 10.24).

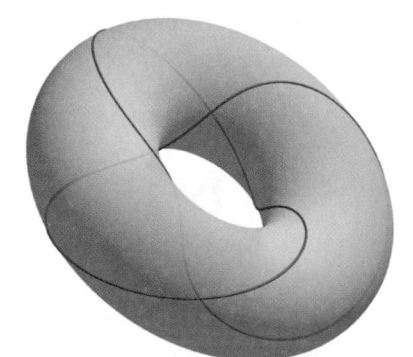

Figure 10.24 : Trefoil knot on the torus

5. It remains only to check that T_r, and hence K_r, lies in a standard three-dimensional space inside $\mathbb{C} \times \mathbb{C}$, so that K_r is really a trefoil knot. Indeed, T_r lies in the three-dimensional space S_r defined by

$$|x|^2 + |y|^2 = r^4 + r^6 \text{ (a constant).}$$

If we let $x = a + ib$ and $y = c + id$ then S_r is the space of 4-tuples $\langle a, b, c, d \rangle \in \mathbb{R}^4$ with

$$a^2 + b^2 + c^2 + d^2 = r^4 + r^6,$$

called the 3-*sphere* of radius $s = \sqrt{r^4 + r^6}$. The 3-sphere minus one point can be mapped to $\mathbb{R}^3$ by stereographic projection, just as the ordinary sphere minus one point can be mapped to the plane. So K_r does lie in a standard space.

Note also that $s = \sqrt{r^4 + r^6}$ increases monotonically with r; hence the 3-spheres S_r fill the space $\mathbb{C} \times \mathbb{C}$ of all complex ordered pairs $\langle x, y \rangle$.

To sum up: *for all sufficiently small values of $s > 0$, the intersection of the complex curve $y^2 = x^3$ with the 3-sphere of points at distance s from $\langle 0, 0 \rangle$ in $\mathbb{C} \times \mathbb{C}$ is a trefoil knot.*

10.7 REIDEMEISTER MOVES

The curve on the torus shown in figure 10.24 seems knotted all right, and indeed, by deforming it slightly and fattening we get the standard picture of the trefoil knot shown in figure 10.25.

Figure 10.25 : The trefoil knot

But how can we *prove* it is knotted? That is, how can we prove that the trefoil knot cannot be transformed into a circle? In particular, what are the appropriate transformations to consider—those that do not destroy "the relative position of the different parts"? The first answers to this question, prompted by previous solutions of topological problems, used the concepts of continuous function and the fundamental group. They led to the first proof that the trefoil knot is indeed knotted, by Tietze (1908). He found the fundamental group of the space ($\mathbb{R}^3$ minus the knot) and showed it is not the same as the group of the space ($\mathbb{R}^3$ minus a circle).

Reidemeister (1927) found a simpler approach, using transformations appropriate to knots. His transformations are now called **Reidemeister moves**. Figure 10.26 shows Reidemeister alongside a poster about him and his moves (courtesy of its designer, Jackie Maldonado). The photograph of Reidemeister is from the Archives of the Mathematisches Forschungsinstitut Oberwolfach and is used with permission.

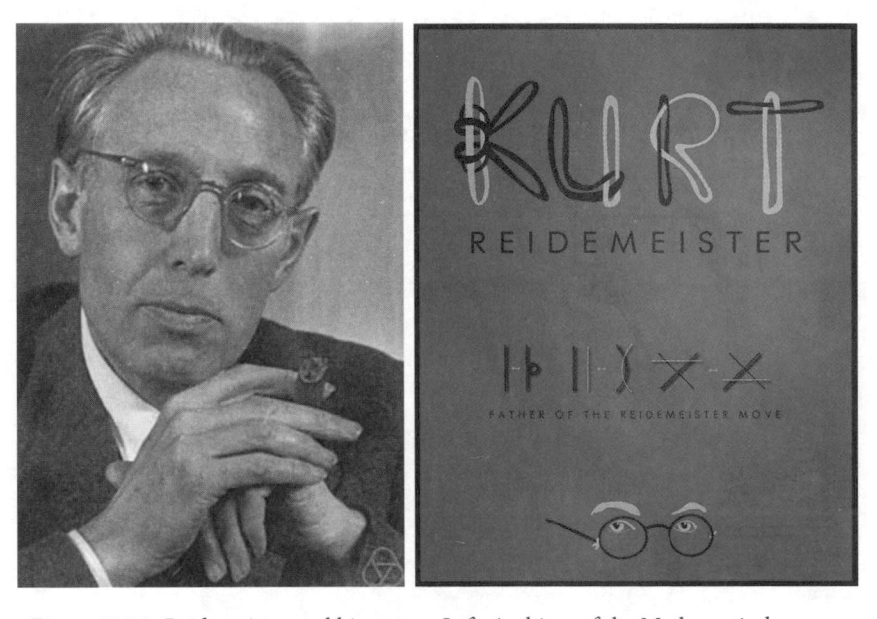

Figure 10.26 : Reidemeister and his moves. Left: Archives of the Mathematisches Forschungsinstitut Oberwolfach. Right: Courtesy of Jackie Maldonado.

The Reidemeister model of a knot is a simple polygon $\mathscr{P}$ in $\mathbb{R}^3$. The knot itself may then be defined as the class $[\mathscr{P}]$ of all polygons obtainable from $\mathscr{P}$ by certain transformations. These transformations are composed from "small" transformations, each of which either replaces an edge AB by a pair of edges AC and CB or the reverse, provided no part of $\mathscr{P}$ meets triangle ABC. The transformations model the idea of "deforming" $\mathscr{P}$ without allowing any edge to pass through another.

Transformations can be "tracked" in a two-dimensional image called the **knot diagram**, a projection of $\mathscr{P}$ onto the plane modified to show which edge is on top at points where the projection of one edge crosses the projection of another. We also assume that the point of projection is chosen so that no three edges cross at the same point. If the polygon $\mathscr{P}$ is first "fattened," then figure 10.25 is an example of a knot diagram for the trefoil. (In this case the polygon has so many edges that it looks like a smooth curve.)

As we make small changes in $\mathscr{P}$, ensuring that no three projected edges meet at the same point, the essential changes in the knot diagram are those that create or destroy crossings or change their relative position. Such changes are of only three kinds, called Reidemeister moves I, II, and III. They are shown in figure 10.27, with the double arrow in each indicating that the portion of the knot diagram on one side can be replaced by the portion on the other side. Any diagram of a knot can be converted to any other by small allowable transformations, and hence the only essential changes in the diagram are by Reidemeister moves.

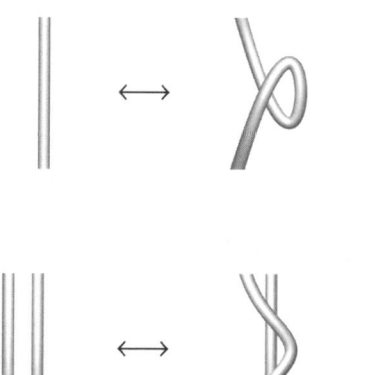

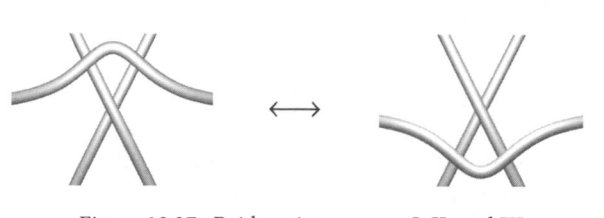

Figure 10.27 : Reidemeister moves I, II, and III

This brings us to **knot invariants**: characteristics of a knot diagram that do not change under the Reidemeister moves. If we can find an invariant with different values for the circle (the "unknot") and a diagram of a knot K, then we have proved that K cannot be transformed to a circle, so K is truly knotted. In the next section we will find some characteristics of knot diagrams that are easily proved not to change under Reidemeister moves. Using these invariants, we can show that the trefoil knot, and indeed, all knots with less than nine crossings, are truly knotted.

10.8 SIMPLE KNOT INVARIANTS

As mentioned in the previous section, the first knot invariants were found using the fundamental group of the knot complement. The fundamental group is a very strong invariant, now known to characterize knots almost completely. Unfortunately, as we also know, it can be hard to extract information from groups.

In the 1920s, more manageable invariants were found (independently) by James Alexander in the United States and Kurt Reidemeister in Germany. Their invariants made it possible to prove not only knottedness but also *distinctness* of all knots with diagrams of fewer than nine crossings.

In the 1950s, Ralph Fox distilled an even simpler invariant, called *p-colorability*, from previous ones. Fox's invariant is weaker than those of Alexander and Reidemeister but still strong enough to prove knottedness for diagrams with up to nine crossings. More important, its invariance is provable directly from the Reidemeister moves, so it can serve as a striking modern example of "proof by pictures." We begin with the simplest case, which is 3-*colorability*.

Figure 10.28 shows a trefoil knot in three colors: red, white, and blue. In this case, there are three segments of the knot diagram between overcrossings, and each has been given a different color. As a result, three colors appear at each crossing: one for the segment above, and a different color on each side of the crossing for the segments below.

Definition. A knot diagram is called 3-*colorable* if each segment between crossings can be assigned one of three colors, in such a way that

Figure 10.28 : The colored trefoil knot

- all three colors are used, and
- the three segments meeting at any crossing are either all the same color or all different colors.

Thus the trefoil knot diagram is 3-colorable, and obviously the circle diagram is not, because it has only one segment. So, to prove that the trefoil cannot be transformed into a circle, it suffices to prove that 3-colorability is preserved by the Reidemeister moves. This is easy to check. We will do so only briefly, because it will also follow from the more general argument for p-colorability we will meet below.

1. For a Reidemeister I move, color the two segments on the right the same as the segment on the left.
2. For a Reidemeister II move there are two cases. If both segments on the left are the same color, use this color for all segments on the right. If they are different colors, use the third color for the new segment on the right.
3. For a Reidemeister III move, if there is only one color on the left, use this color again on the right. If there is more than one color on the left, then there are necessarily three, and they are uniquely determined by the colors a, b, and c at the ends shown. Bearing in mind that segments that extend beyond the diagram must remain the same color, this uniquely determines the coloring on the right.

(Figure 10.29 shows the case where a, b, c are all different; checking what happens when two of them are the same is similar.)

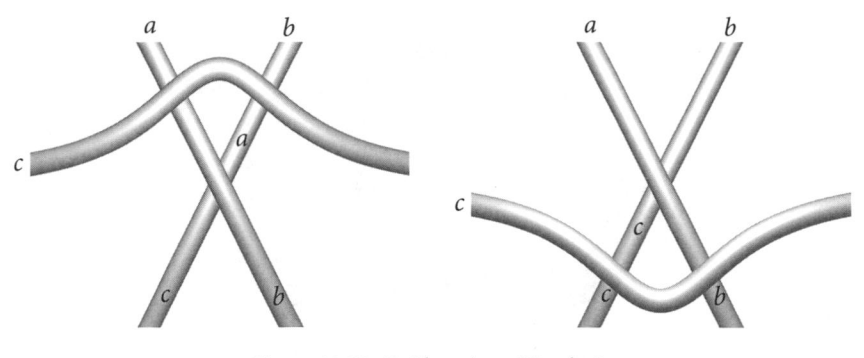

Figure 10.29 : Reidemeister III coloring

Generalization of 3-Coloring

Checking that Reidemeister moves preserve 3-coloring is easy because two colors at a crossing uniquely determine the third. Rather surprisingly, there is an *arithmetic* rule for determining the third color that generalizes to any integer $p \geq 3$: *if we call the three colors 0, 1, and 2 and if a is the top color at a crossing, and b and c are the two others, then*

$$2a - b - c \equiv 0 \quad (\text{mod } 3).$$

For example, if $a = 0$ and $b = 1$ then $c \equiv -1$ (mod 3), which means $c = 2$.

Generalization from 3 to p greatly expands the scope of coloring proofs, because few knots are 3-colorable, but many are p-colorable for primes $p > 3$.

Definition. A knot diagram is called p-**colorable** if each segment between over-crossings may be assigned a number among $0, 1, 2, \ldots, p - 1$ so as to satisfy the **crossing rule**

$$2a - b - c \equiv 0 \quad (\text{mod } p),$$

where a is assigned to the top arc, and b and c to the other two.

As with $p = 3$, we can prove that Reidemeister moves preserve p-coloring for any integer p. So, if a knot diagram is p-colorable, it follows that the corresponding curve is knotted. The proof for Reidemeister moves I and II is the same as for $p = 3$. For Reidemeister III we again observe what happens for input colors a, b, and c. Using the crossing rule

$$2 \times \text{top color} = \text{sum of bottom colors} \quad (\text{mod } p),$$

we deduce the coloration of the "before" picture shown in figure 10.30.

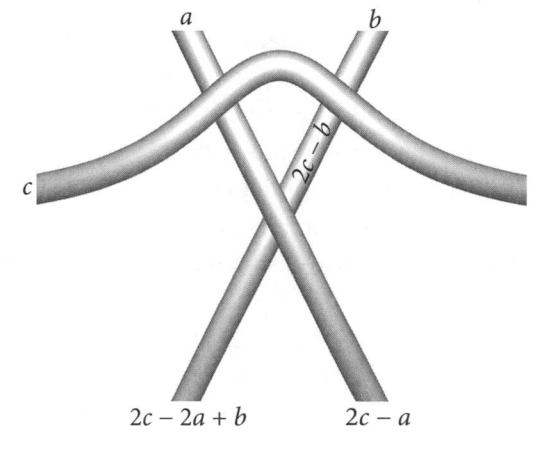

Figure 10.30 : Before Reidemeister III

Now, since the same colors must occur on segments that extend beyond the picture, we get the coloration of the "after" picture shown in figure 10.31. Only the internal segment x remains to be found and it is easy to see that

$$x = 2a - b,$$

because this x satisfies the crossing rule at both its ends.

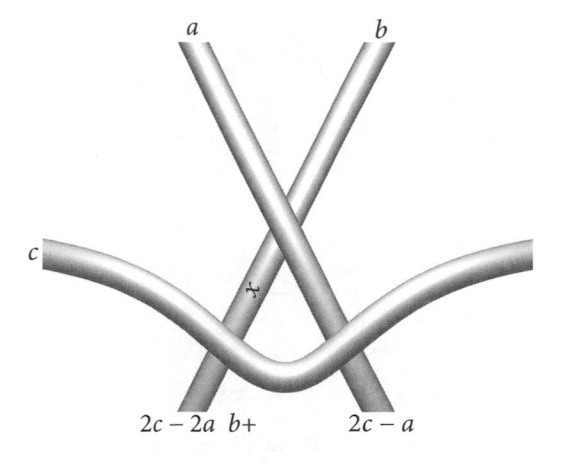

Figure 10.31 : After Reidemeister III

The Knot Determinant

For a p-coloring of a knot diagram we seek a nonzero solution (mod p) of the homogeneous linear equations

$$2a - b - c = 0$$

that hold at each crossing of the diagram. When p is prime this is a situation ripe for linear algebra, where it is well known that the condition for homogeneous linear equations to have a nonzero solution is that their determinant be zero. Some tinkering with this idea is needed, because in this case the equations are not independent, and we have to omit one. But it turns out that the determinant can be defined uniquely, and it is unaffected by Reidemeister moves, so it is an invariant of the knot K. We call it $\det(K)$.

Thus the condition for existence of a p-coloring of K is that $\det(K) = 0$ (mod p) or, equivalently, that p divides $\det(K)$. Turning this result around: K is p-colorable for precisely the primes $p \geq 3$ that divide $\det(K)$.

Of course, if $\det(K) = 1$ we do not get any p-coloring, but this situation is rare. For each knot K whose diagram has less than nine crossings, $\det(K)$ is odd and not equal to 1. So for each knot K with fewer than nine crossings we can find an easy proof, by p-coloring, that K is knotted. For example, the knot in figure 10.32, known as the figure eight knot, has determinant 5, hence a 5-coloring but not a 3-coloring. The figure shows one 5-coloring.

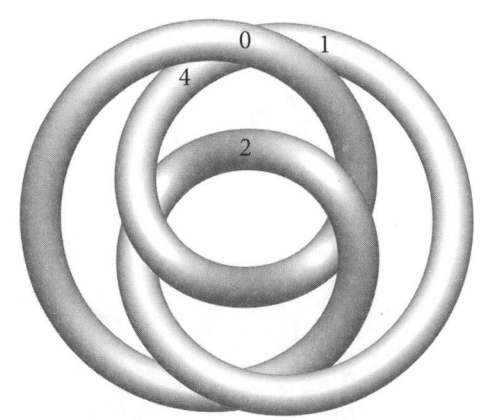

Figure 10.32 : Coloring the figure eight knot

Readers can enjoy finding proofs by coloring for a great variety of knots by looking up the knots and their determinants in the Knot Atlas at

http://katlas.org/wiki/The_Rolfsen_Knot_Table

10.9 REMARKS

The aim of this chapter is to show that topology continues to evolve alongside visual concepts and proofs as it did during the early stages of its development. This is not to say that no other methods of proof have also grown from topology. In fact, two radically different branches of topology—**point set topology** and **algebraic topology**—have developed their own methods of proof, which have spread to other branches of mathematics.

Point set topology began by rigorously analyzing the concepts of limit and continuity, at first in spaces like $\mathbb{R}^n$, but later by defining and working in a general **topological space**. This concept of space was first defined in a book on set theory by Hausdorff (1914)—for more, see section 11.9—and set theory (chapter 12) is now part of the foundations of point set topology.

Algebraic topology began with the paper of Poincaré (1895), which applied group theory (the fundamental group) and a kind of commutative algebra called **homology theory** to find topological invariants of multidimensional objects. One such invariant is the Euler characteristic, which Poincaré generalized to n-dimensional objects. In the 1920s, Emmy Noether saw that the invariant numbers found in Poincaré's homology theory are best packaged in **homology groups**, which exist in each dimension.

The intricate relations between homology groups (and, later, *co*homology groups) were systematized in the 1940s by Eilenberg and MacLane in what is now called **homological algebra**. This form of algebra developed a life of its own in the 1950s, as it was realized that "homology" could be used to probe algebraic objects as well as topological ones. The methodology of homological algebra, called **category theory**, embraces all kinds of mathematics and now competes with set theory as a foundation for mathematical concepts. However, since category theory involves a high level of abstraction, and background too advanced for a book such as this, I refer the reader to Krömer (2007) or Marquis (2009) for more on the story of category theory.

Thus, in a couple of centuries, the role of topology in mathematics has deepened from the superficiality of the Euler polyhedron formula to the very foundations. Recently, topology has been more directly incorporated in the foundations of mathematics with **homotopy type theory**, a system particularly suited to producing computer-verifiable proofs.

Topology has also influenced other branches of mathematics by analogy; for example, the concept of a branch point was applied by Dedekind and Weber (1882) in their theory of algebraic functions, allowing them to define Riemann surfaces and genus without appeal to continuity. This is made possible by the **Riemann-Hurwitz formula**, first observed by Riemann (1857), expressing genus in terms of the degree and the numbers of sheets that fuse together at each branch point. The concept of "ramification" in algebraic number theory carries the analogy of "branching" even further afield.

CHAPTER 11

Arithmetization

By the middle of the nineteenth century there were at least two reasons to seek new foundations for geometry and analysis. Analysis needed to understand the line at the micro level to establish basic properties of continuous functions. Geometry needed a common foundation for both Euclidean and non-Euclidean geometries, in any number of dimensions.

Between 1870 and 1900 mathematicians sought such a foundation in the real numbers, basing them in turn on the rational numbers, and ultimately on the natural numbers. This was the project called **arithmetization**, and it was so successful that Poincaré (1902: 120) could say:

> Today in analysis there remain only natural numbers or finite or infinite systems of natural Numbers. . . . Mathematics, as one says, has been arithmetized.

As we saw in section 8.4, Dedekind built a foundation by defining real numbers in terms of *sets* of rational numbers. Defining the line to be $\mathbb{R}$ then paved the way for a complete arithmetization of geometry and analysis: defining spaces, continuous functions, and geometric objects in terms of natural numbers and sets of natural numbers. For the first time, sets became important mathematical objects.

One virtue of arithmetization was a precise definition of a continuous curve or surface, which embraced some curves and surfaces previously regarded as "pathological." By giving these objects equal status with the common curves, mathematicians were led to sharpen their intuition, with the result that the "pathological" objects—such as **space-filling curves**—became comprehensible to a surprising degree.

11.1 THE COMPLETENESS OF $\mathbb{R}$

Chapter 8 introduced Bolzano's least upper bound property, which describes the **completeness** of $\mathbb{R}$, and showed how it follows from the definition of $\mathbb{R}$ by Dedekind cuts. The least upper bound property is one of several that express completeness by asserting that a certain infinite process leads to a real number. Two of these properties are described here, since they follow from the least upper bound property and are commonly used in analysis. In particular, we will use nested interval completeness in section 11.7.

Nested Interval Completeness

To avoid excessive words, we will denote the least upper bound of a bounded set S by $\text{lub}\,S$, and its greatest lower bound (which is the negative of the lub of the set of negatives of members of S) by $\text{glb}\,S$.

Nested interval property. *If $I_1 \supseteq I_2 \supseteq I_3 \supseteq \cdots$ are closed intervals whose lengths tend to zero, then $I_1, I_2, I_3, \ldots$ have a single common point.*

Proof. Let $I_1 = [a_1, b_1], I_2 = [a_2, b_2], \ldots$, so $a_1 \leq a_2 \leq a_3 \leq \cdots \leq b_3 \leq b_2 \leq b_1$. It follows, by the least upper bound property, that both $\text{lub}\{a_1, a_2, a_3, \ldots\}$ and $\text{glb}\{b_1, b_2, b_3, \ldots\}$ exist. We therefore have

$$a_1 \leq a_2 \leq a_3 \leq \cdots \leq \text{lub}\{a_1, a_2, a_3, \ldots\}$$
$$\leq \text{glb}\{b_1, b_2, b_3, \ldots\} \leq \cdots \leq b_3 \leq b_2 \leq b_1.$$

Thus any x in $[\text{lub}\{a_1, a_2, a_3, \ldots\}, \text{glb}\{b_1, b_2, b_3, \ldots\}]$ is in all of $I_1, I_2, \ldots$. If the length of the intervals $I_1, I_2, \ldots$ tends to zero, then

$$x = \text{lub}\{a_1, a_2, a_3, \ldots\} = \text{glb}\{b_1, b_2, b_3, \ldots\}$$

is the *only* common point. $\qquad\square$

Cauchy Completeness

Definition. A sequence of numbers $c_1, c_2, c_3, \ldots$ **converges** if it has a **limit** c, that is, if c_n tends to c as n increases. More precisely, c is the limit of the sequence $c_1, c_2, c_3, \ldots$ if, for each number $\varepsilon > 0$, there is a natural number N such that

$$n > N \implies |c - c_n| < \varepsilon.$$

We would like to be able to tell whether a sequence converges without knowing in advance what its limit is. The Cauchy convergence criterion of Cauchy (1821) makes this possible. It also provides another definition of real numbers, via convergent sequences of rational numbers.

Cauchy convergence criterion. *A sequence* $c_1, c_2, c_3, \ldots$ *converges if, for each* $\varepsilon > 0$, *there is an* N *such that*

$$m, n > N \implies |c_m - c_n| < \varepsilon.$$

Proof. If the sequence $c_1, c_2, c_3, \ldots$ satisfies the Cauchy convergence criterion, there is a sequence of natural numbers $N_1 < N_2 < N_3 < \cdots$ such that

$$m, n > N_1 \implies |c_m - c_n| < 1/2,$$
$$m, n > N_2 \implies |c_m - c_n| < 1/4,$$
$$m, n > N_3 \implies |c_m - c_n| < 1/6,$$

and so on. Now if $|c_m - c_n| < 1/2$ for all $m, n > N_1$ this means in particular that all c_n stay within distance $1/2$ of c_{N_1+1} for $n > N_1$, and hence within an interval of length 1. Similarly for $N_2, N_3, \ldots$, so we get nested closed intervals

$$I_1, \text{ of length 1, containing all } c_n \text{ with } n > N_1,$$
$$\supseteq I_2, \text{ of length } 1/2, \text{ containing all } c_n \text{ with } n > N_2,$$
$$\supseteq I_3, \text{ of length } 1/3, \text{ containing all } c_n \text{ with } n > N_3,$$
$$\vdots$$

whose length tends to zero. By the nested interval property, these intervals have a single point c, which is clearly the limit of $c_1, c_2, c_3, \ldots$. $\qquad \square$

11.2 THE LINE, THE PLANE, AND SPACE

The completeness of $\mathbb{R}$ assures us that $\mathbb{R}$ can serve as a model of the line in geometry, because $\mathbb{R}$ is "unbroken" or has "no gaps." It follows that $\mathbb{R}^2$ can serve as a model of the plane, and as a model of Euclid's plane geometry when we define the **distance** between points $\langle a_1, a_2 \rangle$ and $\langle b_1, b_2 \rangle$ to be

$$\sqrt{(a_1 - b_1)^2 + (a_2 - b_2)^2}.$$

More generally, we can model n-**dimensional Euclidean geometry** by $\mathbb{R}^n$, with distance between points $\langle a_1, a_2, \ldots, a_n \rangle$ and $\langle b_1, b_2, \ldots, b_n \rangle$ defined by

$$\sqrt{(a_1 - b_1)^2 + (a_2 - b_2)^2 + \cdots + (a_n - b_n)^2}.$$

By changing the definition of "distance" we can also model various non-Euclidean geometries by subsets of $\mathbb{R}^n$. For example, we can define the **sphere** $\mathbb{S}^2$ by

$$\mathbb{S}^2 = \{ \langle x, y, z \rangle \in \mathbb{R}^3 : x^2 + y^2 + z^2 = 1 \}.$$

With "distance" defined by arc length on the sphere as in section 9.2, we then have **spherical geometry**. And we can define the **hyperbolic plane** as the open disk in $\mathbb{R}^2$

$$\mathbb{H}^2 = \{ \langle x, y \rangle \in \mathbb{R}^2 : x^2 + y^2 < 1 \},$$

with "distance" defined by arc length on the pseudosphere as in section 9.5.

As mentioned in section 9.6, Riemann (1854b) defined very general geometries based on $\mathbb{R}^n$, with distance given by a differentiable function of the coordinates. Thus *geometry, in a very broad sense, can be arithmetized* by basing it on $\mathbb{R}$ and calculus. Calculus itself is based on $\mathbb{R}$, as we will now see by investigating the nature of continuity, differentiation, and integration.

11.3 CONTINUOUS FUNCTIONS

Section 8.3 gave Bolzano's definition of continuous function, and section 8.4 proved a basic property of continuous functions, the intermediate value theorem. We needed it there to discuss the fundamental theorem of algebra, but it was perhaps not clear how poorly continuous functions were understood before Bolzano's time. Here is a little background.

In ancient times there was no general concept of function or curve, only particular curves such as the conic sections and a handful of others defined by geometric conditions. With the development of algebraic geometry and calculus in the seventeenth century, *formulas* became available to describe curves: polynomials for the algebraic curves, and formulas of calculus, such as integrals or infinite series, for some transcendental curves. But it was the *formula* that called a curve into being, and no more general way of defining curves was apparent.

Under these circumstances, it is not surprising that mathematicians could be mistaken when they tried to describe general properties of curves, or of "continuous motion." For example, Newton (1671: 71) thought that continuous motion had a well-defined speed at any time, just as motion with a continuous speed had a well-defined distance at any time. It was only in the nineteenth century that the difference between these two situations was sorted out: a continuous function need *not* have a derivative (anywhere!), but a continuous function (on a closed interval) *does* have an integral.

Before continuous functions could be properly understood, mathematicians had to see more examples of them—and also examples of *dis*continuous functions—so as to understand the concept of **function** itself. An important stimulus in the development of the function concept was the problem of the vibrating string.

The Vibrating String

Section 6.2 mentioned in passing the modes of vibration of a stretched string, such as a guitar string. Figure 11.1 shows them again. It is a remarkable talent of the human ear and brain that (with varying degrees of skill) we can hear different modes at the same time, thus recognizing the different notes in a chord. When the string of an instrument is plucked, all modes (in principle) are present to some extent, and

Figure 11.1 : Modes of vibration

the particular mixture of modes creates the characteristic sound of the instrument.

Today, this fact is well known and is used in electronic music to simulate acoustic instruments or to create new ones. But Daniel Bernoulli (1753) already conjectured that any mode of vibration was a sum of the basic ones, which have the shape of sine or cosine curves.

Among other things, this seems to imply that *any continuous function is an infinite sum of terms* $a_n \cos n\pi x + b_n \sin n\pi x$. Fourier (1822) backed up Bernoulli's intuition by showing that

$$f(x) = \frac{1}{2}a_0 + \sum_{n=1}^{\infty}(a_n \cos n\pi x + b_n \sin n\pi x)\, dx,$$

where

$$a_n = \int_{-1}^{1} f(x) \cos n\pi x\, dx \quad \text{and} \quad b_n = \int_{-1}^{1} f(x) \sin n\pi x\, dx.$$

Apparently, $f(x)$ is now "any function that can be integrated," whatever that means. In 1822, mathematics was not ready to decide exactly what Fourier's theorem meant.

11.4 DEFINING "FUNCTION" AND "INTEGRAL"

During the nineteenth and early twentieth centuries, there was a struggle between the function and integral concepts, as mathematicians challenged the integral with highly discontinuous functions, and the integral responded by catching most, if not all, of the functions thrown at it. From the beginning of calculus it was assumed that continuous functions are integrable; for example, the proof of the fundamental theorem of calculus given in section 6.7 assumes this. Admittedly, the Leibniz-style proof given there is questionable, given its use of infinitesimals, but it can be rehabilitated using Bolzano's definition of continuity (section 8.3) and the so-called **Riemann integral**.

Before we turn to the Riemann integral let us review the limit concept for functions implicit in Bolzano's definition of continuity at a point but first explicitly stated (with the $\varepsilon - \delta$ symbolism we use today) by Cauchy (1823).

Definitions. A function $f(x)$ has **limit** l **as** x **tends to** c if, for any $\varepsilon > 0$, there is a $\delta > 0$ such that

$$0 < |x - c| < \delta \quad \text{implies} \quad |f(x) - l| < \varepsilon.$$

We also write this relationship as $f(x) \to l$ as $x \to c$, or $\lim_{x \to c} f(x) = l$. If, in addition, $f(c) = l$ we say that f is **continuous at the point** $x = c$. Finally, if f is continuous for each c with $a \le c \le b$ we say f is **continuous on the interval** $[a, b]$.

The Riemann Integral

The **definite integral** is typically introduced by the idea of approximating the area under the graph of $y = f(x)$ by finitely many rectangles, as in figure 11.2. We subdivide the interval $[a, b]$ at points $a = x_1 < x_2 < \cdots < x_n = b$ and consider the sum of the rectangle areas $(x_{i+1} - x_i) f(x_i)$ shown. If these sums approach a value I as the maximum rectangle width $x_{i+1} - x_i$ approaches zero, then I is the value of the definite integral

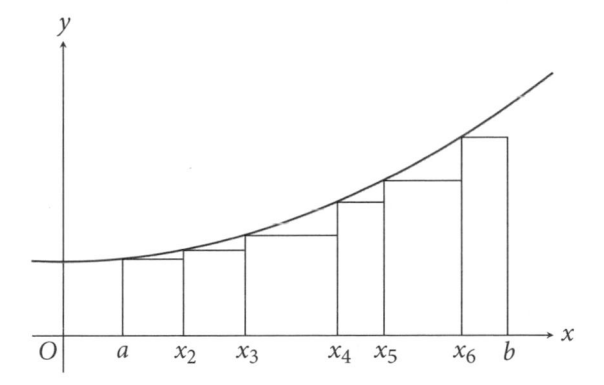

Figure 11.2 : Approximating a curved region by rectangles

$$\int_a^b f(x)\, dx,$$

and f is said to be **Riemann integrable** on $[a, b]$.

This idea goes back to Archimedes, and more formal versions were given by Cauchy (1823) and Riemann (1854a). Their definitions are based on the intuition that the total area of the gaps between the rectangles and the graph approaches zero as the maximum width of rectangles approaches zero. This intuition is correct for continuous functions, though a rigorous proof depends on **uniform continuity**, not mere continuity, as we will see in section 11.6.

The Riemann integral seems tailored to continuous functions, and it obviously fails for some highly discontinuous functions, such as the function

$$d(x) = \begin{cases} 1 & \text{for } x \text{ rational} \\ 0 & \text{for } x \text{ irrational} \end{cases}$$

introduced by Dirichlet (1829).[1] If we choose rational values for the points $x_i \neq a, b$, then all but the end rectangle have height 1 and (whether or not a or b is rational) the sums approach the value $b - a$ as the maximum width $x_{i+1} - x_i$ approaches zero. But if we choose irrational values for $x_i \neq a, b$ the sums approach zero. Thus there is no value I approached by all sums.

However, the Riemann integral actually works on some functions that are discontinuous in much the same way as the Dirichlet function. Take the function of Thomae (1879), for example, which is defined on [0,1] by

$$t(x) = \begin{cases} 1/q & \text{for } x = p/q, \text{ a rational number in lowest terms} \\ 0 & \text{for } x \text{ irrational.} \end{cases}$$

Figure 11.3 (due to Lars Rohwedder on Wikimedia) shows an approximation to the integral of the Thomae function consisting of thin black bars.

Given any natural number n, in the interval [0,1] there are only finitely many rational points $x = p/q$ with $q < n$. At each of these points, erect a rectangle of height $1/q$ but so thin that their total area is at most $1/n$. At all other points of the interval a rectangle of height $1/n$ is tall enough to cover the graph, so these points of the graph can be covered by a single rectangle of height $1/n$ and width 1. Thus the whole graph of the Thomae function can be covered by rectangles of total area at most $2/n$. We can now choose n to make $2/n$ as small as we please, which shows that the integral $\int_0^1 t(x)\,dx = 0$ for the Thomae function t.

It should be emphasized that we are looking only at functions defined on a *closed* interval $[a, b] = \{x \in \mathbb{R} : a \le x \le b\}$. Even a continuous function may not have a Riemann integral over an open, or half-open, interval. The simplest example is the function $1/x$ over the interval

---

1. With this function we approach the idea that the values $f(x) = y$ of a function f can be paired *arbitrarily* with values of x. This means that a **function** f is an arbitrary set of ordered pairs $\langle x, y \rangle$ with the property that there is only one y for each x. But full acceptance of this idea had to wait for acceptance of the set concept, described in the next chapter.

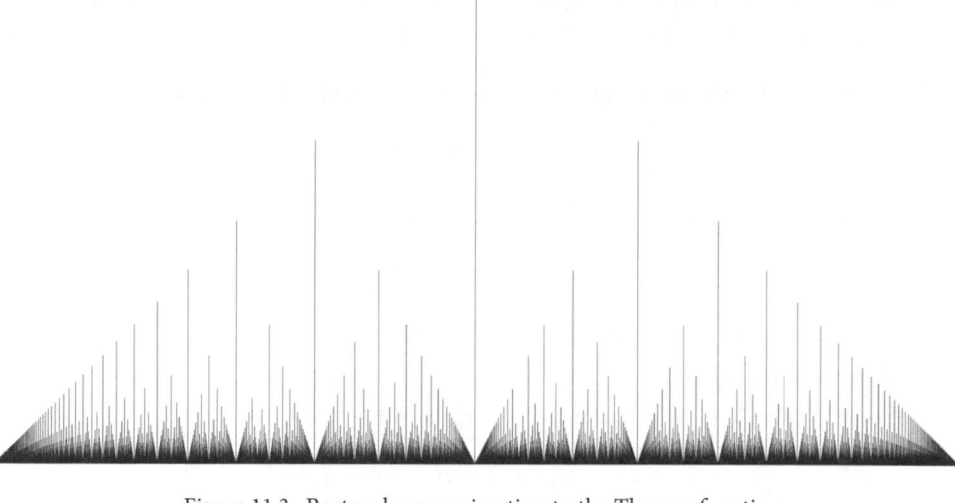

Figure 11.3 : Rectangle approximation to the Thomae function

$(0, 1] = \{x : 0 < x \leq 1\}$. Since

$$\int_{1/n}^{1} \frac{dx}{x} = \ln n,$$

which grows beyond all bounds as $1/n$ approaches 0, the integral on $(0,1]$ does not exist. The problem with the function $1/x$ is of course that it grows beyond all bounds on the interval $(0,1]$. It is continuous, but not *uniformly* so, and problems arise from lack of uniformity, as we will see in section 11.6.

Lebesgue Measure and Integral

Since bounded functions with and without a Riemann integral seem difficult to distinguish, there is need for a more powerful integral concept, able to deal better with discontinuous functions. Such an integral was introduced by Lebesgue (1902) and is called the **Lebesgue integral**.

The Lebesgue integral is based on **Lebesgue measure**, which assigns length to sets on the line, area to sets in the plane, and so forth. Since the integral of a positive-valued function f of one variable may be viewed as the area under the graph $y = f(x)$, Lebesgue measure is the main concept to understand. The Lebesgue measure function μ for sets in the line is defined by the following simple conditions (which generalize to the plane and higher dimensions in the obvious way).

Measure of basic sets. The measure of the interval $[a, b]$ is given by

$$\mu([a, b]) = b - a.$$

Complementation. If $S \subseteq T$ and if $\mu(S)$ and $\mu(T)$ exist, then

$$\mu(T - S) = \mu(T) - \mu(S),$$

where $T - S = \{x \in \mathbb{R} : x \in T \text{ and } x \notin S\}$.

Countable additivity. If S is the union of disjoint sets $S_1, S_2, S_3, \ldots$ with measures $\mu(S_1), \mu(S_2), \mu(S_3), \ldots$, whose sum exists, then

$$\mu(S) = \mu(S_1) + \mu(S_2) + \mu(S_3) + \cdots.$$

Measure zero sets. Any subset of a measure zero set has measure zero.

These conditions have some easy consequences:

1. The measure of a point is 0, because $\{c\} = [c, c]$ and $\mu([c, c]) = c - c = 0$.
2. All the intervals $(a, b], [a, b), (a, b)$ have measure $b - a$. For example,

$$(a, b] = [a, b] - \{a\}, \quad \text{so} \quad \mu((a, b]) = \mu([a, b]) - \mu(\{a\})$$
$$= b - a - 0 = b - a.$$

3. A set $S = \{c_1, c_2, c_3, \ldots\}$, called a **countable** set, has measure 0, because

$$\mu(S) = \mu(\{c_1\}) + \mu(\{c_2\}) + \mu(\{c_3\}) + \cdots = 0 + 0 + 0 + \cdots = 0.$$

The latter result, showing that certain infinite sets have measure zero, can also be supported by the following argument. Given the sequence of points $c_1, c_2, c_3, \ldots$, we can cover c_1 by a line interval of length $\varepsilon/2$, c_2 by a line interval of length $\varepsilon/4$, c_3 by a line interval of length $\varepsilon/8$, and so on. Then the whole set $S = \{c_1, c_2, c_3, \ldots\}$ is covered by intervals of total length at most ε, which can be as small as we please. Therefore, the only measure that can be reasonably assigned to S is zero.

What is surprising, at first, is that *the rational points in* $[0, 1]$ *are a set of the form* $S = \{c_1, c_2, c_3, \ldots\}$, *and hence they have measure zero*. To see why, arrange these rational points in the following order, which groups them according to the denominators of the corresponding fractions:

$$0, \quad 1, \quad \frac{1}{2}, \quad \frac{1}{3}, \frac{2}{3}, \quad \frac{1}{4}, \frac{3}{4}, \quad \frac{1}{5}, \frac{2}{5}, \frac{3}{5}, \frac{4}{5}, \quad \dots$$

This dramatically changes the complexion of the Dirichlet function $d(x)$ in the previous subsection. The points where $d(x) = 1$ form a set of Lebesgue measure zero, so the area under the graph, equal to the Lebesgue integral of $d(x)$ from 0 to 1, is also zero. Thus $d(x)$, which utterly fails to be Riemann integrable, is Lebesgue integrable and has integral zero.

This result raises the hope that virtually all functions—or, at least, the bounded functions on a bounded interval—are Lebesgue integrable. This is a reasonable hope, but to explain whether it comes true or not depends on information from set theory, which will emerge in the next three chapters.

Remark. Finally, it should be added that Lebesgue measure elegantly explains which bounded functions are Riemann integrable: *those that are continuous except on a set of measure zero*. Thus the Riemann integrable functions are actually close to being continuous; they are continuous "almost everywhere" in the sense of Lebesgue measure. This result is due to Lebesgue (1902), and a proof may be seen in Stillwell (2013: 201).

For example, the Thomae function t is continuous at each irrational point $c \in [0, 1]$. This is because, for any positive integer n, all rational points sufficiently close to c have denominator at least n, in which case $t(x)$ has value at most $1/n$ for these points. This means that $0 = t(c)$ is the limit of $t(x)$ as $x \to c$, so t is continuous at $x = c$. Thus $t(x)$ is continuous except at the rational points, and these form a set of measure zero.[2]

11.5 CONTINUITY AND DIFFERENTIABILITY

As mentioned in section 11.3, the founders of calculus did not anticipate the fine distinctions among continuous, differentiable, and integrable functions that would come to light in the nineteenth century. In particular, with their narrow concept of function, they did not realize that

---

2. In contrast, the Dirichlet function is discontinuous at all points. So, despite initial appearances, the Thomae function is *less* discontinuous than the Dirichlet function.

continuous functions could be *nowhere* differentiable. This was apparently first noticed by Bolzano in the 1830s, and specific examples were given by Riemann, Weierstrass, and other analysts. These and many other examples are in Thim (2003).

The examples of the analysts are infinite sums of sines and cosines that look almost "normal" in the analysis of their times. For example, Darboux (1875) showed that the function

$$\sum_{n=1}^{\infty} \frac{\sin(n+1)!x}{n!} \tag{*}$$

is continuous everywhere but differentiable nowhere. The sine functions in (*) have the job of creating "wavy waves." The first term is a sine wave, the second deforms it into a wave of smaller waves, each of which is deformed by the third into many more much smaller waves, and so on. Figure 11.4 shows the first three approximations (black, red, blue) to the function.

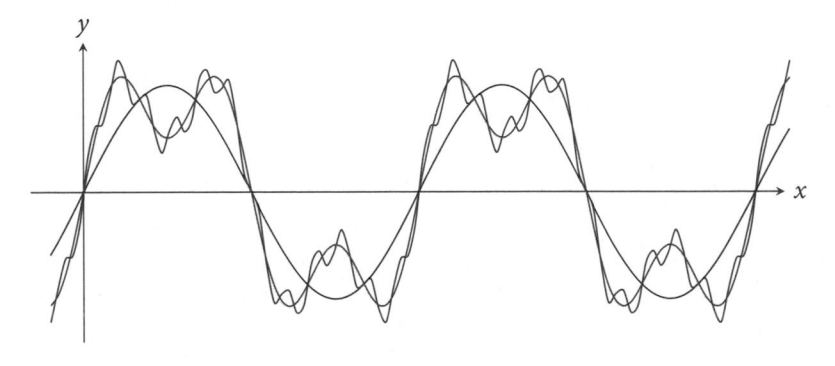

Figure 11.4 : Approximating the Darboux function

Because the amplitude of the waves decreases rapidly, the infinite sum exists and is continuous. But because the wave length decreases equally rapidly, the curve looks "wavy" at any scale and hence cannot have a tangent at any point. This informal argument makes the result plausible, I hope, but some rather delicate analysis is needed to make it rigorous.

More convincing examples can be given when sine waves are replaced by simple sawtooth waves. This is essentially what Bolzano did, but the simplest example was given by von Koch (1904). The Koch curve, also known as the *snowflake curve*, is not the graph of a function but a closed

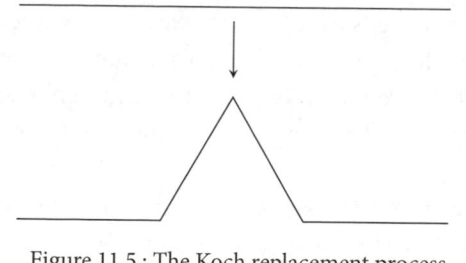

Figure 11.5 : The Koch replacement process

curve in the plane. It is the limit of a process that begins with an equilateral triangle and then replaces each edge with a polygonal path of four edges, as shown in figure 11.5. Each edge in the new polygon is replaced similarly, ad infinitum. Since each new edge is 1/3 of the length of the replaced edge, this process converges, and one can see a continuous limit curve emerging by running the process for a few more steps. (For simplicity, figure 11.6 shows just one third of the curve, which results from replacing a single edge of the triangle.)

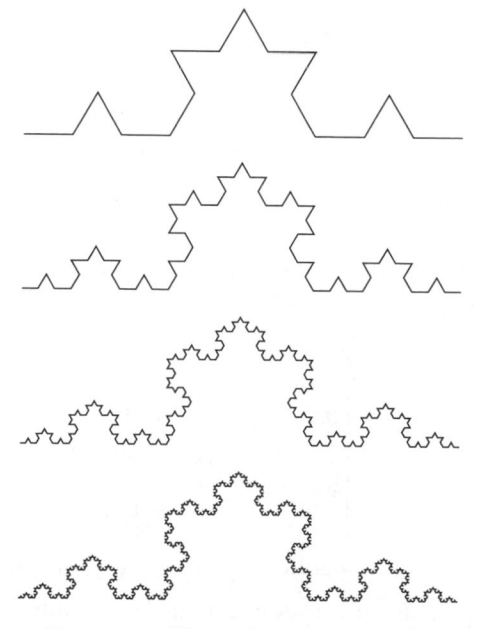

Figure 11.6 : The Koch polygon sequence

The beauty of the snowflake curve is its **self-similarity**. That is, if one takes the leftmost quarter of the limit curve and magnifies it by 3, then the resulting curve is *exactly the same* as the whole. This is because the leftmost of the four edges in the replacement polygon of figure 11.5 is 1/3 the size of the replaced edge, but it undergoes exactly the same replacement process and hence produces the same curve, only at 1/3 of the size.

So, if we zoom in on the snowflake curve by repeatedly magnifying by 3, we will continue to see the same thing. In contrast to the case of the Darboux function (*) above, we have found a precise sense in which the snowflake curve is "wavy at all scales." In particular, there will be no flattening under magnification at any point, as there would be if the curve had tangents.

11.6 UNIFORMITY

While Bolzano was the sharpest thinker about the foundations of calculus in the early nineteenth century, his work was not influential until midcentury because most of it was unpublished or unknown. The first influential works, which rediscovered some of Bolzano's ideas on limits and continuity, were by Cauchy (1821, 1823). Based on lectures at the Ecole Polytechnique in Paris, these books gave the first thorough and rigorous treatment of limits, convergence, continuity, differentiation, and integration. But they also contained a famous mistake: Cauchy (1821) claimed to prove that *the sum of a convergent series of continuous functions is continuous.*

To see why this is false, first consider the *sequence* of continuous "spike" functions (figure 11.7):

$$s_n(x) = \begin{cases} 0 & \text{for } x < -1/n \\ nx + 1 & \text{for } -1/n \leq x \leq 0 \\ -nx + 1 & \text{for } 0 \leq x \leq 1/n \\ 0 & \text{for } x > 1/n. \end{cases}$$

Since $s_n(x) = 0$ for all sufficiently large n when $x \neq 0$, and since $s_n(1) = 1$ for all n, this sequence converges to the discontinuous function

$$s(x) = \begin{cases} 1 & \text{for } x = 0 \\ 0 & \text{for } x \neq 0. \end{cases}$$

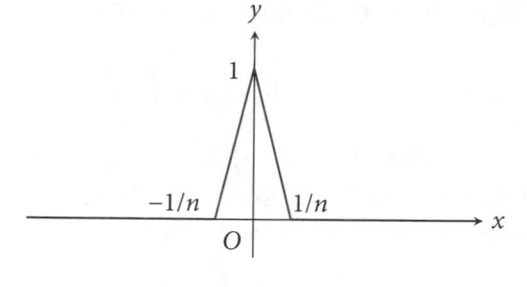

Figure 11.7 : The spike function $s_n(x)$

Now, to get a series rather than a sequence, we let

$$u_n(x) = s_n - s_{n-1}, \quad \text{where} \quad s_0 = 0,$$

so $u_n(x)$ is a continuous function and

$$s_n = u_1 + u_2 + \cdots + u_n.$$

Then, instead of the convergent sequence $s_1(x), s_2(x), s_3(x), \ldots$ we have the convergent *series* of continuous functions

$$u_1 + u_2 + u_3 + \cdots,$$

whose sum, by definition, is $\lim_{n \to \infty} s_n$. As we have seen, this limit is the discontinuous function $s(x)$.

Uniform Convergence

Cauchy's mistake was slowly identified and rectified over the next decades, along with a number of related difficulties involving infinite sequences and series. These included such things as term-by-term integration and differentiation, which had been taken for granted by the early exponents of calculus, such as Newton and Euler. For the full story see Grabiner (1981) or Gray (2015). In brief, the instrument needed to control infinite processes on continuous functions is **uniformity**. What happens with and without uniformity can be illustrated by the above sequence of continuous functions.

The sequence of functions $s_1(x), s_2(x), s_3(x), \ldots$ converges to its limit $s(x)$ in a way that is *non*uniform in the following sense. None of the functions $s_n(x)$ is "close" to the limit function for *all* values of x. In particular,

$$s_n\left(-\frac{1}{2n}\right) = \frac{1}{2}, \quad \text{whereas} \quad s\left(-\frac{1}{2n}\right) = 0.$$

Suppose, on the other hand, that $f_1(x), f_2(x), f_3(x), \ldots$ is a sequence of continuous functions that **converges uniformly** to a function $f(x)$. That is, for each $\varepsilon > 0$ there is an N such that

$$n > N \quad \text{implies} \quad |f(x) - f_n(x)| < \varepsilon \quad \text{for all } x.$$

In this case we can prove that f is a continuous function.[3] The proof is an easy "$\frac{\varepsilon}{2} + \frac{\varepsilon}{2}$" argument: to ensure that $f(x) \to f(a)$ as $x \to a$ we first choose N so that

$$|f(x) - f_n(x)| < \frac{\varepsilon}{2} \quad \text{for all } n > N \text{ and all } x,$$

by uniform convergence. Then use continuity of f_n to choose δ so that

$$|f_n(x) - f_n(a)| < \frac{\varepsilon}{2} \quad \text{for all } x \text{ such that } |x - a| < \delta.$$

This choice of N and δ gives

$$|f(x) - f(a)| \le |f(x) - f_n(x)| + |f_n(x) - f_n(a)| < \frac{\varepsilon}{2} + \frac{\varepsilon}{2} = \varepsilon,$$

so $f(x) \to f(a)$ as required.

This result on sequences transfers to a corresponding result on series. We say that a series

$$u_1(x) + u_2(x) + u_3(x) + \cdots$$

is **uniformly convergent** to $u(x)$ if the sequence $s_1(x), s_2(x), s_3(x), \ldots$ converges uniformly to $u(x)$, where

$$s_n = u_1 + u_2 + \cdots + u_n.$$

Thus the sum, $u(x)$ of a uniformly convergent series $u_1(x) + u_2(x) + u_3(x) + \cdots$ of continuous functions is continuous. In the 1870s Weierstrass developed these results and other basic results on continuous functions in his lectures at the University of Berlin.

---

3. A similar argument proves that the limit of a uniformly convergent sequence of Riemann integrable functions is Riemann integrable. However, the corresponding result for differentiability is false, as the Darboux function $\sum_{n=1}^{\infty} \frac{\sin(n+1)!x}{n!}$ shows. The terms of the series are differentiable and the series is uniformly convergent because its terms rapidly decrease in amplitude, but the sum is not differentiable.

Uniform Continuity

The idea of uniformity can also be applied to a single continuous function. It often depends on the domain whether a continuous function is uniformly continuous or not. The definition is almost the same as the definition of continuity except for the position of the quantifier "for all a."

Definition. A function f is **uniformly continuous** on domain D if, for each $\varepsilon > 0$ there is a $\delta > 0$ such that, *for all $a, x \in D$,*

$$|x - a| < \delta \quad \text{implies} \quad |f(x) - f(a)| < \varepsilon.$$

It follows from this definition that $f(x) \to f(a)$ as $x \to a$ for each $a \in D$, so f is certainly continuous at each point of D, which is what we mean by being continuous *on D*.[4] However, a continuous function need not be uniformly continuous. An example is the function $f(x) = 1/x$ on the open interval $(0, 1) = \{x \in \mathbb{R} : 0 < x < 1\}$. This function is continuous at each $a \in (0, 1)$, but not uniformly so, because as a approaches 0 we need smaller and smaller values of δ to keep $|1/x - 1/a|$ less than a given ε for $|x - a| < \delta$.

However, a continuous f is uniformly continuous on the *closed* interval $[a, b] = \{x \in \mathbb{R} : a \leq x \leq b\}$, as we will see in the next section. This result is the key to showing that every continuous function is Riemann integrable on closed intervals.

11.7 COMPACTNESS

What is it about closed intervals $[a, b]$ that makes continuous functions uniformly continuous on them? The answer is **compactness**, a property that was not clearly recognized in the nineteenth century and first defined by Fréchet (1906). Here we will concentrate on the case of closed intervals and uniform continuity, before elucidating the nature of compact sets more generally.

The first explicit proof that a continuous function is uniformly continuous on a closed interval seems to be in Heine (1872), but Heine was not fully aware of the decisive property of closed intervals that ensures

---

4. Thus the definition of continuity on D reads, "For each $a \in D$ and each $\varepsilon > 0$ there exists a δ such that, for each x, ...," whereas the definition of uniform continuity reads, "For each $\varepsilon > 0$ there is a δ such that, for all a and x, ..." In continuity, δ may depend on a as well as ε; in uniform continuity, δ depends only on ε.

uniformity. The relevant property was rediscovered and made clearer by Borel (1895), leading to a jointly named theorem.

The Heine-Borel Theorem

The theorem depends only on basic properties of open and closed intervals, but it involves an interesting infinite process of repeated bisection that we will see again later.

Heine-Borel theorem. *If the closed interval $I = [a, b]$ is covered by a set of open intervals $J_k = (c_k, d_k)$—that is, if each x in $[a, b]$ is in some (c_k, d_k)—then finitely many of the intervals J_k also cover $[a, b]$.*

Proof. Suppose that there are infinitely many intervals J_k, else there is nothing to prove, and suppose for the sake of contradiction that no finite set of the J_k covers $[a, b]$. In that case, one of the two (closed) halves of I,

$$I' = \left[a, \frac{a+b}{2}\right], \quad I'' = \left[\frac{a+b}{2}, b\right],$$

also cannot be covered by finitely many of the J_k.

Let I_1 be the leftmost of I', I'' that cannot be covered by finitely many J_k, and repeat the argument with I_1. This leads to a closed half I_2 of I_1 that also cannot be covered by finitely many J_k, and so on. Thus we get an infinite sequence of closed intervals

$$I \supset I_1 \supset I_2 \supset I_3 \supset \cdots,$$

each of which is half of the one before, and none of which can be covered by finitely many of the J_k.

It follows, by the nested interval property of section 11.1, that the closed intervals $I, I_1, I_2, I_3, \ldots$ have a single point y in common. Since the intervals J_k cover I, y belongs to some particular J_k. But since J_k is an *open* interval, any sufficiently small closed interval containing y is also in J_k. In particular, some $I_l \subset J_k$, so J_k covers I_l, contrary to what we have just proved.

This means that our initial assumption, that no finite set of J_k covers I, is false. Therefore, finitely many of the intervals J_k cover $[a, b]$. □

The Heine-Borel property of closed intervals is today viewed as the *defining property* of compactness.

Definition. A subset S of $\mathbb{R}$ is called **compact** if any set of open intervals that cover S includes finitely many intervals $I_1, I_2, \ldots, I_n$ that also cover S.

Thus closed intervals are compact, but so are many more complicated sets, some of which we will see later. The concept of compactness extends in a natural way to Euclidean spaces $\mathbb{R}^n$ of any dimension and, indeed, to spaces more general than that—the general **topological spaces** (see section 11.9). However, the germ of general topology is already in $\mathbb{R}$.

Uniform Continuity on Closed Intervals

Now if f on $[a, b]$ is a continuous function, the definition of continuity from section 11.4 says that, for each $c \in [a, b]$ and each ε, there is a δ_c such that

$$|x - c| < \delta_c \quad \text{implies} \quad |f(x) - f(c)| < \varepsilon.$$

As remarked in the footnote to the previous subsection, this does not say f is uniformly continuous on $[a, b]$, because δ_c depends on c and uniform continuity demands a single δ. However, the Heine-Borel theorem allows us to get by with *finitely many* values δ_c, and from them we can get a single δ. The details are in the proof below.

Continuity on a closed interval. *If f is a continuous function defined on a closed interval $[a, b]$, then f is uniformly continuous on $[a, b]$.*

Proof. Given an $\varepsilon > 0$, for each $c \in [a, b]$ there is a $\delta_c > 0$ such that

$$|x - c| < \delta_c \quad \text{implies} \quad |f(x) - f(c)| < \frac{\varepsilon}{2}.$$

(The reason for choosing $\varepsilon/2$ is that we eventually have to add two of them.) If we let $I_c = (c - \delta_c, c + \delta_c)$ then I_c is an open interval, and the set of all I_c covers $[a, b]$ as c ranges over $[a, b]$. Therefore, by the Heine-Borel theorem, there is a finite collection $I_{c_1}, I_{c_2}, \ldots, I_{c_n}$ that also covers $[a, b]$. Since the I_{c_j} are open intervals, any two of them either are disjoint or overlap in a finite open interval. We let

$$\delta = \frac{1}{2} \text{ minimum length of the overlaps.}$$

We are now going to prove uniform continuity by showing that, for any $d \in [a, b]$,

$$|x - d| < \delta \quad \text{implies} \quad |f(x) - f(d)| < \varepsilon.$$

Let I_{c_j} be an interval that covers d. If also $x \in I_{c_j}$ then we have both

$$|f(x) - f(c_j)| < \frac{\varepsilon}{2} \quad \text{and} \quad |f(c_j) - f(d)| < \frac{\varepsilon}{2}$$

by definition of c_j. Therefore,

$$|f(x) - f(d)| = |f(x) - f(c_j) + f(c_j) - f(d)| \le |f(x) - f(c_j)| + |f(c_j) - f(d)|$$

$$< \frac{\varepsilon}{2} + \frac{\varepsilon}{2} = \varepsilon.$$

If x is *not* in I_{c_j} then, since $|x - d| < \delta$, x is in some I_{c_k} that overlaps with I_{c_j}. And since x lies beyond the middle of the overlap, d must also be in I_{c_k}, in which case the above argument applies with c_k in place of c_j. So again we can conclude that $|f(x) - f(d)| < \varepsilon$. $\square$

Riemann Integrability of Continuous Functions

Uniform continuity is the ideal property to guarantee Riemann integrability; hence the theorem in the previous subsection ensures Riemann integrability of continuous functions on closed intervals.

Riemann integrability of continuous functions. *If f is continuous on a closed interval $[a, b]$, then the Riemann integral $\int_a^b f(x)\, dx$ exists.*

Proof. Since f is uniformly continuous on $[a, b]$, for any $\varepsilon > 0$ there is a $\delta > 0$ such that, for any $x_1, x_2 \in [a, b]$, $|x_1 - x_2| < \delta$ implies $|f(x_1) - f(x_2)| < \varepsilon$. So, if we divide $[a, b]$ into intervals of length less than δ, the difference between the lub and glb of $f(x)$ in each interval (which both exist because the values of $f(x)$ in such an interval $[c, d]$ are bounded by $f(c) \pm \varepsilon$) is at most ε.

Therefore, if we erect an *upper rectangle* on each subinterval I of height lub$\{f(x) : x \in I\}$, and a *lower rectangle* of height glb$\{f(x) : x \in I\}$, then the difference between the sums of upper and lower rectangles is at most $(b - a)\varepsilon$.

This means that, as $\varepsilon \to 0$, both the upper and lower sums have a common limit, which is $\int_a^b f(x)\, dx$. $\square$

The Extreme Value Theorem

Another way in which compactness improves the behavior of continuous functions is by ensuring the existence of maximum and minimum values.

Extreme value theorem. *If f is a continuous function on the closed interval $[a, b]$ then f takes a maximum and a minimum value there.*

Proof. It follows from the uniform continuity of f on $[a, b]$, proved above, that f is bounded on $[a, b]$. To see why, divide $[a, b]$ into a finite number, n, of subintervals on which $f(x)$ varies by at most ε. Then $f(x)$ lies between the values $f(a) - n\varepsilon$ and $f(a) + n\varepsilon$ over the whole of $[a, b]$.

The least upper bound principle gives values

$$u = \mathrm{lub}\{f(x) : x \in [a, b]\}, \quad l = \mathrm{glb}\{f(x) : x \in [a, b]\},$$

so it remains to show that $f(x)$ actually attains these values. Suppose on the contrary that $f(x)$ does *not* attain the value u. Then the continuous function $u - f(x)$ is positive for all $x \in [a, b]$, and hence the function $g(x) = 1/(u - f(x))$ is continuous. But $g(x)$ is *unbounded*, since $f(x)$ takes values arbitrarily close to its lub u.

This contradicts the boundedness of continuous functions proved in the first paragraph, so we were wrong to assume that $f(x)$ does not take the value u. Thus u is in fact the maximum value of f on $[a, b]$, and similarly l is its minimum value. □

It is possible to give a more direct proof of the boundedness of f on $[a, b]$ by using the bisection process used to prove the Heine-Borel theorem. Suppose, for the sake of contradiction, that f is *unbounded* on $[a, b]$. Then f is also unbounded on a closed half of $[a, b]$, call it I_1. Likewise, f is unbounded on a closed half I_2 of I_1, on a closed half I_3 of I_2, and so on.

The intervals $I_1 \supset I_2 \supset I_3 \supset \cdots$ have a single common point, c. Since f is continuous it is bounded, by the values $f(c) \pm \varepsilon$, in some open interval $I = (c - \delta, c + \delta)$. But since I is open, it contains one of the I_n, and we have our contradiction. □

The extreme value theorem also applies to continuous functions on the plane, such as the function $|p(z)|$ for a polynomial p, which we saw in the d'Alembert-Argand proof of the fundamental theorem of algebra (section 8.2). In this case the compact set, on which we seek a minimum of $|f(z)|$, is the closed disk $\{z : |z| \le R\}$.

This is close to what the above version of the extreme value theorem says, except that the domain of our continuous, real-valued function

$|f(z)|$ is a disk and not a line interval. We can get around this difficulty by a slight modification of the proof:

- Instead of looking for a minimum on the disk of radius R, look for a minimum on the square of width $2R$ that contains it.
- Instead of dividing intervals into halves, divide squares into quarters.
- Instead of determining a point by nested intervals whose lengths tend to 0, determine a point by nested squares whose widths tend to 0.

Then if f is unbounded on the square it is unbounded on some quarter of the square, and the argument can proceed as in the proof of the extreme value theorem above.

Thus when Weierstrass finally gave rigorous proofs of the intermediate and extreme value theorems in the 1870s, the long march to the fundamental theorem of algebra was finally over.

11.8 ENCODING CONTINUOUS FUNCTIONS

Having seen the role of the real numbers and continuous functions in analysis, we can now come back to the quotation from Poincaré at the beginning of this chapter. What did he mean when he said analysis had been reduced to natural numbers and finite or *infinite systems of* natural numbers?

From what we have seen, analysis reduces to real numbers and continuous functions, but how are these reduced to natural numbers and infinite systems of natural numbers? The main steps from natural numbers to continuous functions are these:

1. From the natural numbers $0, 1, 2, 3, \ldots$ we get **integers** from ordered pairs of natural numbers when we interpret $\langle m, n \rangle$ as $m - n$. For example, -1 is represented by the pairs $\langle 0, 1 \rangle, \langle 1, 2 \rangle,$ $\langle 2, 3 \rangle, \ldots$ So an integer is really an *equivalence class* of pairs $\langle a, b \rangle$, where $\langle a, b \rangle$ is equivalent to $\langle c, d \rangle$ if and only if $b - a = d - c$ (or $a + d = b + c$, writing the condition in terms of addition).

2. From integers we get **rational numbers** as fractions m/n, where m, n are integers and $n \neq 0$. Again, a rational number is really an equivalence class of fractions, where

$\dfrac{m}{n}$ and $\dfrac{p}{q}$ represent the same rational number if and only if $mq = np$.

So if we interpret each fraction m/n as an ordered pair $\langle m, n \rangle$ of integers, each rational number is a class of such pairs.

3. Each **real number** α can be represented by an infinite set of rational numbers, for example, by the lower set in the Dedekind cut that defines α. It is here that infinite sets enter in a really essential way. The infinite sets involved in the definitions of integers and rational numbers can each be replaced by a *representative*, such as the reduced fraction for a rational number. But there is no way to represent the lower set for an irrational number by a single member.[5]

Another way to represent a real number, which is particularly useful when we want to represent infinite sequences of real numbers, is by **decimal expansion**. In particular, each number between 0 and 1 can be represented by a decimal expansion between $0.0000\ldots$ and $0.9999\ldots$. Each such expansion can be viewed as an infinite set of ordered pairs $\langle n, d_n \rangle$, where n is a positive integer and d_n, the nth digit of α, is an integer between 0 and 9.

4. An infinite sequence $\alpha_1, \alpha_2, \alpha_3, \ldots$ of numbers in $[0,1]$ can then be viewed as an array of decimal digits, such as the array shown in figure 11.8.

This array can be encoded by a single infinite decimal α whose digits are those encountered in succession on the zigzag path shown. The same idea can be adapted to encode any sequence of real numbers by a single real number, but we skip the details.

5. Finally, a continuous function f is determined by its values on the rational numbers. This is because any real number x is the limit of a sequence of rational numbers r, and then $f(x) = \lim_{r \to x} f(r)$. Also, the rational numbers can be arranged in sequence $r_1, r_2, r_3, \ldots$ by a slight variation of the idea used to list the rational numbers in $[0,1]$ in section 11.4. Namely, first list the fractions m/n where m, n are positive integers for which $m + n = 2$, then those for which $m + n = 3$, and so on. Finally, alternate the fractions m/n on this list with 0 and their negatives $-m/n$.

Thus any continuous function f is determined by the sequence of numbers $\alpha_n = f(r_n)$, which finally can be encoded by a single number α by the zigzag process above.

---

5. Not just no *obvious* way, but really no way at all. This is because the members of a lower Dedekind set are rational, and there are *more* irrational numbers than rational numbers, as we will see in the next chapter.

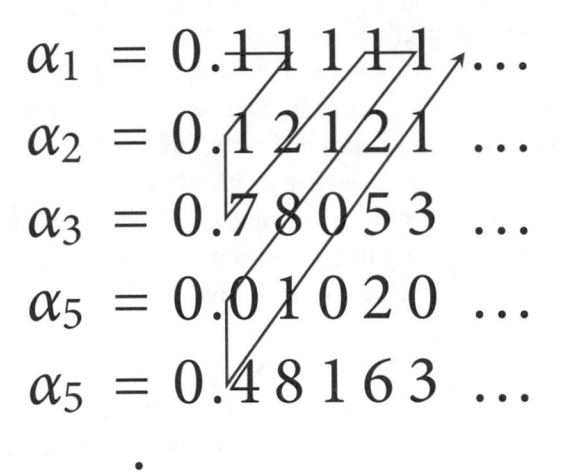

$$\alpha_1 = 0.\cancel{1}\,1\,1\,\cancel{1}\,\cancel{1}\,\ldots$$
$$\alpha_2 = 0.1\,2\,1\,2\,1\,\ldots$$
$$\alpha_3 = 0.7\,8\,0\,5\,3\,\ldots$$
$$\alpha_5 = 0.0\,1\,0\,2\,0\,\ldots$$
$$\alpha_5 = 0.4\,8\,1\,6\,3\,\ldots$$

$$\vdots$$

$$\alpha = 0.111721118041021\ldots$$

Figure 11.8 : Encoding an array by a single decimal

In this account, I have interpreted Poincaré's "systems" of natural numbers, somewhat liberally, to include ordered pairs of natural numbers, ordered pairs *of* ordered pairs, and so on. However, by encoding ordered pairs of natural numbers by single natural numbers, a few times, we can encode everything up to continuous functions by either natural numbers or infinite sets of natural numbers. A simple way to do this is to encode the ordered pair $\langle m, n \rangle$ of natural numbers by the number $2^m 3^n$.

11.9 REMARKS

The program of arithmetization was essentially completed when Borel (1898: 109) observed that each continuous function may be encoded by a real number. This accounts for Poincaré's 1902 statement that "everything in analysis" had been arithmetized. But of course this was not really the end of geometry, nor had absolutely everything in analysis been arithmetized. Many questions remained about sets of real numbers, as we will see in the next chapter.

Geometry Lives On

As described in sections 3.1–3.4, Hilbert (1899) derived the structure of $\mathbb{R}$, as a complete, ordered, Archimedean field, from his axioms for Euclidean plane geometry. In fact, it appears that Hilbert was largely motivated by the search for a geometric basis for $\mathbb{R}$, since he includes a completeness axiom. This axiom is not essential for Euclid's geometry, where constructible numbers suffice, but it is needed (naturally) for the completeness of $\mathbb{R}$.

Reinforcing the impression that $\mathbb{R}$ was Hilbert's real goal, he also gave a derivation of $\mathbb{R}$ from his axioms for hyperbolic *non*-Euclidean geometry. Here he found quite a different model of $\mathbb{R}$ in the **line at infinity**, which is the real axis in the half-plane model. The operations of addition and multiplication are also modeled quite differently in non-Euclidean geometry.

Another way in which geometry proved helpful was in motivating constructions of interesting but "counterintuitive" functions. We saw one example in von Koch's construction of a continuous but nondifferentiable curve in section 11.5. While arithmetization is useful in defining this curve precisely, the geometric definition makes its nondifferentiability *clearer* than is the case for functions defined by analytic means, such as infinite series. It is fair to say that the Koch curve sharpens our intuition by making visible what was previously thought counterintuitive.

Another example where geometry ultimately convinces us of a "counterintuitive" truth is the **space-filling curve** of Peano (1890). The intuition here is of a point moving through the unit square, from the bottom left corner at time $t = 0$ to the top right corner at time $t = 1$. The path of the moving point is obtained by a process of successive refinements, beginning with its positions at $t = 0$ and $t = 1$ and then repeatedly subdividing according to a simple geometric scheme. The first subdivision is shown in figure 11.9.

The unit square is divided into nine subsquares, each of which the moving point traverses from one corner to its opposite in a time interval of length 1/9. For example, at time 1/9 the point reaches the corner opposite to 0 in the first subsquare, and it revisits the same point at time 5/9. Since the path in each subsquare is from one corner to its opposite, we can similarly divide each subsquare into nine subsubsquares, each of which the point traverses from one corner to its opposite in a time interval of length 1/81.

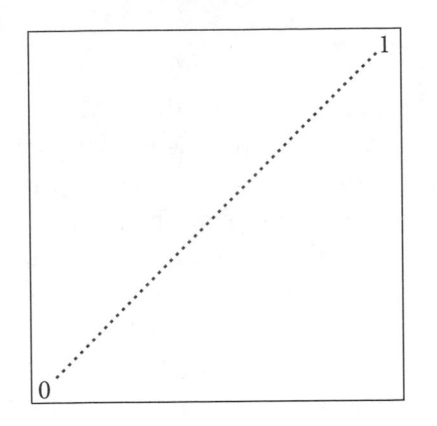 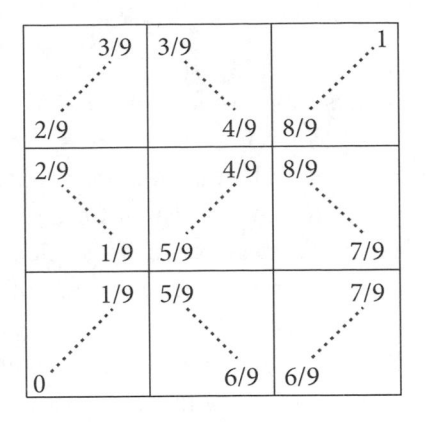

Figure 11.9 : Refining an approximation to the Peano curve

The limit of this process is a curve that

- is continuous, because at times close together the moving point is at places close together, and
- includes each point of the square, because each point P is the limit of a nested sequence of squares, and the moving point reaches P at the instant of time that is the limit of the nested time intervals in which the moving point traverses the corresponding squares.

Thus, even when arithmetization was at its height, there was room for visual arguments in analysis. Hyperbolic geometry, too, lent itself to visible but infinite constructions. We saw examples in section 9.7, and a whole book devoted to visualizing infinite constructions is Vilenkin (1995).

From the 1970s, thanks to computer graphics, many previously unimaginable objects have become visible and the subject of new investigations. The Mandelbrot set is a well-known example. Other eye-catching examples, which update the classical illustrations from Fricke and Klein, are in Mumford et al. (2002). In the English translations, Klein and Fricke (2017) and Fricke and Klein (2017), the original illustrations have been painstakingly recreated.

Point Set Topology

The fundamental concepts of point set topology are those of **open set** and the complementary concept of **closed set**. The motivating examples of each are those of open interval $(a, b) = \{x \in \mathbb{R} : a < x < b\}$ and closed interval $[a, b] = \{x \in \mathbb{R} : a \le x \le b\}$ in $\mathbb{R}$. An open interval (a, b) is "open" in the sense that it does not include the endpoints a, b that "close" the closed interval $[a, b]$. We can also say that $[a, b]$ is "closed" under the operation of taking limit points, whereas (a, b) is not closed under this operation, because a and b are the limits of sequences of points belonging to (a, b).

More generally, we define an **open subset** of $\mathbb{R}$ to be any union of open intervals, and a **closed subset** of $\mathbb{R}$ to be the complement of an open set. It follows that both the empty set and $\mathbb{R}$ are open, and consequently, they are also both closed. This example shows that "open" and "closed" are not mutually exclusive. Nor are they exhaustive; for example, the set of rational numbers is neither open nor closed. (Sets are not like doors!)

The basic property of an open set $\mathcal{O} \subseteq \mathbb{R}$ is that for any $P \in \mathcal{O}$ there is also a **neighborhood** of P in $\mathcal{O}$, that is, an open interval $(a, b) \subseteq \mathcal{O}$ with $P \in (a, b)$. The basic property of a closed set $\mathcal{K} \subseteq \mathbb{R}$ is that all limit points of sequences from $\mathcal{K}$ are also in $\mathcal{K}$. Because if $P_1, P_2, P_3, \ldots \in \mathcal{K}$ and $P_n \to P$, then $P \in \mathcal{K}$ —otherwise P is in the open complement of $\mathcal{K}$, in which case some neighborhood of P is also in the complement, contrary to the assumption $P_n \to P$, which means that any neighborhood of P contains some $P_n \in \mathcal{K}$.

The argument just given about limit points suggests that the concept of continuous function may be definable in terms of open sets, and indeed it is. A function $f : \mathbb{R} \to \mathbb{R}$ is **continuous** if and only if $f^{-1}(\mathcal{O})$ is open for any open set $\mathcal{O}$. This definition, incidentally, bypasses the concept of "continuity at a point" that we originally used to define continuity in section 8.3. In fact, the definition can be vastly generalized. We can define an **open subset** $\mathcal{O}$ **of** $\mathbb{R}^n$ to be any set with the property that, for any $P \in \mathcal{O}$, there is an **open ball** $\mathcal{B} = \{Q \in \mathbb{R}^n : |P - Q| < \varepsilon\} \subseteq \mathbb{R}^n$. Then $f : \mathbb{R}^m \to \mathbb{R}^n$ is continuous if and only if $f^{-1}(\mathcal{O})$ is open for any open set $\mathcal{O}$.

The ultimate generalization was given by Hausdorff (1914) when he defined a **topological space** $\mathcal{T}$ as an arbitrary set with a family $\mathcal{F}$ of subsets $\mathcal{O} \subseteq \mathcal{T}$, called **open**, and subject only to these axioms:

1. The empty set and the whole space $\mathcal{T}$ are open.
2. The union of any collection of open sets is open.
3. The intersection of two open sets is open.

It is clear that $\mathbb{R}$ with its open subsets is a topological space according to this definition, as is $\mathbb{R}^n$. But the definition covers much more, in fact, any space where we can imagine talking about "neighborhoods" of points. It also covers any conceivable type of continuity, because a **continuous function** f can be defined as one such that $f^{-1}(\mathcal{O})$ is open for any open set $\mathcal{O}$.

Another noteworthy feature of Hausdorff's concept of topological space is that Heine-Borel is no longer a theorem but rather the definition of compactness: a subset $\mathcal{K}$ of a topological space $\mathcal{T}$ is **compact** if, for any open sets $\mathcal{O}_i$ whose union contains $\mathcal{K}$, there are finitely many $\mathcal{O}_1, \mathcal{O}_2, \ldots, \mathcal{O}_k$ whose union contains $\mathcal{K}$. From these definitions we can prove, for example, that the continuous image of a compact set is compact. We can also prove, as did Fréchet (1906), that a nested sequence of closed subsets of a compact set has a common point. This is the ultimate generalization of the result from section 11.1 that a nested sequence of closed intervals has a common point.

This kind of topology seems a far cry from the kind of topology we looked at in chapter 10—and so it is! It is one end of the wide spectrum of topology, the analysis end, opposite to the combinatorial/geometric end that began with questions like the Euler polyhedron formula. The two ends eventually were connected when it became clear that invariants such as the Euler characteristic are actually invariant under certain continuous functions (**homeomorphisms**). So continuity and related ideas, such as compactness, now underlie the precise study of surfaces and knots. We downplayed them in chapter 10 because that part of topology was originally, and still can be, developed largely in a combinatorial/geometric fashion.

CHAPTER 12

.....

Set Theory

Analysis in the nineteenth century made cautious moves toward infinity, by accepting certain infinite collections as legitimate mathematical objects. But it was hoped that, by "arithmetizing," the infinite collections could be viewed as merely "potentially" infinite, on a par with the set of natural numbers.

In the 1870s Cantor made a series of groundbreaking discoveries about infinite sets that forced a complete rethink of the ancient dichotomy between potential and actual infinity. First, Cantor (1874) showed that the set $\mathbb{R}$ is **uncountable,** which means there is no way to finesse $\mathbb{R}$ as a potential infinity. The **continuum** $\mathbb{R}$, the foundation of analysis and mathematical physics, is an *actual* infinity. Similarly uncountable is the collection of sets of natural numbers, so the arithmetization of analysis cannot be about natural numbers alone. It is unavoidably about *sets* of natural numbers as well, since these sets are vastly more numerous than the natural numbers themselves.

Cantor (1891) noticed that his argument about natural numbers extended to any set: any set has more subsets than members, and hence *there is no largest set*. This discovery (and some related ones) provoked a new "crisis of foundations." Since there is no largest set, there is no "set of all sets," and we have a problem: which properties define sets, if the property of being a set is not one of them? Is there a sensible set of axioms for set theory? Before we see the answer to that question, we should look at what was known about sets and real numbers in the 1890s. Then, in the next chapter, we will look at some axiom systems known in the 1890s and return to the question of axioms for sets.

12.1 A VERY BRIEF HISTORY OF INFINITY

From ancient times until the early nineteenth century there was a consensus among mathematicians that infinity was acceptable in mathematics only in the *potential* sense. That is, an endless collection or process was acceptable if it was finite at any stage, though growing, but at no stage complete.

Examples of this kind of infinity are the collection of natural numbers, which corresponds to the process of starting with 0 and repeatedly adding 1; Archimedes's decomposition of the parabolic segment into a triangle, then two triangles next to it, then two triangles next to each of them, and so on; and the set of rational numbers in [0,1], beginning with 0 and 1, then the fractions with denominator 2, then those with denominator 3, and so on.

Yet another infinity like the natural numbers is suggested by the pavement in figure 12.1. In this painting by Francesco di Giorgio Martini the pavement actually stops after about 20 tiles, but one can *imagine* it continuing indefinitely. If this were done, there would be room for infinitely many paving stones before the horizon—a potential infinity. Thus, in perspective drawing the horizon suggests itself as *the place at*

Figure 12.1 : Potentially infinite pavement

which infinity is completed. Indeed, as we saw in chapter 3, geometers aptly called the horizon the **line at infinity**, so geometers and artists perhaps embraced the completed, or **actual**, infinity before other mathematicians.

Not only that, I believe that a few artists even embraced the idea of going *beyond infinity* by showing objects beyond the horizon. Arguably, this occurs in the famous frontispiece to Thomas Hobbes's *Leviathan* of 1651, drawn by Abraham Bosse (figure 12.2).[1]

Figure 12.2 : The Leviathan beyond the horizon

If one looks closely at the Leviathan, who represents the body politic, one sees that he is composed of many small figures. Thus, the Leviathan is the *embodiment of a set*. The idea of viewing a set as a single entity is another groundbreaking feature of the picture, possibly suggested to Bosse by Hobbes. This particular idea took hold in mathematics only in

---

1. Bosse, as we know from section 3.7, was an authority on perspective drawing. Therefore, I think he would have been well aware that his Leviathan figure was too big to fit in the landscape. But, perhaps to avoid the shock of showing the figure rising from beyond the horizon, he has concealed that part of the horizon behind some low hills.

the nineteenth century, and we will see in section 12.4 that it became the standard way to go "beyond infinity."

12.2 EQUINUMEROUS SETS

The potentially infinite sets in the previous section are **equinumerous**, or of the same **cardinality**, in the sense that there is a bijective correspondence between any two of them. When also, as here, they are equinumerous with $\mathbb{N}$, such sets are called **countable**. Another example is the set $\mathbb{Z}$ of integers, which is in the following correspondence with $\mathbb{N}$:

$$
\begin{array}{ccccccccccc}
0 & 1 & 2 & 3 & 4 & 5 & 6 & 7 & 8 & 9 & \cdots \\
\updownarrow & \updownarrow & \updownarrow & \updownarrow & \updownarrow & \updownarrow & \updownarrow & \updownarrow & \updownarrow & \updownarrow & \cdots \\
0 & 1 & -1 & 2 & -2 & 3 & -3 & 4 & -4 & 5 & \cdots .
\end{array}
$$

We have also seen a correspondence with rational numbers in $[0,1]$:

$$
\begin{array}{ccccccccccc}
0 & 1 & 2 & 3 & 4 & 5 & 6 & 7 & 8 & 9 & \cdots \\
\updownarrow & \updownarrow & \updownarrow & \updownarrow & \updownarrow & \updownarrow & \updownarrow & \updownarrow & \updownarrow & \updownarrow & \cdots \\
0 & 1 & \frac{1}{2} & \frac{1}{3} & \frac{2}{3} & \frac{1}{4} & \frac{3}{4} & \frac{1}{5} & \frac{2}{5} & \frac{3}{5} & \cdots .
\end{array}
$$

Bijective correspondence is the simplest and best way to compare "size" for infinite sets, though it has consequences that seem paradoxical at first. In particular, a set can be equinumerous with a *part* of itself, as $\mathbb{Z}$ is with $\mathbb{N}$.

Dedekind (1888: definition 64), made a virtue of this necessity by making it the *definition* of an infinite set: a set S is infinite if it is in bijective correspondence with a proper subset of itself. The sets with this property certainly cannot be finite, but there is some difficulty establishing the converse, as we will see in section 14.1. Dedekind, even more boldly, ventured to *prove* that infinite sets exist, following an idea similar to that of Bolzano (1851: §13). Their argument concerns the "realm of thoughts," in which every thought t may be paired with t', the thought that "t is thinkable." Since not every thought is of the form t', this is a bijective correspondence between the realm of thoughts and a proper subset; hence the realm is infinite.

This "proof" did not look like mathematics to most people, though Russell (1903: sec. 339) accepted it. To be on the safe side, most mathematicians take existence of infinite sets to be an axiom (see section 13.4 for details).

Countable Sets

Not all infinite sets are countable, as will soon be clear, but it is worth seeing a few ways of "counting" infinite sets, as the methods will be useful when we come to uncountable sets. Here are some important examples of countable sets.

Any infinite subset of $\mathbb{N}$. If $S \subseteq \mathbb{N}$ is infinite we can list its members as $s_0, s_1, s_2, \ldots$, thereby pairing them with the members $0, 1, 2, \ldots$ of $\mathbb{N}$, by letting

$$s_0 = \text{least member of } S$$
$$s_1 = \text{least member of } S - \{s_0\}$$
$$s_2 = \text{least member of } S - \{s_0, s_1\}$$
$$\vdots$$

This simple observation is used in each of the following examples, where we map certain infinite sets to infinite subsets of $\mathbb{N}$.

The set $\mathbb{N} \times \mathbb{N}$. This is the set ordered pairs $\langle a, b \rangle$ where $a, b \in \mathbb{N}$. We can arrange these pairs in a planar array and count them by following a zigzag path through the array, as in figure 11.8. Another way, which extends to arbitrary finite sequences, is to use unique prime factorization: map $\langle a, b \rangle$ to the natural number $2^a 3^b$.

Finite sequences of natural numbers. An ordered n-tuple $\langle k_1, k_2, \ldots, k_n \rangle$ of natural numbers may be encoded by $2^{k_1} 3^{k_2} \cdots p_n^{k_n}$, where p_n is the nth prime. Then each n-tuple is mapped to a distinct natural number, because of unique prime factorization, and n can be of any size, because there are infinitely many primes.

Finite subsets of $\mathbb{N}$. Each nonempty finite subset $F \subseteq \mathbb{N}$, when arranged in increasing order, corresponds to the ordered n-tuple of members of $\mathbb{N}$. This n-tuple is then encoded by a positive integer as above. This leaves the natural number 0 available to represent the empty set.

The next question is whether all infinite sets are equinumerous with $\mathbb{N}$. It is interesting that $\mathbb{N}$ is equinumerous with some dense subsets of $\mathbb{R}$, such as the set $\mathbb{Q}$ of rationals, as we saw in step 5 of section 11.8. An even more extensive set that is equinumerous with $\mathbb{N}$ is the set of algebraic numbers. This was observed by Dedekind in the 1870s.

Algebraic and Transcendental Numbers

It is worth sketching Dedekind's idea for pairing algebraic numbers with the members of $\mathbb{N}$, because it led to a remarkable proof by Cantor (1874) that **transcendental** (that is, nonalgebraic) numbers exist.

By the results in the previous subsection, it suffices to pair each algebraic number α with a unique finite sequence of natural numbers. Certainly, α is one of up to m solutions of an equation

$$a_m x^m + a_{m-1} x^{m-1} + \cdots + a_1 x + a_0 = 0,$$

where $a_0, \ldots, a_{m-1}, a_m \neq 0$ are integers, by the definition of algebraic number from section 7.6. Since any a_i could be negative, we will encode its sign by another natural number s_i, equal to 1 for positive a_i and 0 for negative a_i. Then the equation for α is encoded by the $(2m+2)$-tuple of natural numbers

$$\langle s_0, |a_0|, \ldots, s_{m-1}, |a_{m-1}|, s_m, |a_m| \rangle,$$

and finally α itself is encoded by a $(2m+3)$-tuple,

$$\langle s_0, |a_0|, \ldots, s_{m-1}, |a_{m-1}|, s_m, |a_m|, k \rangle,$$

where α is the kth root of the equation, in some arbitrary ordering of the roots.

Thus the algebraic numbers are countable, by the third example in the previous subsection.

Dedekind's friend Cantor was surprised by this result but quick to exploit it by showing that the set $\mathbb{R}$ of real numbers is *not* countable. We will see some proofs of Cantor's theorem below, but first we will use it, as Cantor did, to prove that transcendental numbers exist.[2] The proof is just one line: *since the algebraic numbers are countable, and the real numbers are not, some real numbers are not algebraic.*

Although the existence of transcendental numbers was already known—first proved by Liouville (1844), and for the number e specifically by Hermite (1873)—Cantor's proof was radically simpler, because it involved no algebra or analysis.

---

2. It seems that Cantor was persuaded by Weierstrass to downplay his uncountability theorem in favor of this application, which was more palatable to the mathematicians of his time. For more about this, see Ferreirós (2007: 183).

The Uncountability of $\mathbb{R}$

The best-known proof that $\mathbb{R}$ is not equinumerous with $\mathbb{N}$ takes a sequence of real numbers $\alpha_0, \alpha_1, \alpha_2, \ldots$, given by decimal expansions, and finds a real number α unequal to each α_n. Thus any set of real numbers equinumerous with $\mathbb{N}$, via the pairing $n \leftrightarrow \alpha_n$, is *not all* of $\mathbb{R}$. We also say $\mathbb{R}$ is **uncountable** because to "count" a set is to pair its members with $0, 1, 2, 3, \ldots$.

$\mathbb{R}$ **is uncountable.** *If $\alpha_0, \alpha_1, \alpha_2, \ldots$ is any sequence of real numbers, then there is a real number $\alpha \neq$ each α_n.*

Proof. Given $\alpha_0, \alpha_1, \alpha_2, \ldots$ we tabulate their decimal expansions (ignoring their integer parts) in an infinite array like that shown in figure 12.3.

$$\alpha_0 = 0.2\ 7\ 5\ 1\ 3\ \cdots$$
$$\alpha_1 = 0.1\ 1\ 1\ 1\ 1\ \cdots$$
$$\alpha_2 = 0.7\ 1\ 7\ 1\ 7\ \cdots$$
$$\alpha_3 = 0.1\ 4\ 1\ 5\ 9\ \cdots$$
$$\alpha_4 = 0.5\ 4\ 3\ 2\ 1\ \cdots$$
$$\vdots$$
$$\alpha = 0.1\ 2\ 1\ 1\ 2\ \cdots$$

Figure 12.3 : Diagonalizing a sequence of real numbers

Then α is defined to be different from each α_n in the nth decimal place. Specifically, if the nth digit of α_n is not 1, make it 1 in α, and if the nth digit of α_n is 1, make it 2 in α. This certainly gives α a decimal expansion different from that of α_n, in the nth place, but α is also a *number* different from α_n. This is because we avoided the digits 0 and 9 in α, which prevents α from having another decimal expansion.

Thus the sequence $\alpha_0, \alpha_1, \alpha_2, \ldots$ does not include all the real numbers. $\qquad \square$

This proof is called the **diagonal argument** because it uses the digits along the diagonal of the array. There are many variations of the argument, also called "diagonal" after this prototype version. For example, one can prove that $\mathbb{N}$ has more subsets than members by tabulating a sequence $S_0, S_1, S_2, \ldots$ of subsets of $\mathbb{N}$ and using the diagonal of the table to produce a subset $S \neq$ each S_n. The way to tabulate sets is shown in figure 12.4.

$$
\begin{array}{c|ccccccc}
 & 0 & 1 & 2 & 3 & 4 & \cdots \\
\hline
S_0 & 1 & 0 & 1 & 0 & 1 & \cdots \\
S_1 & 0 & 0 & 0 & 0 & 0 & \cdots \\
S_2 & 1 & 1 & 1 & 1 & 1 & \cdots \\
S_3 & 1 & 1 & 0 & 0 & 1 & \cdots \\
S_4 & 0 & 1 & 0 & 1 & 0 & \cdots \\
 & \vdots \\
S & 0 & 1 & 0 & 1 & 1 & \cdots
\end{array}
$$

Figure 12.4 : Diagonalizing a sequence of subsets of $\mathbb{N}$

Each S_n is given by a sequence of 0s and 1s, with 1 in the kth place if and only if $k \in S_n$. Thus we would get the table shown if S_0 is the even numbers, S_1 is the empty set, and S_2 is $\mathbb{N}$. The set S, obtained by *switching* all the digits on the diagonal, differs from each S_n with respect to the number n. Hence the list $S_0, S_1, S_2, \ldots$ does not include all subsets of $\mathbb{N}$.

The sequence of 0s and 1s describing S_n is the sequence of values of the **characteristic function** χ_n of S_n, defined by

$$
\chi_n(k) = \begin{cases} 1 & \text{if } k \in S_n \\ 0 & \text{if } k \notin S_n. \end{cases}
$$

More generally, we can tabulate any list of functions $f_0, f_1, f_2, \ldots$ from $\mathbb{N}$ to $\mathbb{N}$ by placing the value of $f_n(k)$ in the table beneath k in the top row. Then there are many obvious ways to define a function $f \neq$ each f_n, for example, by letting $f(k) = f_k(k) + 1$. It is even possible to make f *grow faster* than each f_n, in the sense that

$$f(k)/f_n(k) \to \infty \quad \text{as} \quad k \to \infty.$$

The diagonal argument was actually introduced in this more complicated context by du Bois-Reymond (1875). The argument of Cantor (1874) was not so obviously a diagonal argument, though in some sense it is. To make a real number x different from each x_n, one needs to make x unequal to x_0 in some way, then unequal to x_1, then unequal to x_2, and so on—and this cannot help looking sort of "diagonal." Cantor gave the first really clear version of the diagonal argument in 1891, for characteristic functions.

Cantor's Generalization of the Diagonal Argument

Cantor (1891) generalized the diagonal argument to any set X.

Definition. The set of all subsets of a set X is called the **power set** of X and is denoted by $\mathscr{P}(X)$, or sometimes 2^X.

Cardinality of the power set. *For any set X, there is no bijection between X and $\mathscr{P}(X)$.*

Proof. Suppose there is a bijection $x \leftrightarrow S_x$ between elements $x \in X$ and subsets $S_x \subseteq X$. But then the sets S_x do not include all subsets of X. In particular, they do not include the set S defined by

$$x \in S \quad \text{if and only if} \quad x \notin S_x,$$

because S differs from each S_x with respect to the element x. □

This theorem smashed through the ceiling that was thought to exist in analysis: it is *not* true that all objects can be encoded by natural numbers and sets of natural numbers, or (equivalently) by natural numbers and real numbers. Specifically, there are more *sets* of real numbers than there are real numbers, so any attempt to encode all sets of real numbers *by* real numbers will fail. Likewise, despite the promising discovery that all continuous functions can be encoded by real numbers (section 11.8), it is not possible to encode all real functions by real numbers.

Indeed, there is no ceiling on the size of infinity, since any set X is exceeded in size by $\mathscr{P}(X)$. This discovery has many implications, as we will see in section 12.8.

12.3　SETS EQUINUMEROUS WITH $\mathbb{R}$

Just as there are bijections between $\mathbb{N}$ and some of its subsets, and also between $\mathbb{N}$ and certain sets that contain it, there are also bijections between $\mathbb{R}$ and some of its subsets, and with some sets that contain it.

To start with the simplest, there is a bijection $P' \leftrightarrow P$ between the line and an open semicircle, given by projection from the center O of the semicircle as shown in figure 12.5.

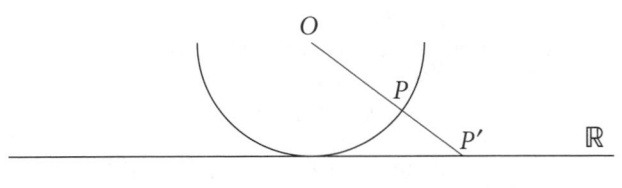

Figure 12.5 : Projecting the semicircle to the line

By projecting the semicircle from a point higher than O, it, and hence also $\mathbb{R}$, can be mapped bijectively onto an open interval, say, $(-1, 1)$. This open interval can be mapped bijectively onto any open interval by a linear function. Thus $\mathbb{R}$ *is equinumerous with any open interval on the line.*

All of these bijections so far are homeomorphisms, but by abandoning continuity we can get bijections between nonhomeomorphic sets, such as open and closed intervals.

For example, to map $[0,1]$ onto $(0,1)$, consider the subset of both,

$$S = \{1/2, 2/3, 3/4, 4/5, \ldots\}.$$

To map $[0,1]$ onto $(0, 1)$, first send 0 and 1 into S at $1/2$ and $2/3$, respectively, then move the rest of S to the right by sending $1/2$ to $3/4$, $2/3$ to $4/5$, $3/4$ to $5/6$, and so on. The remaining points in $[0,1]$ can be sent to themselves.

A similar trick—using a countable subset of a set B as a container for some countable subset of a set A, plus some extra elements—can be used to make a bijective map from the subsets of $\mathbb{N}$ to points in $[0,1]$.

Cardinality of $\mathscr{P}(\mathbb{N})$. $\mathscr{P}(\mathbb{N})$ *is equinumerous with* $[0,1]$*, and hence with* $\mathbb{R}$.

Proof. The subsets of $\mathbb{N}$, when represented by sequences of 0s and 1s as above, are already "almost" the same as points in $[0,1]$, if we represent the latter by binary expansions. The trouble with binary expansions is that, like decimal expansions, there are countably many exceptional numbers with two different expansions. For example, 1/4 is represented by both

$$0.01000000\ldots \quad \text{and} \quad 0.00111111\ldots,$$

and of course the sequences $01000000\ldots$ and $00111111\ldots$ represent different subsets of $\mathbb{N}$, the first finite and the second **cofinite** (complement of a finite set). This trouble can be fixed by mapping the countable collections of finite and cofinite subsets of $\mathbb{N}$ onto the countable collection of numbers with terminating binary expansions.

To be specific, suppose that $\alpha_0, \alpha_1, \alpha_2, \alpha_3, \ldots$ are the numbers in $[0,1]$ with terminating binary expansions. Then let the finite subsets of $\mathbb{N}$ correspond to $\alpha_0, \alpha_2, \alpha_4, \ldots$ and let the cofinite subsets correspond to $\alpha_1, \alpha_3, \alpha_5, \ldots$. □

A nice way to visualize $\mathscr{P}(\mathbb{N})$ is to interpret each sequence of 0s and 1s as a *path in a binary tree*: starting from a single trunk, the tree splits into two infinitely often, and the path corresponding to a given sequence turns left for each 0 and right for each 1. Figure 12.6 gives an idea of the intricate structure of the infinite tree. (I made this using Toby Schachman's online drawing tool at recursivedrawing.com.)

Finally, here is a result of Cantor (1878) that Cantor himself found hard to believe. He wrote to Dedekind in 1877, "I see it but I do not believe it" (quoted in Ferreirós 2007: 188).

Cardinality of the unit square. *The unit square* $[0,1] \times [0,1]$ *is equinumerous with the unit interval* $[0,1]$*, and hence with* $\mathbb{R}$.

Proof. We use the bijection between points in $[0,1]$ and infinite binary sequences just found. Each point in the unit square is an ordered

Figure 12.6 : The tree of infinite binary sequences

pair $\langle \alpha, \beta \rangle$ of real numbers $\alpha, \beta \in [0, 1]$, which correspond to infinite sequences, say,

$$\alpha \leftrightarrow 10010111\ldots$$
$$\beta \leftrightarrow 00111001\ldots.$$

Since these expansions are unique, their digits may be interwoven into a unique single binary sequence γ. With the α and β above,

$$\gamma = 1000011101101011\ldots.$$

Conversely, given a number $\gamma \in [0, 1]$, we split the corresponding binary sequence into two: α, whose digits are in the even-numbered places; and β, whose digits are in the odd-numbered places. Thus we have a bijection between points $\langle \alpha, \beta \rangle \in [0, 1] \times [0, 1]$ and points $\gamma \in [0, 1]$. □

This astonishing result caused alarm at the time, because it challenged the concept of **dimension**. What does "dimension" mean if there is a bijection between the square, of dimension 2, and the line segment, of dimension 1? The bijection is highly discontinuous, so Dedekind realized that the problem might be resolved by requiring dimension to be preserved by *continuous* bijections. But a rigorous result along these lines had to wait until Brouwer (1912) proved that there is no homeomorphism $\mathbb{R}^m \to \mathbb{R}^n$ when $m \neq n$.

12.4 ORDINAL NUMBERS

In section 12.1 we saw how artists have depicted infinity within a finite space by introducing a "horizon" that lies beyond the finite. If one places the natural numbers $0, 1, 2, 3, \ldots$ in the picture, as in figure 12.1, then one can see a point on the horizon that lies beyond them all. We might name this point ω, after the last letter of the Greek alphabet, and indeed ω is the usual name for the "number" that comes after all the natural numbers.

Cantor, in the 1870s, was the first to see a need for numbers beyond the natural numbers, now called **ordinal numbers** or **ordinals**. Cantor needed ordinals to measure the complexity of sets of real numbers. He encountered complicated sets while investigating the discontinuities of integrable functions, and he introduced an operation ′ that simplifies a set S by removing each of its isolated points, that is, each point, of S that has a neighborhood free from other points of S.

For example, if $S = \{0, 1/2, 2/3, 3/4, 4/5, \ldots, 1\}$ then $S' = \{1\}$, and $S'' = \{\}$ (the empty set). Cantor quickly realized that, by inserting infinite sequences between isolated points, he could construct sets to which the ′ operation can be applied any finite number of times. If none of $S, S', S'', S''', \ldots$ is empty, then it is appropriate to define $S^{(\omega)}$ as what remains after any finite number of ′ operations, that is, as

$$S^{(\omega)} = S \cap S' \cap S'' \cap S''' \cap \cdots.$$

Then, if $S^{(\omega)}$ has isolated points, we can apply ′ again, obtaining

$$S^{(\omega)'} = S^{(\omega+1)}.$$

Thus ω is not the last number, by any means! Examples show that we need infinitely many further numbers, $\omega + 1, \omega + 2, \omega + 3, \ldots$, and a number after all of them, called $\omega \cdot 2$.

In fact, any infinite increasing sequence of ordinals has a "horizon," or **limit**, with more ordinals beyond it. Figure 12.7, which is from Wikimedia, shows about as many ordinals as can be easily visualized in this way.

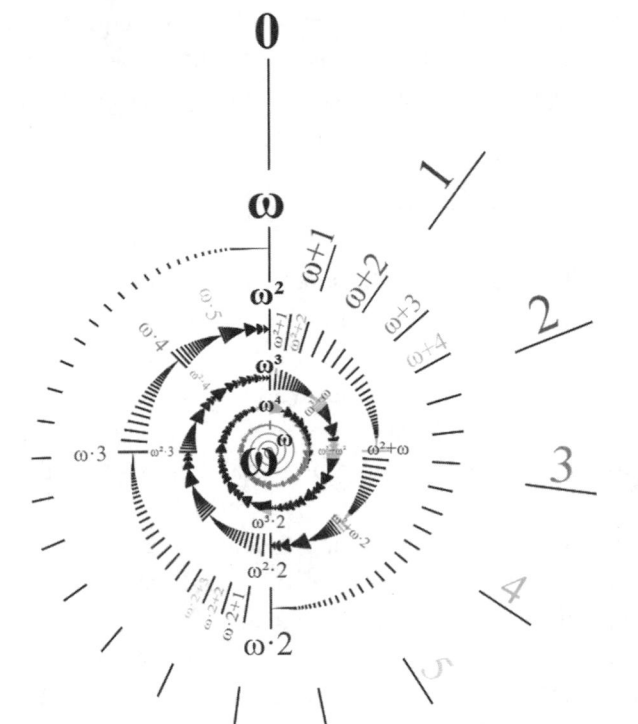

Figure 12.7 : View of the ordinal numbers up to ω^{ω}

Before describing some mathematical realizations of ordinal numbers, we introduce a concept that goes hand in hand with ordinal numbers: the concept of **well-ordering**. This is the strictest among several concepts of ordering.

Definitions. A **partial** ordering is a binary relation, written $a < b$, with the properties that

- $a \nless a$ (irreflexivity) and
- $a < b$ and $b < c$ implies $a < c$ (transitivity).

A partial ordering is a **linear**, or **total**, ordering if also

- $a < b$ or $b < a$ for any a, b in the ordering (symmetry).

Finally, a linear ordering is a **well**-ordering if also

- any set S of elements in the ordering has a least member, that is, an element $l <$ any other member of S (**well-foundedness**).

The loosest concept, partial ordering, is exemplified by the ordering of sets under the inclusion relation ⊂. It is often useful in modern algebra, as section 14.6 will show. Linear ordering, as the name suggests, is exemplified by the left-to-right order of points on a line, and well-ordering by the ordering of $\mathbb{N}$ by the $<$ relation. Well-ordering is the all-important property that underlies the principle of proof by **induction**, introduced in section 2.6. Many sets of numbers are well-ordered by the $<$ relation, some of them being **order isomorphic** to $\mathbb{N}$. For example, the set $0, 1/2, 3/4, 7/8, 15/16, \ldots$ is ordered

$$0 < \frac{1}{2} < \frac{3}{4} < \frac{7}{8} < \frac{15}{16} < \cdots,$$

which is order isomorphic to

$$0 < 1 < 2 < 3 < 4 < 5 < \cdots$$

in the sense that there is the bijective correspondence $n \leftrightarrow 1 - 2^{-n} = f(n)$ such that $a < b$ if and only if $f(a) < f(b)$.

12.5 REALIZING ORDINALS BY SETS

The concept of well-ordering developed before the concept of ordinal, and an ordinal was at first considered the common property of a class of isomorphic well-orderings. In particular, ω was viewed as characteristic of the set $\mathbb{N}$ and any other set order isomorphic to it, such as $\{0, 1/2, 3/4, 7/8, 15/16, \ldots\}$. Sets of rational numbers are particularly good at modeling ordinal numbers in this way; in particular, they can model all the ordinals in figure 12.7.

In fact, sets of rational numbers can model all ordinals generated from 0 by the following two operations:

Successor. From α form its successor $\alpha + 1$.
Sum of an infinite sequence. From a sequence $\alpha_0, \alpha_1, \alpha_2, \ldots$ of ordinals form the sum $\alpha_0 + \alpha_1 + \alpha_2 + \ldots$.

We model these two ordinal operations using well-ordered sets of rationals in the open interval $[0,1)$. We can model the ordinal 1 by any single point, so it remains to show how to model the successor and infinite sum operations.

Successor. Suppose that α is modeled by a set A of rationals in $[0,1)$. Then α is also modeled by the isomorphic set of rationals

$$A' = \left\{ \frac{a}{2} : a \in A \right\} \subset \left[0, \frac{1}{2} \right),$$

and $\alpha + 1$ is modeled by the set $A' \cup \left\{ \frac{3}{4} \right\} \subset [0, 1)$.

Sum of an infinite sequence. Suppose that $\alpha_0, \alpha_1, \alpha_2, \ldots$, respectively, are modeled by sets of rationals $A_0, A_1, A_2, \ldots \subset [0, 1)$. Then $\alpha_0, \alpha_1, \alpha_2, \ldots$ are also modeled by sets $A'_0, A'_1, A'_2, \ldots$, respectively, in the disjoint intervals

$$I_0 = \left[0, \frac{1}{2} \right), \quad I_1 = \left[\frac{1}{2}, \frac{3}{4} \right), \quad I_2 = \left[\frac{3}{4}, \frac{7}{8} \right), \quad \ldots.$$

A'_n is obtained by shrinking the members of A_n by a factor equal to the length of I_n and then translating them into I_n by adding $1 - 2^{-n}$. Then the union

$$A = A'_0 \cup A'_1 \cup A'_2 \cup \cdots$$

is a well-ordered set, because any subset S has a member in an A'_k with minimal k, and $S \cap A'_k$ has a least member by the well-ordering of A'_k. This well-ordered set A represents the sum $\alpha = \alpha_0 + \alpha_1 + \alpha_2 + \cdots$.

Figure 12.8 shows how $\omega^2 = \omega + \omega + \omega + \cdots$ is represented using this construction. The corresponding rational numbers are emphasized by drawing vertical line segments through them. A similar idea is used in figure 12.7, except that the line interval is bent into an infinite spiral to give an additional sense of approaching a limit.[3]

The von Neumann Definition of Ordinals

So far we have modeled ordinals with familiar objects, such as the natural numbers and the rationals. But we need *sets* of these objects, so it would be more economical to use sets from the outset. A supremely economical way to employ sets was introduced by von Neumann (1923): it starts with

---

3. The ordinal number depicted in figure 12.7, ω^ω, actually occurs elsewhere in mathematics. It occurs in three-dimensional hyperbolic geometry: the volumes of the so-called hyperbolic three-manifolds form a set of numbers of order type ω^ω. This is a theorem of Thurston and Jørgensen first published in Gromov (1981).

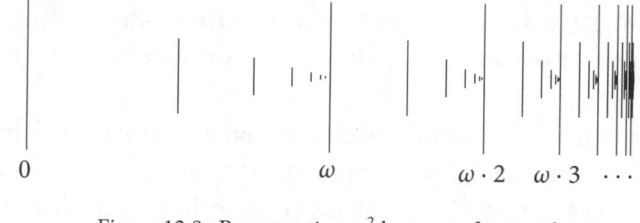

Figure 12.8 : Representing ω^2 by a set of rationals

the empty set $\{\}$ and uses the set formation process itself to build all the ordinals. First, the finite ordinals $0, 1, 2, 3, \ldots$ are defined by

$$0 = \{\}$$
$$1 = \{0\}$$
$$2 = \{0, 1\}$$
$$3 = \{0, 1, 2\}$$
$$\vdots$$

In other words, the **finite ordinal** $n + 1$ *is the set* $\{0, 1, 2, \ldots, n\}$ *of all the preceding ordinals.* This definition has the following bonus properties:

- The "less than" relation is set membership: $m < n$ if and only if $m \in n$.
- The successor function is easily defined: $n + 1 = n \cup \{n\}$.

The step to infinite ordinals is almost as easy, and it maintains the properties of $<$ and successor:

$$\omega = \{0, 1, 2, 3, \ldots\}$$
$$\omega + 1 = \{0, 1, 2, 3, \ldots, \omega\}$$
$$\omega + 2 = \{0, 1, 2, 3, \ldots, \omega, \omega + 1\}$$
$$\vdots$$
$$\omega \cdot 2 = \{0, 1, 2, 3, \ldots, \omega, \omega + 1, \omega + 2, \ldots\},$$

and so on. Ordinals such as ω and $\omega \cdot 2$, which are not successors, are called **limit ordinals**. Like other ordinals, they are the sets of all their predecessors. Thus ω, the first infinite ordinal, is the set of all the finite ordinals. The general definition of ordinal, which embraces all the preceding examples, is the following:

Definition. An **ordinal** is a set that is **transitive**—that is, any member of a member is itself a member—and linearly ordered by the $\in$ relation.

It follows from this definition that members of an ordinal are themselves ordinals. It also follows, by the foundation axiom of set theory we will meet later (section 13.4), that any set of ordinals has a least member and hence that any ordinal is *well*-ordered by the $\in$ relation. Thus ordinals represent well-ordered sets, and it turns out that each class of isomorphic well-orderings contains a unique ordinal. Another convenience of von Neumann ordinals is their **least upper bound** operation, which is simply union (like the least upper bound operation for lower Dedekind cuts). For example, ω is the least upper bound of $0, 1, 2, 3, \ldots$, and it is also their union.

It is an axiom of set theory that any set of sets has a union, so any set of ordinals has a least upper bound. This idea crystallizes an intuition from the genesis of the ordinal concept in Cantor (1883). He had no precise definition, but he insisted that every ordinal have a successor, and that every set of ordinals have a least upper bound. In particular, with von Neumann ordinals it becomes clear that Cantor's concept of a **countable ordinal** is meaningful, and that there is a set of all countable ordinals. Then the least upper bound of this set is the least *uncountable* ordinal, which is called ω_1.

The long road to the uncountable we have just followed, via ordinals, is very different from the short road via the diagonal argument. However, it has the advantage of showing us the *smallest* uncountable set, ω_1. Whether ω_1 is equinumerous with $\mathbb{R}$ is the most famous problem in set theory. First raised by Cantor (1883), it is known as the **continuum problem**. When Hilbert (1900) posed mathematical problems for the twentieth century, the continuum problem was first on the list (see Hilbert 1902: 445).

12.6 ORDERING SETS BY RANK

As we know from Cantor's general diagonal argument (section 12.2), larger and larger sets can be generated by applying the power set operation $\mathscr{P}$. Ordinals can "count" the number of applications of the power set operation, and thereby assign an ordinal number to each set, called its **rank**.

Definitions. Sets V_α, whose members are called the sets of **rank** less than α, are defined inductively for all ordinals α as follows:

- $V_0 = \{\}$ (the empty set).
- $V_{\alpha+1} = \mathscr{P}(V_\alpha)$.
- $V_\lambda = \cup_{\beta<\lambda} V_\lambda$, for each limit ordinal λ.

We say that $\{\}$ has rank 0, and otherwise the rank of a set is the least ordinal greater than the ranks of its members.

Thus 1 has rank 1, any finite ordinal n has rank n, and ω has rank ω. V_ω includes the finite ordinals, but also all finite sets of ordinals, all finite sets of such sets, and so on. The sets in V_ω are called **hereditarily finite**. A step beyond them is ω, which calls to mind Abraham Bosse's picture of the Leviathan again: ω is the set of finite sets that "count," namely, the finite ordinals (see figure 12.9).

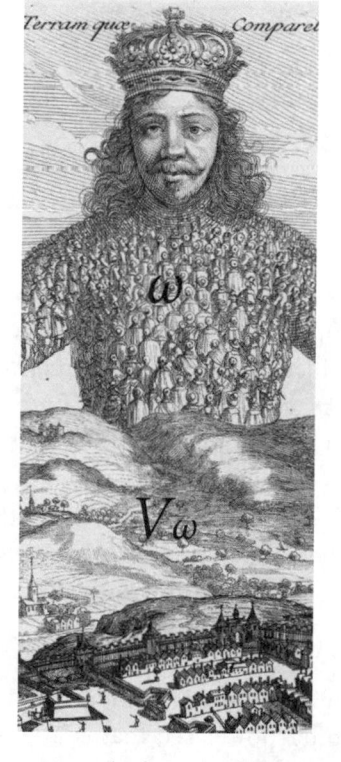

Figure 12.9 : The finite world and beyond

12.7 INACCESSIBILITY

In our informal description of the sets V_α we have tied the existence of these sets to the existence of ordinals, which themselves depend on the existence of certain sets. In particular, we have assumed the existence of the infinite set $\omega = \{0, 1, 2, \ldots\}$, the existence of a power set $\mathscr{P}(X)$ for any set X, and the existence of the union of a set of sets. All of these are natural assumptions, and they are part of the usual axiom system for set theory that we will discuss more fully in section 13.4. A more subtle assumption, which was overlooked in the first axiom systems for set theory, is *existence of the range of a function*, which is called the **replacement** axiom.

The replacement axiom comes into play when we try to prove existence of the ordinal

$$\omega \cdot 2 = \{0, 1, 2, \ldots, \omega, \omega + 1, \omega + 2, \ldots\}.$$

We have assumed that $\omega = \{0, 1, 2, \ldots\}$ exists, but here we also need the set

$$\{\omega, \omega + 1, \omega + 2, \ldots\},$$

which is the range of the function $f(n) = \omega + n$. So $\omega \cdot 2$, which is the union of the sets in the range of f, exists by virtue of the replacement axiom. Larger ordinals α, such as

$$\omega \cdot 3, \omega \cdot 4, \ldots, \omega^2, \omega^3, \ldots, \omega^\omega, \ldots,$$

exist for similar reasons, and with them the corresponding sets V_α.

The replacement axiom, in conjunction with the axioms of infinity, power set, and union, gives a dizzying universe of sets, large enough to model any of the objects usually considered in mathematics: natural numbers, real numbers, complex numbers, functions, sets of real numbers or functions, and much more. But we might ask, Is there a V_α that satisfies all the axioms of set theory?

I know, we haven't seen all the axioms of set theory yet, but some of them are automatically satisfied by most of the sets V_α. The hard axioms to satisfy simultaneously are **infinity**, **power set**, and **replacement**.

Definition. A set V_α (and with it the ordinal α) is called **inaccessible** if

- V_α has an infinite member (which means $\alpha > \omega$);
- V_α is closed under power set; that is, if $X \in V_\alpha$ then $\mathscr{P}(X) \in V_\alpha$; and

- V_α is closed under replacement; that is, for any function f and $X \in V_\alpha, f(X) \in V_\alpha$.

Thus V_α is "inaccessible" in the sense that we cannot exhaust it by repeated application of the power set and replacement operations. It seems difficult to put our hands on any inaccessible set, and indeed Zermelo in 1928 observed that *if an inaccessible set exists, its existence is **not** provable from the axioms of set theory!* This result was reported in Baer (1928).

Zermelo's argument went like this. Suppose that V_α is inaccessible, so V_α satisfies the axioms of set theory. We can also suppose that α is the *least* ordinal with this property, because any set of ordinals has a least member. Then *no $V_\beta \in V_\alpha$ is inaccessible, because $\beta < \alpha$ for any $V_\beta \in V_\alpha$.* But then V_α satisfies not just the axioms of set theory; V_α *also satisfies the sentence "there is no inaccessible set,"* because V_α has no inaccessible member. Therefore, assuming the axioms of set theory are consistent, they do not prove the existence of an inaccessible set.   □

This result of Zermelo was stunning, because one expects the axioms of set theory to be consistent, and hence to have a model, which should be an inaccessible set. Thus we *expect* inaccessible sets to exist, and it is a shock not to be able to prove it. Zermelo's result was the first theorem of this kind, but many more examples were to follow, mainly in the work of Gödel we will study in chapter 16.

One branch of mathematics that not only expects inaccessible sets to exist, but actually *demands* them is **category theory**, mentioned in section 10.9. For more on this, see Krömer (2007) and Marquis (2009).

12.8 PARADOXES OF THE INFINITE

In a book called *Paradoxien des Unendlichen (Paradoxes of the Infinite)*, Bolzano (1851) collected many of the facts about the infinite then considered paradoxical. Like most of us today, he took the view that these properties—such as equinumerosity of a set and a subset—are not "paradoxical" so much as characteristic of infinity. They are not a bug but a feature.

The "easy" paradoxes, such as the equinumerosity of $\mathbb{N}$ and the set of squares, and of the open interval $(0,1)$ and the whole line, are resolved by a better understanding of the concept of "size" for infinite sets. In particular, it is natural for an infinite set to be in one-to-one correspondence with a proper subset, so infinite sets demand a more elastic concept of "size," under which sets in one-to-one correspondence are deemed to have the *same* size.

Another class of paradoxes involve infinite constructions in geometry, where a solid of infinite length may have finite volume, a region of infinite length may have finite area, and a continuous curve may have no tangents. These intuitively surprising facts are explained by clarifying the definitions of "length," "volume," "continuity," and "tangent." As we saw in chapter 11, these definitions ultimately rest on the definition of real numbers. There is not really a crisis of intuition, as some mathematicians thought, but rather a need to *hone our intuition* on more complicated examples.

Harder paradoxes arise from the diagonal argument. Among other things, it shows that there is no "set of all sets" because no set S is the largest set—its power set $\mathscr{P}(S)$ is always larger. But this situation, again, is not so much paradoxical as characteristic of the nature of sets. Because of the power set operation, there is no ceiling to the size of sets. There is no more reason to have a largest set than there is to have a largest natural number.

While all the paradoxes arising so far have been resolved by clarifying the concepts of set and infinity, potential problems remain with the existence of very large sets, such as inaccessibles. As we have just seen, such sets cannot be proved to exist, and their existence is entangled with the **consistency** of set theory—or what you might call its freedom from paradox. This question leads us deep into logic, as we will see in the next four chapters.

12.9 REMARKS

Cantor's discovery of uncountable sets led him to the concept of **cardinal number**, which is a common characteristic of a class of equinumerous sets. The **finite cardinal numbers** are simply the natural numbers $0, 1, 2, 3, \ldots$, which measure the size of finite sets, and their arithmetic reflects certain operations on finite sets. For example:

- The sum $m + n$ is the cardinal number of the union of disjoint sets of size m and n.

- The product mn of $m \times n$ is the cardinal number of the **Cartesian product,**

$$S \times T = \{\langle s, t \rangle : s \in S, t \in T\},$$

of a set S of size m with a set T of size n.
- The power 2^n is the cardinal number of the power set $\mathscr{P}(S)$ of a set S with cardinal number n. More generally, m^n is the cardinal number of the set of functions $f : S \to T$, where S has cardinal number n and T has cardinal number m.

Motivated by these operations on finite sets, Cantor assigned the cardinal number $\aleph_0$ ("aleph null" or "aleph nought" or "aleph zero") to the infinite set $\mathbb{N}$, and investigated its arithmetic. The equinumerosity results of section 12.2 are easily adapted to show that

$$\aleph_0 + \aleph_0 = \aleph_0 \quad \text{and} \quad \aleph_0 \times \aleph_0 = \aleph_0,$$

so cardinal sums and products do not lead to anything new. However, the uncountability of $\mathscr{P}(\mathbb{N})$ shows that

$$2^{\aleph_0} > \aleph_0.$$

This raised the **continuum problem,** mentioned in section 12.4 in connection with the first uncountable ordinal ω_1. Cantor gave the set ω_1 the cardinal number $\aleph_1$, and his **continuum hypothesis** can be written as the equation

$$2^{\aleph_0} = \aleph_1.$$

However, this is where cardinal arithmetic runs into trouble. The first problem is how to extend the sequence of alephs beyond $\aleph_1$. Cantor (1883) tried to solve this problem by assuming that every set can be **well-ordered** or, equivalently, that every set is equinumerous with an ordinal. Under this assumption, there is an aleph, $\aleph_\alpha$, for every ordinal α, and every set is equinumerous with an aleph. But even with this assumption, which is equivalent to the **axiom of choice** studied in the next chapter, one does not know *which* aleph equals $2^{\aleph_0}$. The best we can do, using the equinumerosity arguments of section 12.3, is prove results like

$$2^{\aleph_0} + 2^{\aleph_0} = 2^{\aleph_0} \times 2^{\aleph_0} = \aleph_0^{\aleph_0} = 2^{\aleph_0}.$$

Many equinumerosity arguments can be simplified with the help of the following theorem.[4]

---

4. The names of Dedekind and Schröder are often attached to this theorem as well, but Ferreirós (2007) settles on Cantor and Bernstein after a detailed study of its history.

Cantor-Bernstein theorem. *If A and B are infinite sets and there are injections from A into B and B into A, then there is a bijection of A onto B.*

Proof. Suppose that $f : A \to B$ and $g : B \to A$ are injective functions. Without loss of generality we can assume that A and B are disjoint. We take the members a of A and members b of B as vertices of a graph $\mathcal{G}$, with an edge from each $a \in A$ to $f(a) \in B$ and an edge from each $b \in B$ to $g(b) \in A$.

Since f and g are injective, each $a \in A$ equals $g(b)$ for at most one $b \in B$, and each $b \in B$ equals $f(a)$ for at most one $a \in A$. It follows that the vertices of $\mathcal{G}$ fall into disjoint subsets lying along zigzag paths like that shown in figure 12.10. Since A and B are infinite, each such path has no end in the direction of the arrows, but there are three possibilities for its beginning.

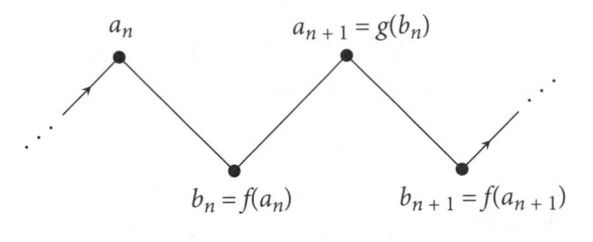

Figure 12.10 : A zigzag path

- The path has an initial vertex in A. In this case we define $h(a) = f(a)$ for each $a \in A$ in the path. Then h is a bijection of the subset of A in the path onto the subset of B in the path.
- The path has an initial vertex in B. In this case we let $h(a) = g^{-1}(a)$ for each $a \in A$ in the path, so h is again a bijection of the subset of A in the path onto the subset of B in the path.
- The path has no initial vertex. In this case we let $h(a) = f(a)$ for each $a \in A$ in the path, so h is again a bijection of the subset of A in the path onto the subset of B in the path.

It follows, since the paths are disjoint and together they include all elements of A and B as vertices, that $h : A \to B$ is a bijection of A onto B. □

The idea of using graph theory to prove the theorem seems to stem from Kőnig, and it led him to his infinity lemma in Kőnig (1927); see Franchella (1997).

As an application of the Cantor-Bernstein theorem, here is another proof of the result from section 12.3 that $(0,1)$ is equinumerous with $\mathscr{P}(\mathbb{N})$.

We define an injection of $(0,1)$ into $\mathscr{P}(\mathbb{N})$ by taking, for each $x \in (0, 1)$, the binary expansion of x with infinitely many zeros. For example, for $x = 1/2$ we choose the expansion $0.10000\ldots$ rather than $0.01111\ldots$. Then interpret the chosen expansion as the characteristic function of a subset of $\mathbb{N}$ (in this case, the set $S = \{0\}$).

We define an injection of $\mathscr{P}(\mathbb{N})$ into $(0, 1)$ by taking, for each $S \in \mathscr{P}(\mathbb{N})$, the number x whose decimal has 1 in the nth place if $n \in S$, and 2 in the nth place if $n \notin S$. This map is an injection because it avoids the digits 0 and 9 occurring in numbers x with nonunique decimal expansions.

The Dimensions of $\mathbb{R}$ and $\mathbb{R}^2$

As mentioned at the end of section 12.3, Brouwer (1912) proved that there is no *continuous* bijection between $\mathbb{R}^m$ and $\mathbb{R}^n$ when $m \neq n$. For general m and n this theorem on **invariance of dimension** is quite difficult, but we can prove there is no continuous bijection between $\mathbb{R}$ and $\mathbb{R}^2$ quite simply, using the intermediate value theorem.

Suppose on the contrary that $h : \mathbb{R}^2 \to \mathbb{R}$ is a continuous bijection and that $h(\langle 0,0 \rangle) = c$. Then h is also a continuous bijection $\mathbb{R}^2 - \{\langle 0,0 \rangle\} \to \mathbb{R} - \{c\}$. However, we can see that $\mathbb{R}^2 - \{\langle 0,0 \rangle\}$ has a property that $\mathbb{R} - \{c\}$ does not have: it is **path-connected**. That is, any two points in $\mathbb{R}^2 - \{\langle 0,0 \rangle\}$ may be connected by a continuous path in $\mathbb{R}^2 - \{\langle 0,0 \rangle\}$. On the other hand, $\mathbb{R} - \{c\}$ is *not* path-connected; for example, there is no continuous path from a point $a < c$ to a point $b > c$ by the intermediate value theorem.

Since a continuous bijection obviously sends paths to paths, this is a contradiction. It follows that there is no continuous bijection from $\mathbb{R}^2 - \{\langle 0,0 \rangle\}$ to $\mathbb{R} - \{c\}$, and hence also none from $\mathbb{R}^2$ to $\mathbb{R}$.

■ ■ ■ ■ ■

CHAPTER 13

Axioms for Numbers, Geometry, and Sets

Until the late nineteenth century, Euclid's *Elements* remained the only axiomatic development of mathematics in common use. After Beltrami found models of non-Euclidean geometry in 1868, Euclid's axioms lost their privileged position, and mathematicians sought new foundations of mathematics, first in arithmetic and later in set theory. We saw this in chapters 11 and 12.

The change of viewpoint led to several new axiom systems between 1888 and 1908. First were axioms for arithmetic, proposed independently by Peano and Dedekind in 1888. At about the same time, Peano proposed a new foundation of Euclidean geometry by means of axioms for a real vector space with an inner product. Peano's vector space axioms took the real numbers as given, but Hilbert (1899) gave a remarkable *geometric* foundation for the real numbers in his axioms for geometry.

Meanwhile, Cantor's development of set theory revealed unexpected complexity in the real numbers. There were also paradoxes about such things as the "set of all sets." This raised a demand for set theory axioms, which were given by Zermelo (1908) and supplemented by Fraenkel (1922).

All of these axiom systems were intended to *completely characterize* numbers, geometry, and sets, encapsulating each by a finite set of sentences. In contrast, the axioms for groups, rings, and fields—which also developed in this period—are *defining properties* of the concepts in question. The question of *consistency* does not arise with the group, ring, and field axioms, and in fact they all have trivial models. In the case of axioms for natural numbers and sets, however, consistency is a serious question.

13.1 PEANO ARITHMETIC

Sections 2.6 and 4.6 touched on the role of induction in number theory, and mentioned that Grassmann (1861) used it to prove the commutative, associative, and distributive properties of addition and multiplication in $\mathbb{N}$. In doing so he saw deeper than the most eminent number theorists. Even Dirichlet, whose work was considered the last word in rigor by his contemporaries, was content to "prove" that $ab = ba$ on the first page of his textbook (Dirichlet 1863) by considering a rectangular array of width a and height b.

Apparently the only mathematicians to appreciate the depth of Grassmann's thinking were Peano and Dedekind, who almost simultaneously built arithmetic on the foundation of induction in 1888. First we will discuss the axioms, now known as the **Peano axioms**, and then look at Dedekind's investigation of the nature of induction itself. Peano sees the natural numbers generated from 0 by the **successor** function, and the values of sum and product functions generated from their values for smaller natural numbers, ultimately from their values at 0. Using s to denote the successor function, and + and · for sum and product, the Peano axioms are:

1. 0 is not a successor.
2. Numbers with the same successor are equal.
3. For all m and n, $m + 0 = m$ and $m + s(n) = s(m + n)$ (inductive definition of sum).
4. For all m and n, $m \cdot 0 = 0$ and $m \cdot s(n) = m \cdot n + m$ (inductive definition of product).
5. If a set X is such that $0 \in X$ and $n \in X$ implies $s(n) \in X$, then X includes all natural numbers (induction axiom).

The induction axiom is the basis for **proof by induction**, whereby one proves a property P is true for all m by proving that the *set* $\{m : P(m)\}$ includes all numbers. Thanks to the induction axiom, it suffices to prove that

- P holds for 0 (the **base step**), and
- if P holds for n then it holds for $s(n)$ (the **induction step**).

Essentially the Peano axioms were used by Grassmann (1861) to give inductive proofs of the basic rules of arithmetic, such as $a + b = b + a$ and $ab = ba$. The proofs are not difficult, though they form quite a long sequence and some care is needed to get them in the right order. To give an idea how induction is involved in even the simplest properties, here is one of the early steps in the sequence (Grassmann's proposition 20):

Adding a unit. *For any natural number m, $m + 1 = 1 + m$, where $1 = s(0)$.*

Proof. By the definition of sum, $m + 1 = m + s(0) = s(m + 0) = s(m)$.

We now use induction on m to prove that $1 + m = s(m)$ also. This statement is true for $m = 0$ because $1 + 0 = 1 = s(0)$ by the definition of the sum function. Now suppose that the statement is true for $m = n$, in other words, that $1 + n = s(n)$. We wish to prove that the statement is also true for $m = s(n)$, that is, that $1 + s(n) = s(s(n))$.

Well, $1 + s(n) = s(1 + n)$ by definition of sum, and $1 + n = s(n)$ by our induction hypothesis. So $1 + s(n) = s(s(n))$, as required. This completes the induction step, and hence the proof. □

Induction is often a stumbling block for mathematics students, perhaps because a complicated induction hypothesis is often pulled out of the air, and the student wonders, How could I think of that? A mild example is the formula, encountered in section 6.6,

$$1 + 2 + 3 + \cdots + n = \frac{n(n+1)}{2}.$$

This formula can be proved by induction on n, but one would prefer to know where the expression $n(n + 1)/2$ comes from. This is done by writing down the sum $1 + 2 + 3 + \cdots + n$ and, beneath it, the same sum in reverse:

$$
\begin{array}{ccccccccccccc}
1 & + & 2 & + & 3 & + & \cdots & + & (n-2) & + & (n-1) & + & n \\
n & + & (n-1) & + & (n-2) & + & \cdots & + & 3 & + & 2 & + & 1.
\end{array}
$$

This new sum can be viewed as n columns, each of which sums to $n + 1$, so the new sum is $n(n + 1)$ and the old sum $1 + 2 + 3 + \cdots + n$ is half of it: $n(n + 1)/2$. This proof is certainly clearer, and it explains where the answer comes from. It also seems, at first glance, to avoid induction—but it doesn't! It has actually replaced a complicated induction hypothesis by one so simple it seems obvious: that the sum of the ith column is $n + 1$ for $i = 1, 2, \ldots, n$. Our belief that the ith column sums to $n + 1$ for each i is really an unconscious induction on i.

So, the moral of this example is not that induction can be avoided but, rather, that it can sometimes be drastically simplified.

Dedekind on Induction

The definitions of sum and product in the Peano axioms are "by induction" in the sense that the induction axiom proves that sum and product are defined uniquely for all natural numbers m and n. Thus the first clause $m + 0 = m$ defines the sum $m + k$ uniquely for all m and for $k = 0$. Then the second clause $m + s(n) = s(m + n)$ defines $m + k$ uniquely for $k = s(n)$, assuming it is already defined uniquely for $k = n$. A similar argument then applies to the definition of product, which presupposes the definition of sum.

Dedekind (1888) recognized the importance of definition by induction, or **definition by recursion** as it is generally called today, and sought to ground it in more fundamental concepts. He asked in a letter (Dedekind 1890), How can we be sure that any interpretation of the Peano axioms is essentially the same as our intuitive conception of the natural numbers?

Certainly, any interpretation must include objects denoted by

$$0, s(0), s(s(0)), \ldots,$$

and these objects are distinct by axioms 1 and 2. The object denoted by 0 can be anything, and the function denoted by s can be any injective function that does not take the value 0, but apart from that, can there be what Dedekind called "alien intruders"? These are objects *not* in the set obtained from 0 by finitely many applications of the function s. The induction axiom is precisely what rules out "alien intruders": the set $X = \{0, s(0), s(s(0)), \ldots\}$ of "nonaliens" satisfies the hypothesis of the induction axiom, and therefore it includes everything.

Thus any model of the Peano axioms must include an object denoted by 0, distinct nonzero objects denoted by $s(0), s(s(0)), s(s(s(0))), \ldots$, and *no other* elements ("alien intruders"). This means the model is essentially the same as $\mathbb{N}$, as we hoped. We say that the Peano axioms are **categorical**, meaning that all their models are essentially the same. However, this conclusion depends on the induction axiom being **second order**, that is, a statement about all *sets* X of numbers. In this respect, the induction axiom differs from the other axioms, which are about all *numbers*. Such axioms are called **first order**.

For many purposes, it is convenient to replace Peano's induction axiom by first-order axioms asserted to hold only for all numbers. To do so, we use a collection of infinitely many axioms (called a **schema**), one for each property $\Phi(n)$ that can be written in the language of arithmetic. We skip the details of the language and simply state what the first-order induction schema looks like:

If Φ holds for 0, and if Φ holds for $s(n)$ whenever it holds for n, then Φ holds for all numbers.

Mathematicians generally prefer first-order axioms, because the logic of first-order statements is better behaved and better understood. We explain why in chapter 15. However, we will see that the first-order induction schema allows "alien intruders" to creep back in! Luckily, this is not necessarily a bad thing, as the "alien intruders" can serve as "infinite numbers," and their reciprocals as "infinitesimals." We will say a little more about this in the Remarks at the end of this chapter.

13.2 GEOMETRY AXIOMS

Chapter 3 described the development of geometric axioms from Euclid to Hilbert. The system of Hilbert (1899) was virtually the last word on traditional axiom systems for both Euclidean and hyperbolic geometry. By including enough axioms to ensure that the line was essentially the same as $\mathbb{R}$, Hilbert was able to show that his axioms were categorical. In particular, any model of his axioms for the Euclidean plane is essentially the same as $\mathbb{R}^2$ with the Pythagorean distance function, as we saw in section 3.5.

Chapter 5 also described the linear algebra approach to geometry, which began with Grassmann (1844) and took axiomatic form with the Peano (1888) axioms for Euclidean vector spaces (section 4.8). These axioms also have essentially only one model—$\mathbb{R}^n$ with the Pythagorean distance function—so they are categorical, like Hilbert's. However, they have the advantage of applying to any dimension n, whereas Hilbert's axioms are special for dimension 2.

Another advantage of the vector space approach is that it is open to different kinds of inner product, an option that turned out to be valuable in the early twentieth century.

Inner Product Spaces

Grassmann's inner product, defined for n-dimensional vectors

$$\boldsymbol{u} = \langle u_1, u_2, \ldots, u_n \rangle \quad \text{and} \quad \boldsymbol{v} = \langle v_1, v_2, \ldots, v_n \rangle$$

by

$$\boldsymbol{u} \cdot \boldsymbol{v} = u_1 v_1 + u_2 v_2 + \cdots + u_n v_n,$$

is natural for Euclidean geometry because it gives the Pythagorean length

$$|\boldsymbol{u}| = \sqrt{\boldsymbol{u} \cdot \boldsymbol{u}} = \sqrt{u_1^2 + u_2^2 + \cdots + u_n^2}.$$

Grassmann (1847) made this point. The option of a *different* inner product, allowing some lengths to be negative, arose early in the twentieth century with Einstein's special relativity theory. Minkowski (1908) noticed that in Einstein's four-dimensional space-time of vectors $\langle x, y, z, t \rangle$, where x, y, z are space coordinates and t is time, it is natural to assign length by the formula

$$|\langle x, y, z, t \rangle|^2 = x^2 + y^2 + z^2 - t^2,$$

which corresponds to an inner product defined by

$$\langle x_1, y_1, z_1, t_1 \rangle \cdot \langle x_2, y_2, z_2, t_2 \rangle = x_1 x_2 + y_1 y_2 + z_1 z_2 - t_1 t_2.$$

In **Minkowski space**, as it is called, it is obviously possible for $|\langle x, y, z, t \rangle|^2$ to be negative, in which case $|\langle x, y, z, t \rangle|$ is imaginary.

The same is true in the three-dimensional space of vectors $\langle x, y, t \rangle$ with

$$\langle x_1, y_1, t_1 \rangle \cdot \langle x_2, y_2, t_2 \rangle = x_1 x_2 + y_1 y_2 - t_1 t_2.$$

This space is not so relevant to physics, but it is remarkably relevant to geometry, because it contains an object first dreamed of by Lambert (1766): a *sphere of imaginary radius*. Lambert speculated, purely by analogy with the ordinary sphere and formulas of spherical trigonometry, that non-Euclidean (hyperbolic) geometry might hold on such a sphere. This turns out to be true!

The sphere of radius $\sqrt{-1}$ in this space has equation

$$x^2 + y^2 - t^2 = -1 \quad \text{or} \quad t^2 - x^2 - y^2 = 1,$$

which is what we would normally call a *hyperboloid*. When the Minkowski distance is restricted to the surface of the hyperboloid (the same way we get distance on the ordinary sphere from the Pythagorean distance function in $\mathbb{R}^3$, using arc length), then the hyperboloid becomes a model of the hyperbolic plane. Indeed, this model was already known to Poincaré (1881).

Figure 13.1, by Konrad Polthier at the Freie Universität Berlin, shows how geometry on the hyperboloid is related to the conformal disk model of hyperbolic geometry. It shows a non-Euclidean tiling of the disk, by triangles with angles $\frac{\pi}{2}$, $\frac{\pi}{3}$, and $\frac{\pi}{7}$, projected to a tiling of the hyperboloid by triangles that are congruent under the Minkowski distance function.

Figure 13.1 : The hyperboloid model

13.3 AXIOMS FOR REAL NUMBERS

Grassmann's and Peano's setting for geometry as a real vector space with an inner product takes the real numbers as given. Hilbert was the first to realize that the real numbers themselves needed an axiomatic foundation. As we have just seen, he gave one such foundation, built on geometry with the help of the Archimedean and completeness axioms. An equivalent, more algebraic foundation is provided by the concept of a complete ordered field, as we saw in section 3.4.

However, Hilbert also realized that writing down axioms for $\mathbb{R}$ is not the end of the story. $\mathbb{R}$ is a highly infinite structure, and we should not be content simply to assume that it exists. Rather, we should attempt to *guarantee* the existence of $\mathbb{R}$ by proving that its axioms are consistent. He posed this as problem 2 in a list of mathematical problems for the twentieth century he presented to the mathematical community in Hilbert (1900). In principle, having axioms for $\mathbb{R}$ seems like half the battle. We have a finite set of axioms, each of which is a finite string of symbols, and the rules of logic (as we will see in section 15.5) are basically computations on finite strings of symbols. To prove consistency we have only to prove that such a computation does not produce a contradictory output, such as the string "$0 = 1$." This was the **Hilbert program** for proving the consistency of reasoning about infinity.

Asking whether a computation produces a certain output sounds like a *much* easier question than the existence of the uncountable structure $\mathbb{R}$. And perhaps it is, but nevertheless there is no general procedure for answering such questions, as we will see in section 15.5. Indeed, the consistency of axioms for $\mathbb{R}$ still does not have a proof that Hilbert would have accepted.

This raises the question, Given the difficulties around axioms for $\mathbb{R}$, is it a good idea to make $\mathbb{R}$ part of the foundations of geometry? As already mentioned in section 11.9, Euclid's geometry really needs only the constructible numbers. A careful development of Euclidean geometry based on the constructible numbers may be found in Hartshorne (2000). They form a more tractable, countable set and in fact it is possible to prove the consistency of Euclidean geometry if we drop the Archimedean and completeness axioms.

Tarski (1948) gave a general procedure for answering questions in a system that includes traditional Euclidean geometry in a certain sense, and he thereby showed the consistency of geometry without the Archimedean and completeness axioms. His procedure is able to decide whether a polynomial equation or inequality in any number of variables has a real solution. It therefore can answer questions not only about lines and circles (linear and quadratic equations), as required for Euclidean geometry, but also about real algebraic curves of any degree.

The catch is that Tarski's system cannot distinguish natural numbers from numbers in general, as Hilbert's system does to even *state* the Archimedean axiom. So Tarski's system cannot express some traditional geometric questions, such as: for which n is the regular n-gon constructible?

As we will see in section 15.5, questions about the natural numbers are the core of all difficulties in mathematics.

13.4 AXIOMS FOR SET THEORY

Broadly speaking, the axioms of set theory are like the Peano axioms for arithmetic. The Peano axioms have a starting object, 0, the successor function for generating new objects, and induction. The **Zermelo-Fraenkel axioms** for set theory have a starting object, the empty set $\{\}$, plus power set and replacement for generating new objects, and an induction axiom called **foundation**. The crucial difference—without which set theory would be essentially the same as Peano arithmetic—is the **axiom of infinity**, which asserts the existence of an infinite set.

Here is the precise list of Zermelo-Fraenkel axioms, ZF, with some comments on how they capture some informal ideas about sets. In particular, they formalize the ideas about ordinals sketched in section 12.4. One of the axioms is a schema of infinitely many formulas, one for each function that can be defined in the language of ZF. We will not go into detail about the language, except to say that it has just one type of variable, for sets, and symbols for just two relations between sets: equality, =, and membership, $\in$.

Extensionality. Sets are equal if they have the same members.
> This implies in particular that $\{1, 2\} = \{2, 1\} = \{1, 1, 2\}$ because all of these sets have the same members: 1 and 2.

Empty set. There is a set $\{\}$ with no members.
> It follows from extensionality that there is only one empty set. As we saw in section 12.4, $\{\}$ can serve as the number 0.

Pairing. For any sets x, y there is a set whose members are x and y.
> This is the set we denote by $\{x, y\}$. If $x = y$ it is the singleton set $\{x\}$ by extensionality. Since $\{x, y\} = \{y, x\}$, by extensionality again, $\{x, y\}$ is not an *ordered* pair. However, $\{\{x\}, \{x, y\}\}$ can serve as the ordered pair $\langle x, y \rangle$ of x and y, because $\{\{x\}, \{x, y\}\} = \{\{y\}, \{y, x\}\}$ only if $x = y$.

Union. For any set x, there is a set whose members are the members of members of x.

If $x = \{a, b\}$ then the members of members of x make up the set we call $a \cup b$, the "union of a and b." A particular union of two sets is the set $n \cup \{n\}$ we used in section 12.4 as the successor of the finite ordinal n. More generally, if α is any ordinal, then $\alpha + 1 = \alpha \cup \{\alpha\}$.

It is also useful to be able to form unions of infinitely many sets. We did this in section 8.4 when we formed the union L of the lower Dedekind cuts L_i to create the least upper bound of the corresponding real numbers. This process is validated by applying the union axiom to the set whose members are the L_i. We also used infinite unions in section 12.4 to form the least upper bound of a set of ordinals.

Notice also that pairing and union enable us to build finite sets with three or more members. For example, to build the set $\{a, b, c\}$ we use pairing to build $x = \{a, b\}$ and $y = \{c\}$ and then use union to form $x \cup y = \{a, b, c\}$.

Infinity. There is an infinite set; specifically, there is a set x whose members include $\{\}$ and, along with any member y, the member $s(y)$.

Here $s(y)$ is the successor set $y \cup \{y\}$, so this axiom says there is a set that includes the natural numbers. To get the set $\mathbb{N}$ whose members are *exactly* the natural numbers, we need an axiom that allows us to collect sets with a particular *defining property* into a set. For technical reasons, this axiom is stated in terms of function definitions, under "Replacement" below.

Power set. For any set x there is a set $\mathscr{P}(x)$ whose members are the subsets of x.

As we saw in section 12.3, this axiom creates a surprisingly large set $\mathscr{P}(x)$ when $x = \mathbb{N}$. In fact, because $\mathscr{P}(\mathbb{N})$ is not countable, there are not enough formulas in the language of ZF to define all its members. This is why we need an axiom to guarantee existence of the power set for $\mathbb{N}$ and other infinite sets.

Replacement (schema). If $\varphi(u, v)$ is a formula defining v as a function $f(u)$, then the range of f for u in a set x is itself a set.

Replacement is a schema generalizing the "definable subset" schema used by Zermelo (1908). Zermelo's axiom was that the elements u of a set x satisfying a formula $\varphi(u)$ form a set. Fraenkel (1922) pointed out that the replacement schema is needed to obtain sets such as $\{\mathbb{N}, \mathscr{P}(\mathbb{N}), \mathscr{P}(\mathscr{P}(\mathbb{N})), \ldots\}$. We also saw that it gives ordinals such as $\omega \cdot 2$.

Foundation. Any set has an $\in$-least member.

When applied to the set $\mathbb{N}$, foundation gives ordinary induction, because it is equivalent to saying each subset of $\mathbb{N}$ has a least member, since $\in$ is the usual ordering relation on $\mathbb{N}$, as mentioned in section 12.4.

The last two axioms enable us to generalize proof by induction from the natural numbers to ordinals in general and to define functions by recursion over the ordinals. Foundation implies that *any ordinal is a well-ordered set*, since it says that the $\in$ relation is well-founded, and the definition of ordinal (section 12.4) already says that an ordinal is linearly ordered by $\in$. Just as the well-ordering of $\mathbb{N}$ yields the method of proof by induction, the well-ordering of ordinals yields a more general proof method called **transfinite induction**. We will say more about transfinite induction in section 14.5. Replacement allows functions on the ordinals, such as extensions of addition and multiplication, to be defined by recursion.

ZF Minus the Infinity Axiom

The set V_ω of hereditarily finite sets, introduced in section 12.6, is a model of ZF minus the infinity axiom, which we will abbreviate by ZF−Infinity. It includes the empty set, as part of its definition, and it is easily seen to be closed under pairs, unions, power set, and replacement, because all members of V_ω are finite. And, *because* all of its members are finite, V_ω does not satisfy the axiom of infinity. This confirms that the axiom of infinity does not follow from the other ZF axioms. Conversely, V_ω cannot be proved to exist in ZF−Infinity.

V_ω is a good model for the mathematics of finite objects, since it includes the natural numbers, finite sets of natural numbers, finite sets of finite sets, and so on. Indeed, the natural numbers are precisely the ordinals of V_ω, as we saw in sections 12.4 and 12.6. Their successor operation $s(n) = n \cup \{n\}$ is defined in V_ω, hence also the sum and product operations, by their inductive definitions in the Peano axioms.

Finally, the induction axiom holds in V_ω, because it is equivalent to the axiom of foundation. Thus all the Peano axioms hold in ZF−Infinity. We say that Peano arithmetic is **interpretable** in ZF−Infinity.

Conversely, we can prove that ZF−Infinity is interpretable in Peano arithmetic. This takes more work, because each hereditarily finite set has to be encoded by a natural number, and because operations on sets, such as union and power set, have to be encoded by operations on numbers, ultimately definable in terms of addition and multiplication. This is a big project, called the **arithmetization of syntax**, which was not even contemplated until Gödel (1931) carried it out as part of his famous incompleteness proofs. We will say more about this kind of arithmetization, which took the nineteenth-century arithmetization project in a new direction, in section 15.5.

At any rate, the upshot of Gödel's arithmetization is that ZF−Infinity is equivalent to Peano arithmetic. Or, to put it more dramatically, set theory equals arithmetic plus infinity!

13.5 REMARKS

Here is how the first-order axioms for Peano arithmetic allow the presence of Dedekind's "alien intruders." Suppose we introduce a new constant N and add to the Peano axioms the infinitely many new axioms

$$N > 0, \quad N > 1, \quad N > 2, \quad N > 3, \quad \ldots,$$

where $a > b$ is an abbreviation for $\exists c [a = b + s(c)]$. These axioms say that N is an infinite number. Yet, assuming that the Peano axioms do not lead to a contradiction, neither do the new axioms. Why not? Well, suppose on the contrary that we can prove a contradiction, say, $0 = 1$, from the new axioms and the Peano axioms. *Any proof is finite,* so this hypothetical proof of $0 = 1$ uses only finitely many of the new axioms $N > k$. But any finite number of axioms of this form can be satisfied by interpreting N as the successor of the maximum k, so finitely many such axioms do *not* lead to a contradiction.

Moreover, we will see in section 15.4 that if a first-order axiom A does not lead to a contradiction then A has a model. The idea generalizes to give a model of the first-order Peano axioms plus the infinitely many axioms

$$N > 0, \quad N > 1, \quad N > 2, \quad N > 3, \quad \ldots.$$

The object interpreting N in this model may be regarded as an "infinite number." Then, if we expand Peano arithmetic to include the division operation, $1/N$ may be regarded as an "infinitesimal."

These results show how it can be possible to use infinitesimals without contradiction. The idea can be extended from arithmetic to analysis, so as to obtain various systems of **nonstandard analysis** in which Leibniz's arguments using infinitesimals can be put on a sound footing. Nonstandard analysis originally came from logic in Robinson (1966), but it has sometimes been used by mainstream mathematicians, such as Tao (2013). An especially interesting use of infinite numbers is their application to probability theory by Nelson (1987).

CHAPTER 14

The Axiom of Choice

Before Zermelo (1908) introduced his axioms for set theory that eventually became the Zermelo-Fraenkel axioms (ZF), he had used another axiom, the **axiom of choice** (AC), to prove a conjecture of Cantor known as the **well-ordering theorem** (Zermelo (1904). The well-ordering theorem states that every set has a well-ordering—a controversial claim, because no *explicit* well-ordering is known for many sets, such as $\mathbb{R}$. Since AC is easily seen to be equivalent to the well-ordering theorem, this put AC under suspicion.

However, it was not feasible to ban the use of AC in mathematics. In fact some of its opponents had unconsciously used it themselves, since AC is involved in highly plausible statements such as the existence of a countable subset of any infinite set. Moreover, AC greatly simplifies both the world of sets and many areas of mathematics, such as algebra and analysis.

In set theory, AC correlates each set with an ordinal, enabling us to say of any two sets that they are either equal in cardinality or else one is strictly larger than the other. In algebra, AC enables us to say that any vector space has a basis, any field has an algebraic closure, and that any nonzero ring has a maximal ideal.

On the other hand, AC also has "undesirable" consequences, particularly in measure theory. Vitali (1905) first discovered that AC implies the existence of nonmeasurable sets, and his result was intensified by Hausdorff (1914) and Banach and Tarski (1924) into the so-called **Banach-Tarski paradox.**

This led to the study of weaker axioms of choice, which were strong enough to prove the plausible consequences of AC but not strong enough to prove the "paradoxical" ones.

14.1 AC AND INFINITY

Let us begin with a formal statement of AC.

Axiom of choice. *For any set X whose members are nonempty sets x there is a function f, called a **choice function**, such that $f(x) \in x$ for each $x \in X$.*

The mere statement of the axiom seems highly plausible, but bear in mind that AC is needed only when there is no known way to *define* a choice function. To some mathematicians, it is unacceptable to claim existence of a function without having the means to define it, and unfortunately in many cases we do not have the means. A simple example is where X is the set of nonempty sets of real numbers.

On the other hand, it is difficult to manage without AC, because the very concept of infinite set is connected with it. Without some form of AC we cannot prove that each infinite set S has the property Dedekind used to define infinitude: the existence of a bijection between S and a proper subset $S' \subset S$.

If there is a bijection $g : S \to S'$ then there is an element a in S but not in S', in which case $a, g(a), g(g(a)), \ldots$ are distinct members of S. Thus a set that is infinite according to Dedekind's definition has a countably infinite subset. Conversely, if a set S has a countably infinite subset $\{a_0, a_1, a_2, a_3, \ldots\}$ then there is a bijection g from S to its proper subset $S' = S - \{a_0\}$. Namely, if $s \in S$ is not in $\{a_0, a_1, a_2, a_3, \ldots\}$ we let $g(s) = s$. If, on the contrary, $s = a_n$, we let $g(s) = a_{n+1}$. Then g is clearly a bijection of S onto S'.

Thus S *has the Dedekind property if and only if S has a countably infinite subset.* But to prove outright that an infinite set S has a countably infinite subset we seem to need AC. The obvious proof goes as follows. Choose any element $a_0 \in S$, and then choose

$$a_1 \in S - \{a_0\},$$
$$a_2 \in S - \{a_0, a_1\},$$
$$a_3 \in S - \{a_0, a_1, a_2\},$$
$$\vdots$$

and so on. Since S is infinite, there is no end to this sequence of choices, so we get a countably infinite subset $\{a_0, a_1, a_2, \ldots\}$ of S.

But we are making an infinite sequence of choices, and in fact this theorem is not provable in ZF alone. The existence of a countably infinite subset is one of many theorems known not to be provable from the ZF axioms alone. Others worth mentioning in this context are the theorem that a countable union of countable sets is countable and—even more astonishing—that $\mathbb{R}$ is not a countable union of countable sets. We know most of these results about unprovability in ZF from the work of Paul Cohen in 1963, which is explained in Cohen (1966). Cohen constructed models of ZF in which various consequences of AC do not hold. One of his models contains an infinite set of reals with no countable subset. Feferman and Levy (1963) constructed a model of ZF in which $\mathbb{R}$ is a countable union of countable sets.

In view of these weaknesses of ZF, it is useful to know which important statements are *equivalent* to AC, since any equivalents will not be provable in ZF. We discuss some examples in the next few sections, along with some consequences of AC that, while they do not imply AC, are still not provable in ZF. The latter statements may be viewed as weaker choice principles.

14.2 AC AND GRAPH THEORY

Section 10.1 gave formal definitions of graph, connectedness, and tree, and we proved the existence of a spanning tree in any connected graph. Section 10.1 also (briefly) pointed out the use of AC in the proof. In this section we will prove, conversely, that the existence of a spanning tree in any connected graph implies AC. Thus existence of a spanning tree is an equivalent of AC, and it is probably one of the easiest equivalents to state and to prove.

Existence of spanning trees implies AC. *If every connected graph has a spanning tree, then AC holds.*

Proof. Given a set X of nonempty sets x, we wish to obtain a choice function for X. It will help to use the notation x_i for a general member of X, where i is an "index" or name for the set x. (The name could be, for example, the set $\{x\}$. The indices i, which we will call *set indices*, are just separate names for the members of X, and there could be uncountably many of them.)

Then we define a graph $\mathcal{G}$ whose vertices are

- A vertex i for each set index i.
- A vertex x_{ij} for each member of x_i, where j indexes the members of x_i.

The edges of $\mathcal{G}$ are of two kinds:

- An edge from each vertex i to each vertex x_{ij}. These edges connect the set index i to all members of the set x_i.
- An edge from each member x_{ij} of x_i to each member x_{kl} of x_k, when $i \neq k$. These edges make $\mathcal{G}$ connected.

Now a choice function for X assigns exactly one x_{ij} to each i. This assignment can be made from a spanning tree $\mathcal{T}$ of $\mathcal{G}$. Here is how: Fix a vertex i_0, and for any vertex $i \neq i_0$, consider the unique simple path in the tree $\mathcal{T}$ from i_0 to i. The last edge in this path must be from some x_{ij} to i, since all edges out of i are of this form. This x_{ij} is the one we associate with i. We still have to associate an $x_{i_0 j}$ with i_0, but this can be done arbitrarily.[1] □

14.3 AC AND ANALYSIS

The existence of a countable subset of every infinite set is an important issue in analysis. This issue arises in two theorems: the strong Bolzano-Weierstrass theorem and continuity from sequential continuity.

Strong Bolzano-Weierstrass theorem. *Assuming AC, if S is a bounded infinite set of real numbers, then S contains a countable subset* $\{x_0, x_1, x_2, \ldots\}$ *for which the* sequence $x_0, x_1, x_2, \ldots$ *is convergent.*

Proof. We first prove that S contains an **accumulation point** x, that is, a point whose neighborhoods $(x - \delta, x + \delta)$ all include members of S. This part of the proof does *not* need AC. Then, using AC, we find a sequence of members $x_0, x_1, x_2, \ldots$ of members of S that converge to x.

---

1. AC is not needed to make *one* choice. Logic always allows us to make finitely many choices, for example, when we say "Let A, B, C be any points in the plane."

Without loss of generality we can assume that $S \subset [0, 1] = I_0$. Since S is infinite, one of the half-intervals $[0, 1/2]$ or $[1/2, 1]$ includes infinitely many points of S. Let the leftmost of these two intervals that includes infinitely many points of S be I_1.

We similarly define a half I_2 of I_1 that includes infinitely many points of S, then a half I_3 of I_2 likewise, and so on. Thus we get intervals

$$I_1 \supset I_2 \supset I_3 \supset \cdots,$$

each half the length of the one before and each containing points of S. Let x be the unique common point of these intervals, which exists by the completeness of $\mathbb{R}$ (section 11.1). This x is an accumulation point of S, because any neighborhood of x contains some I_k.

By our construction, for any n there are points of S within distance 2^{-n} of x, namely, the points of S in I_n. By AC we can choose such a point x_n for each n, and then the sequence $x_0, x_1, x_2, \ldots$ converges to x. □

For many purposes, it is sufficient to know that an accumulation point of S exists, which is why I separated its proof from the part that uses AC to get the stronger conclusion of a convergent subsequence of S. Obviously, for a convergent subsequence to exist it is necessary for the infinite set S to have a countable subset. But, as mentioned in section 14.1, Cohen found a model showing that the latter statement is not provable in ZF for all infinite sets $S \subset \mathbb{R}$. The same model causes trouble for another basic theorem, involving the concept of **sequential continuity**.

Definition. A function f is **sequentially continuous** at $x = a$ if, for each sequence $a_0, a_1, a_2, \ldots$ with limit a, $f(a_n) \to f(a)$ as $a_n \to a$.

Continuity from sequential continuity. *Assuming AC, if f is sequentially continuous at $x = a$ then f is continuous at $x = a$.*

Proof. Suppose that f is sequentially continuous at $x = a$, but suppose, for the sake of contradiction, that f is *not* continuous at $x = a$. This means that there is an $\varepsilon > 0$ such that, for *any* $\delta > 0$, there is an x in $(a - \delta, a + \delta)$ with $|f(x) - f(a)| \geq \varepsilon$.

In particular, by AC we can choose $a_0, a_1, a_2, \ldots$ such that

$$a_0 \in (a-1, a+1) \quad \text{and} \quad |f(a_0) - f(a)| \geq \varepsilon,$$
$$a_1 \in (a-1/2, a+1/2) \quad \text{and} \quad |f(a_1) - f(a)| \geq \varepsilon,$$
$$a_2 \in (a-1/2^2, a+1/2^2) \quad \text{and} \quad |f(a_2) - f(a)| \geq \varepsilon,$$
$$\vdots$$

Then $a_n \to a$ but $f(a_n)$ does *not* tend to $f(a)$, so f is *not* sequentially continuous at $x = a$. This contradiction shows that in fact f is continuous at $x = a$. □

The implication in this theorem cannot be proved in ZF because of Cohen's model, mentioned above, in which there is an infinite set of reals with no countably infinite subset. By mapping this set into $[0,1]$ we get an infinite bounded set S with no countably infinite subset, yet it has an accumulation point a, by the first part of the Bolzano-Weierstrass theorem. We can use this set S to define a function that is sequentially continuous at a but not continuous there, as follows.

Without loss of generality we can assume that $a \notin S$ (otherwise, remove a from S and the set that remains still has a as an accumulation point). Now define f to be the characteristic function of S; that is,

$$f(x) = \begin{cases} 1 & \text{if } x \in S \\ 0 & \text{if } x \notin S. \end{cases}$$

Since $f(a) = 0$ but $f(x) = 1$ for points x arbitrarily close to a, f is certainly not continuous at $x = a$. Now suppose $a_0, a_1, a_2, \dots$ is a sequence with limit a. Since S contains no countably infinite subset, only finitely many of the points a_n are in S. From some term a_N onward, the a_n are not in S, and hence $f(a_n) = 0$. It follows that $f(a_n) \to 0 = f(a)$ as $a_n \to a$, so f is sequentially continuous at $x = a$.

14.4 AC AND MEASURE THEORY

Recall from section 11.4 the properties of the Lebesgue measure μ, which implicitly define the Lebesgue measurable subsets of $\mathbb{R}$:

Measure of basic sets. The measure of the interval $[a, b]$ is given by

$$\mu([a, b]) = b - a.$$

Complementation. If $S \subseteq T$ and $\mu(S)$ and $\mu(T)$ exist, then

$$\mu(T - S) = \mu(T) - \mu(S),$$

where $T - S = \{x \in \mathbb{R} : x \in T \text{ and } x \notin S\}$.

Countable additivity. If S is the union of disjoint sets $S_1, S_2, S_3, \ldots$ with measures $\mu(S_1), \mu(S_2), \mu(S_3), \ldots$, then

$$\mu(S) = \mu(S_1) + \mu(S_2) + \mu(S_3) + \cdots,$$

provided the sum exists.

Measure zero sets. Any subset of a measure zero set has measure zero.

An important consequence of this definition is that the Lebesgue measure μ is **translation invariant**. That is, if S is Lebesgue measurable and

$$S + r = \{x + r : x \in S\}$$

is the translate of S by r, then $\mu(S + r) = \mu(S)$. This property obviously holds for intervals, and from them it propagates to any measurable set, since any measurable set may be approximated within measure ε by a finite union of intervals.

Translation invariance was used by Vitali (1905) to give the first example of a set that is not Lebesgue measurable, using AC. Vitali's example is most naturally viewed as a subset of the circle with circumference 1, on which Lebesgue measure is defined the same way as on the line, except that angle θ takes the place of distance x. Thus a "translation" in this context is rotation, and the Lebesgue measure is **rotation invariant**.

Nonmeasurable set. *Assuming AC, there is a subset of the circle that is not Lebesgue measurable.*

Proof. For each point θ on the circle of circumference 1, consider the set

$$X_\theta = \{\theta + \varphi : \varphi \text{ is rational in } [0,1)\}$$

of "rational translates of θ." Then any two of the sets X_θ, $X_{\theta'}$ are either identical (when $\theta - \theta'$ is rational) or disjoint (when $\theta - \theta'$ is irrational). By AC, there is a set X including exactly one member from each X_θ.

Now consider the rational translates of X,

$$X + \varphi = \{\theta + \varphi : \theta \in X, \varphi \text{ rational in } [0,1)\}.$$

From the definitions of the X_θ and X we draw the following conclusions:

1. The sets $X + \varphi, X + \varphi'$ are disjoint for distinct rationals φ and φ' in $[0, 1)$, because if χ is a common member of $X + \varphi$ and $X + \varphi'$ we have sets X_θ and $X_{\theta'}$ with

$$\chi = \theta + \varphi = \theta' + \varphi', \quad \text{so} \quad \theta - \theta' = \varphi' - \varphi.$$

   But if $X_\theta \neq X_{\theta'}$ then $\theta - \theta'$ is irrational, so the only possibility is that $\theta = \theta'$ and therefore $\varphi = \varphi'$.

2. On the other hand, the *union* of these disjoint sets $X + \varphi$ is the whole circle, because each θ is in some X_θ; hence some rational translate $\theta + \varphi \in X$, and then $\theta \in X + (-\varphi)$.

3. Finally, there are countably many sets $X + \varphi$, because there are countably many rational numbers φ.

Now, if X is Lebesgue measurable, rotation invariance of μ implies that $\mu(X) = \mu(X + \varphi)$ for each of the countably many rational $\varphi \in [0, 1)$. Then there are two possibilities: either $\mu(X) = 0$ or $\mu(X) = m > 0$. In the first case, countable additivity implies that the measure of the circle is $0 + 0 + 0 + \cdots = 0$. In the second case, countable additivity implies that the measure of the circle is $m + m + m + \cdots$, which is infinite. Since neither of these values is correct, we conclude that X is *not* Lebesgue measurable. □

Hausdorff (1914) found a worse kind of nonmeasurability in subsets of the sphere in $\mathbb{R}^3$, using AC to partition the sphere minus a countable set into three sets $\mathscr{A}, \mathscr{B}, \mathscr{C}$ that are congruent to each other and to $\mathscr{B} \cup \mathscr{C}$. Thus, if measurable, $\mathscr{A}, \mathscr{B}, \mathscr{C}$ are each 1/3 of the surface area of the sphere, yet also 2/3 of the area. This contradiction shows that $\mathscr{A}, \mathscr{B}, \mathscr{C}$ are *not* measurable. In fact, it shows that there is not even a *finitely* additive measure on the sphere that assigns equal measure to congruent subsets.

Banach and Tarski (1924) leveraged Hausdorff's proof to get an even more paradoxical result about the ball in $\mathbb{R}^3$: there is a partition of the unit ball into five subsets that can be reassembled, by translations and rotations, into *two* unit balls. This consequence of AC is known as the **Banach-Tarski paradox**. It has been the subject of at least two books: Wapner (2005) and Tomkowicz and Wagon (2016). Needless to say, the partitioning of the ball is not one that could be made with a knife.

The five subsets are highly disconnected clouds of points, like Vitali's nonmeasurable sets.

14.5 AC AND SET THEORY

As mentioned at the start of this chapter, Zermelo (1904) used AC to prove that every set has a well-ordering. It is also true, and easier to prove, that the well-ordering theorem implies AC, so we will do that first. We take the well-ordering theorem to say that the members of every nonempty set X can be written as $x_0, x_1, x_2, \ldots, x_\alpha, \ldots$ for all the ordinals α less than some ordinal β. (For a refresher on ordinals, see section 12.4.)

Well-ordering implies AC. *If every set has a well-ordering, then every set Y of nonempty sets y has a choice function.*

Proof. Given the set Y, let X be the union of all the sets y in Y. Also let $x_0, x_1, x_2, \ldots, x_\alpha, \ldots$ be a well-ordering of the elements of X. Then, for each set $y \in Y$ we define

$$f(y) = \text{the member } x_\alpha \text{ of } y \cap X \text{ for which } \alpha \text{ is least.}$$

Then $f(y)$ is well defined because each set of ordinals has a least member, and $f(y) \in y$, so f is a choice function for Y. □

The idea for proving that AC implies well-ordering is also simple, but it involves an inductively defined function, which requires some justification.

AC implies well-ordering. *If AC holds, then the members of any nonempty set X may be written $x_0, x_1, x_2, \ldots, x_\alpha, \ldots$ for all the ordinals α less than some ordinal ν.*

Proof. Given a nonempty set X we write the members of X as follows, with the help of AC. We suppose that f is a choice function for the set of nonempty subsets of X. Then define a function $g(\alpha) = x_\alpha$ "inductively" as follows:

$$g(0) = x_0 = f(X),$$
$$g(1) = x_1 = f(X - \{x_0\}),$$
$$\vdots$$
$$g(\alpha) = x_\alpha = f(X - \{x_\gamma : \gamma < \alpha\}),$$
$$\vdots$$

The main problem with this proof is showing that it really defines a function $g(\alpha) = x_\alpha$. To prove the validity of "definition by induction" we first consider functions g_β that *partially* achieve what g is supposed to do, namely, up to the ordinal β. That is,

- $g_\beta(\alpha)$ is defined for $\alpha < \beta$,
- $g_\beta(0) = f(X)$, and
- $g_\beta(\alpha) = f(X - \{g_\beta(\gamma) : \gamma < \alpha\})$ for each $\alpha < \beta$. (*)

The range of each g_β is a subset of X, so the ordinals β that index these partial functions correspond to some subset of $\mathscr{P}(X)$, and hence form a set themselves, by replacement. Union then gives the least upper bound of the β for which g_β exists, and there is a least ordinal $\nu >$ all these β.

It is clear that $g_\beta(\alpha) = g_{\beta'}(\alpha)$ for any α for which both $g_\beta(\alpha)$ and $g_{\beta'}(\alpha)$ are defined; if not, consider the least α for which $g_\beta(\alpha) \neq g_{\beta'}(\alpha)$, and a contradiction is immediate from the last clause of (*).

We can therefore form the union, g_ν, of the g_β for $\beta < \nu$, and g_ν is a function satisfying the conditions (*), with ν instead of β. Also, g_ν is onto the whole of X. If not, consider the nonempty set

$$X - \{g_\nu(\alpha) : \alpha < \nu\}.$$

Then we can extend g_ν to $g_{\nu+1}$ by setting $g_{\nu+1}(\nu) = f(X - \{g_\nu(\alpha) : \alpha < \nu\})$, but no such extension exists, by definition of ν.

Thus we can take $g = g_\nu$, and $g(0), g(1), g(2), \ldots, g(\alpha), \ldots$, for $\alpha < \nu$, is a well-ordering of X. □

The idea used in this theorem, building an inductively defined function g from the partial functions that satisfy the defining conditions of g, was introduced by Dedekind (1888: theorems 125 and 126) to prove the validity of inductive definitions in Peano arithmetic. Of course, in Dedekind's situation one does not need a choice function f because one can always choose from a set of natural numbers by taking the least. But when a choice function f is *given*, by AC, then the problem of justifying "definition by induction" is much the same.

14.6 AC AND ALGEBRA

The applications of AC in algebra are often about the existence of *maximal* elements. For example, a basis of a vector space is a set B of linearly independent vectors that is maximal in the sense that any set consisting of B and some other vectors is *not* independent. The concept of a maximal ideal in a ring is similar. Because of this, algebraists since the 1930s have preferred to use AC in an equivalent form that states the existence of a maximal element. This form is known as **Zorn's lemma**.

However, some of the early proofs of these theorems were along the lines of Zermelo's proof of the well-ordering theorem, defining a structure by transfinite induction. Here is the first example, due to Hamel (1905):

Hamel basis. *Assuming AC, $\mathbb{R}$ has a basis B as a vector space over $\mathbb{Q}$.*

Proof. Let $x_0, x_1, \ldots, x_\alpha, \ldots$ be a well-ordering of the nonzero members of $\mathbb{R}$, given by the well-ordering theorem. Also let γ be the least upper bound of the ordinals α occurring in this well-ordering. Then construct the following subset B of $\mathbb{R}$ by induction on the ordinals less than $\beta < \gamma$.

Stage $\beta + 1$. If x_β is not a linear combination, with rational coefficients, of elements in B at this stage, put x_β in B. (Thus x_0 enters B at stage 1; see below.)

Stage λ, when λ is a limit ordinal. The members of B at this stage are all x_α put in B at stages $\alpha < \lambda$. (Note that x_λ is considered at stage $\lambda + 1$.)

It is clear from this process that there are enough x_α in B to ensure that every $x \in \mathbb{R}$ is a linear combination of members of B with rational coefficients, because the only x_α omitted are already such linear combinations. Also, no linear combination of members of B with nonzero coefficients is 0, because in any such (hypothetical) combination there is a member x_α with largest index, and all such x_α are omitted. Therefore, B is a basis. $\square$

Notice that the same well-ordering argument applies with any vector space in place of $\mathbb{R}$, since we assume nothing specific about the well-ordering of $\mathbb{R}$. Thus it is not unreasonable to credit Hamel with the theorem that any vector space has a basis. Much later, Blass (1984) proved that the existence of a basis for any vector space implies AC, so AC has a natural algebraic equivalent: existence of a basis for any vector space.

Another important proof by well-ordering was the proof of Krull (1929) that any nonzero ring has a maximal ideal. Krull's proof occurred before Zorn's lemma became the algebraist's preferred method for deploying AC, and his theorem actually lends itself better to a proof by Zorn's lemma. However, Grothendieck (1957) also used transfinite induction for a key theorem, long after Zorn's lemma became well known.

Zorn's Lemma

The usual statement of Zorn's lemma is:

Zorn's lemma. *If P is a partially ordered set in which each linearly ordered subset has an upper bound, then P has a maximal element.*

Before we see how AC implies Zorn's lemma, let us see how Zorn's lemma affects the proof of algebraic theorems such as existence of a Hamel basis. Typically, the partially ordered set P is a collection of sets, ordered by the $\subset$ relation, and the upper bound of a linearly ordered subcollection is its union.

Hamel basis. *Zorn's lemma implies $\mathbb{R}$ has a basis as a vector space over $\mathbb{Q}$.*

Proof. Consider the set P of independent subsets B_i of $\mathbb{R}$, partially ordered by set inclusion. Thus any finite subset of each B_i is a set of real numbers that are linearly independent over $\mathbb{Q}$. It follows that any linearly ordered set of B_i has an upper bound, namely, its union, B. This is because any finite subset F of B is a finite subset of some B_i, and hence F is independent.

Since ordering by $\subset$ is a partial ordering, P satisfies the premises of Zorn's lemma, and therefore it has a maximal element M. In other words, M is an independent set, but $M \cup \{r\}$ is not independent, for any real number r not in M.

This means r is a linear combination of elements of M, with coefficients in $\mathbb{Q}$. Of course, this is also true for any element of M, so M is an independent set that spans $\mathbb{R}$; that is, M is a basis. $\square$

Krull's theorem that any nonzero ring R has a maximal ideal has a similar proof, taking P to be the set of ideals of R, ordered by set inclusion. An even easier theorem to prove by Zorn's lemma is existence of a spanning tree $\mathcal{T}$ in a connected graph $\mathcal{G}$. In this case we get $\mathcal{T}$ as a maximal member of the set of subtrees of $\mathcal{G}$, ordered by graph inclusion.

Now let us see why Zorn's lemma follows from AC. In fact, we will show that Zorn's lemma follows from the well-ordering theorem.

Well-ordering implies Zorn's lemma. *If every set can be well-ordered, then a partially ordered set in which every linearly ordered subset has an upper bound has a maximal element.*

Proof. Suppose that P is partially ordered by $<$ in such a way that every linearly ordered subset of P has an upper bound. Also suppose, for the sake of contradiction, that P does *not* have a maximal element.

Now, given a well-ordering of P,

$$x_0, x_1, x_2, \ldots, x_\alpha, \ldots \quad \text{for} \quad \alpha < \gamma,$$

we construct by transfinite induction a linearly ordered subset C,

$$y_0 < y_1 < \cdots < y_\beta < \cdots.$$

Stage 0. Let $y_0 = x_0$.

Stage 1. Since y_0 is not maximal, there is an $x_\alpha > y_0$. Let y_1 be the one with least index α.

Stage β. Suppose that $y_0 < y_1 < \cdots < y_\gamma < \cdots$ have been selected from P before stage β. By hypothesis, this linearly ordered set has an upper bound y. Also by hypothesis, y is not maximal. Let y_β be the x_α with least index α such that $x_\alpha > y$.

Now the linearly ordered subset $y_0 < y_1 < \cdots < y_\beta < \cdots$ with this inductive definition exists by an argument like that in the well-ordering theorem. And, as in the proof of the well-ordering theorem, there is a least upper bound ν to the stages in the construction of C. But then C cannot have an upper bound in P; otherwise stage ν would take place.

This contradiction shows that P does in fact have a maximal element, so Zorn's lemma is proved. $\square$

So now we have proved that AC implies the well-ordering theorem and that the well-ordering theorem implies Zorn's lemma. We complete the circle by showing that Zorn's lemma implies AC, so all three statements are equivalent:

Zorn's lemma implies AC. *If Zorn's lemma holds, then every set X of nonempty sets has a choice function.*

Proof. Given a set X of nonempty sets x, we consider the set P of *partial* choice functions, that is, functions f such that $f(x) \in x$ for x in some subset $S \subseteq X$. P is not empty, because X has some member x_0 and x_0 has some member x_{01} we can take to be $f(x_0)$.

When each partial choice function f is viewed as a set of ordered pairs $\langle x, f(x) \rangle$, the set P is partially ordered by set inclusion. Also, if $f \subseteq f'$ then f and f' have the same value on any common argument x, so any linearly ordered subset of P has an upper bound: its union. Thus P satisfies the conditions of Zorn's lemma, and therefore P has a maximal member $f_{\max}$.

This maximal function $f_{\max}$ is defined for all $x \in X$. If not, suppose $f_{\max}$ is not defined on $x_0 \in X$. In that case we can extend $f_{\max}$ by adding an ordered pair $\langle x_0, x_{01} \rangle$ for some $x_{01} \in x_0$. This contradicts maximality, so $f_{\max}$ is in fact a choice function for X. $\square$

14.7 WEAKER AXIOMS OF CHOICE

Probably the weakest axiom of choice in common use is this:

Countable choice (CC). Any countable set X of nonempty sets $x_0, x_1, x_2, \ldots$ has a choice function, that is, a function f such that $f(x_n) \in x_n$.

A theorem that seems obvious but actually depends on CC is:

Countable union of countable sets. *If $x_0, x_1, x_2, \ldots$ are disjoint countable sets then the union $x_0 \cup x_1 \cup x_2 \cup \cdots$ is countable.*

Proof. For each natural number n consider the set

$$y_n = \{\text{enumerations of } x_n\},$$

where an **enumeration** of the countable set x_n is a bijection $f : \mathbb{N} \to x_n$. We can write $f(k) = x_{nk}$, so an enumeration looks like a list of members of x_n:

$$x_n : \quad x_{n0} \quad x_{n1} \quad x_{n2} \quad \ldots.$$

By CC we can choose an enumeration for each x_n, which gives a table

$$
\begin{array}{llllll}
x_0: & x_{00} & x_{01} & x_{02} & x_{03} & \cdots \\
x_1: & x_{10} & x_{11} & x_{12} & \cdots & \cdots \\
x_2: & x_{20} & x_{21} & \cdots & \cdots & \cdots \\
x_3: & x_{30} & \cdots & \cdots & \cdots & \cdots \\
\vdots &
\end{array}
$$

The table gives the following enumeration of $x_0 \cup x_1 \cup x_2 \cup \ldots$, in which each block lists the members of the table on diagonals sloping from top right to bottom left:

$$x_{00}; \quad x_{01}, x_{10}; \quad x_{02}, x_{11}, x_{20}; \quad x_{03}, x_{12}, x_{21}, x_{30}; \quad \ldots.$$

Hence $x_0 \cup x_1 \cup x_2 \cup \ldots$ is countable. $\qquad\square$

We conclude from this theorem that CC implies $\mathbb{R}$ is not a countable union of countable sets, because $\mathbb{R}$ is uncountable (and this can be proved in ZF). However, as mentioned in section 14.1, Feferman and Levy (1963) found a model of ZF in which $\mathbb{R}$ *is* a countable union of countable sets. It follows that CC *is not provable in* ZF. One can also find models of ZF in which CC holds but not the full AC. Thus CC is a weaker axiom of choice than AC.

Another commonly used choice axiom is this:

Dependent choice (DC). If R is relation on a nonempty set X such that for each $x \in X$ there is a $y \in X$ such that xRy, then there is a sequence $x_0, x_2, x_2, \ldots \in X$ such that

$$x_0 R x_1, \quad x_1 R x_2, \quad x_2 R x_3, \quad \ldots.$$

Intuitively speaking, we first choose x_1 related to x_0, then x_2 related to x_1, then x_3 related to x_2, and so on. Thus each choice depends on the one before. We used DC in section 14.1 when we proved that any infinite set X has a countably infinite subset (though in fact this theorem can be proved using CC, with a little more work). In that case the relation $x_n R x_{n+1}$ was $x_{n+1} \in X - \{x_0, x_1, \ldots, x_n\}$.

An important property of ordinals depends on DC:

Infinite descent theorem. *A linear ordering $<$ on a set P is a well-ordering if and only if there is no infinite descending sequence*

$$x_0 > x_1 > x_2 > x_3 > \cdots.$$

Proof. An infinite descending sequence has no least member, so P is not well-ordered by $<$ if it has such a sequence.

Conversely, if P is not well-ordered then it has a subset S with no least member. If x_0 is any member of S, DC allows us to choose $x_1 \in S$ with $x_0 > x_1$, then $x_2 \in S$ with $x_1 > x_2$, and so on. Thus $x_0 > x_1 > x_2 > x_3 > \cdots$ is an infinite descending sequence. $\qquad \square$

Like CC, DC holds in some models of ZF in which the AC does not hold. A famous example is the model of Solovay (1970), in which DC holds but *all subsets of* $[0, 1]$ *are Lebesgue measurable.* As we saw in section 14.4, there are non-(Lebesgue-)measurable sets whenever AC holds, so AC cannot hold in the Solovay model.

A remarkable feature of Solovay's model is that it is constructed with the help of an *inaccessible* set, and Shelah (1984) showed that the inaccessible is necessary. Thus inaccessible sets, whose existence we are not able to prove (section 12.7), reach down to influence the behavior of sets at the level of real numbers.

Figure 14.1 is a picture by William Blake that captures this phenomenon, I like to think. It shows an inaccessible being, which Blake called Urizen, who reaches down to impose measure on the world. Inaccessibles can reach us, even though we cannot reach them!

14.8 REMARKS

Gödel (1938) proved that AC is consistent with ZF, so AC does not lead to contradiction, unless ZF itself does. Gödel's proof is based on a model L of ZF consisting of the so-called **constructible sets**, which are roughly the sets that must exist in any model that contains all the ordinals. L is the union of sets L_α defined by an induction on the ordinals α, analogous to the induction defining the sets V_α in section 12.7.

The finite sets $L_0, L_1, L_2, \ldots$ and their union L_ω are exactly the same as $V_0, V_1, V_2, \ldots$ and V_ω, respectively. But $L_{\omega+1}$ consists of the subsets of L_ω *definable* by formulas of ZF with symbols for ordinals $< \omega$ and quantifiers ranging over L_ω, which is not all subsets of L_ω. In fact, further subsets of L_ω (in particular, further subsets of $\mathbb{N}$) will occur in L_α for values of $\alpha > \omega + 1$.

The higher L_β are defined like $L_{\omega+1}$ for successor ordinals β, and at a limit ordinal λ we simply let L_λ be the union of all preceding L_α. It turns

Figure 14.1 : The inaccessible reaches down

out that L is a model of ZF. L is also a model of AC, essentially because every set in L has a *definition*, which can be written in the language of ZF supplemented by symbols for the ordinals. Since each definition is a finite string of symbols, and the ordinal symbols are well-ordered, it is possible to well-order the definitions, and hence all the sets in L. It follows that L satisfies the well-ordering theorem, and hence AC.

Since L is a model of ZF + AC, AC cannot lead to a contradiction, unless ZF itself does. In particular, the "paradoxical" consequences of AC, such as Banach-Tarski, are not actually contradictory, unless ZF itself is.

On the other hand, the negation of AC cannot lead to a contradiction either, thanks to the models of ZF + ¬AC found in Cohen (1963). Cohen used Gödel's model as a starting point but modified it by a subtle technique called **forcing**, which allows very flexible changes to the original model. Forcing is beyond the scope of this book, but the book Cohen (1966) can still be recommended in the spirit of learning from the master.

The Continuum Hypothesis

We cannot leave Gödel's model L without pointing out that it also satisfies the **continuum hypothesis**, which can be stated as follows: $\mathscr{P}(\mathbb{N})$ *is equinumerous with the first uncountable ordinal* ω_1.

The brief description of L above pointed out that $\mathbb{N}$ occurs as a subset of L_ω. It follows that $\mathbb{N}$ and some of its subsets are members of $L_{\omega+1}$, due to their definitions in the language of ZF. However, there are only countably many such definitions, so further subsets of $\mathbb{N}$ appear in higher L_α as the language expands with constants for higher ordinals, and we can quantify over larger domains L_α. The interesting thing, which Gödel discovered, is that *all the constructible subsets of* $\mathbb{N}$ occur in the L_α for *countable* α.

This technical fact implies that the members $\mathscr{P}(\mathbb{N})$ in L all have definitions that can be written in the language of ZF together with symbols for the ordinals $< \omega_1$. It follows, for much the same reason that there are countably many finite strings of symbols in a language with countably many symbols, that there are ω_1 finite strings of symbols in a language with ω_1 symbols. Therefore, there are ω_1 members of $\mathscr{P}(\mathbb{N})$ in Gödel's model L.

This means that the continuum hypothesis is consistent with ZF + AC. Cohen (1963) showed, in contrast, that there is a model for ZF + AC in which the continuum hypothesis is false. Therefore the continuum hypothesis, like AC, is independent of the usual axioms of set theory. Despite a huge amount of research on the continuum since the 1960s, we are still at a loss for new axioms of set theory that can settle the continuum problem. Hilbert's first problem from 1900 still does not have a satisfactory answer.

CHAPTER 15

Logic and Computation

Long ago, Llull in the thirteenth century and Leibniz in the seventeenth century suggested that logic might be reducible to computation. The first substantial step toward this goal was the "algebra of logic" of Boole (1847). **Boolean algebra** covers propositional logic, where it is equivalent to solving multivariable polynomial equations in mod 2 arithmetic.

Thus propositional logic does indeed reduce to a simple kind of computation. However, propositional logic is not all of logic, and certainly not enough to cover most proofs in mathematics. A logic strong enough for mathematics is the **predicate logic** of Frege (1879), which does not reduce to algebra or computation in any straightforward way. However, Frege gave **axioms** and **rules of inference** for deriving valid formulas of logic, and later Gödel (1930) showed that all valid formulas were obtainable in this way. It is also possible to prove, fairly easily, that Frege's system is consistent in the sense that it produces *only* valid formulas.

However, since predicate logic is strong enough to contain proofs of all mathematical implications, its complexity equals that of mathematics itself. This led Hilbert in the 1920s to pose the *Entscheidungsproblem* (decision problem): the problem of finding an algorithm to decide the validity of any formula of predicate logic. In the same decade, attempts by Post to mechanize the proof process in the strong system *Principia Mathematica* of Whitehead and Russell (1910) led him to suspect that algorithms for certain problems were nonexistent.

This conclusion was confirmed when Turing (1936) formalized the concept of algorithm and, as a corollary, showed that there is no algorithm for the solution of the *Entscheidungsproblem*.

15.1 PROPOSITIONAL LOGIC

Propositional logic is a very modest analysis of the concept of truth. It aims only to compute the **truth value** (1 for "true" or 0 for "false") of compound propositions built from **propositions** $p, q, r, \ldots$ by means of **connectives** such as AND, OR, and NOT. Nothing is assumed about the propositions $p, q, r, \ldots$ except that they can take the values 0 or 1. The value of any compound proposition can then be computed using the **truth tables**:

p	q	p AND q
0	0	0
0	1	0
1	0	0
1	1	1

p	q	p OR q
0	0	0
0	1	1
1	0	1
1	1	1

p	NOT p
0	1
1	0

The connectives AND, OR, and NOT are not the only connectives—another important one is p IMPLIES q—but they are convenient because any other connective can be expressed in terms of them. For example, p IMPLIES q is the same as (NOT p) OR q, so its truth table can be computed as follows:

p	q	NOT p	(NOT p) OR q
0	0	1	1
0	1	1	1
1	0	0	0
1	1	0	1

Any function of $p, q, r, \ldots$ taking the values 0 or 1 is called a **Boolean function** and may be represented by a truth table. Here is a random example:

p	q	r	$f(p,q,r)$
0	0	0	0
0	0	1	0
0	1	0	1
0	1	1	0
1	0	0	1
1	0	1	0
1	1	0	0
1	1	1	1

For this table we read that $f(p,q,r)$ is true if and only if

$$[(\text{NOT } p) \text{ AND } q \text{ AND } (\text{NOT } r)] \text{ is true,}$$
$$\text{OR } [p \text{ AND } (\text{NOT } q) \text{ AND } (\text{NOT } r)] \text{ is true,}$$
$$\text{OR } [p \text{ AND } q \text{ AND } r] \text{ is true,}$$

so $f(p,q,r)$ is the same Boolean function as

$$[(\text{NOT } p) \text{ AND } q \text{ AND } (\text{NOT } r)]$$
$$\text{OR } [p \text{ AND } (\text{NOT } q) \text{ AND } (\text{NOT } r)]$$
$$\text{OR } [p \text{ AND } q \text{ AND } r].$$

Thus our everyday understanding of the words "and," "or," and "not" gives the less-than-obvious theorem that any Boolean function, in any number of variables, is a compound of the three Boolean functions AND, OR, and NOT.

Mod 2 Arithmetic

Boolean functions get their name from George Boole, who introduced a computational approach to logic by studying the Boolean functions AND, OR, and NOT (Boole 1847). He observed that AND and OR have some properties in common with ordinary sum and product, and his analysis led to a set of axioms defining a class of algebraic structures now known as **Boolean algebras**. An important example studied by Boole is the algebra of *subsets* of an arbitrary set X, where $\cap$ plays the role of AND, $\cup$ plays the role of OR, and complement (with respect to X) plays the role of NOT.

However, the Boolean algebra studied here, which has only the elements 0 and 1, is more simply viewed as **mod 2 arithmetic**. Certainly, p AND q is the same as the function pq on the numbers 0 and 1, but p OR q does not resemble + as much as the EXCLUSIVE OR function ("p or q but not both"), which has the truth table

p	q	p EXCLUSIVE OR q
0	0	0
0	1	1
1	0	1
1	1	0

The EXCLUSIVE OR function is precisely the same as the **mod 2 sum**, the sum in the field $\mathbb{Z}/2\mathbb{Z}$ defined in section 7.1. In $\mathbb{Z}/2\mathbb{Z}$ the ordinary OR function p OR q is the polynomial function $p + q + pq$. Moreover, the function NOT p is simply $p + 1$, so all Boolean functions can be expressed in terms of the mod 2 sum and product—another less-than-obvious theorem.

Validity and Satisfiability

By the method of truth tables or, equivalently, by mod 2 arithmetic, we can now calculate the value of any Boolean function $f(p, q, r, \ldots)$ for any values of the variables. This solves (in principle) the two main problems of propositional logic: validity and satisfiability.

Validity. Decide whether $f(p, q, r, \ldots) = 1$ for all values of $p, q, r, \ldots$. If so, f is called **valid** because it is true for any sentences $p, q, r, \ldots$. Examples are the functions

$$p \text{ OR (NOT } p) \text{ and } p \text{ IMPLIES } (q \text{ IMPLIES } p).$$

Satisfiability. Decide whether $f(p, q, r, \ldots) = 1$ for some values of $p, q, r, \ldots$.

However, if $f(p_1, p_2, \ldots, p_n)$ is a Boolean function of n variables, to solve these problems we may have to compute $f(p_1, p_2, \ldots, p_n)$ for 2^n sequences of values of $p_1, p_2, \ldots, p_n$, since each p_i can take two values. Thus the computational difficulty of establishing either validity or satisfiability appears to grow exponentially with the number of variables.

This seemingly small cloud over propositional logic has never been dispersed. In fact, it has grown to become the main problem in logic and the theory of computation today. No matter which approach one takes to propositional logic—and there are several, as we will see below—the problem of exponential growth reappears.

15.2 AXIOMS FOR PROPOSITIONAL LOGIC

The validity and satisfiability problems for propositional logic are inherently finite, and we should not expect a finite procedure like truth tables to apply to any logic that is adequate for mathematics. To see what is in store when we explore such a logic, we first look at other approaches to propositional logic, with an eye toward generalizing them.

Generating Valid Formulas from Axioms

Another way of approaching valid formulas is to view them as *theorems* that might follow from some special formulas that we take as *axioms*. As we have seen, this has been the method of mathematics since Euclid. However, mathematics took logic for granted, and we now want to explain logic itself, so we need to be more precise about what it means for one statement to "follow" from others. The precise means by which a statement follows from statements proved previously are called **rules of inference**.

We now introduce some symbolism to shorten the statement of axioms and rules of inference for logic:

$$\wedge \text{ for AND, } \quad \vee \text{ for OR, } \quad \neg \text{ for NOT, } \quad \Rightarrow \text{ for IMPLIES.}$$

(The symbols for AND and OR are analogous to the symbols $\cap$ and $\cup$ for set intersection and union).

Many axiom systems for propositional logic are known, and typically they are written with a restricted set of connectives, such as $\{\neg, \vee\}$ or $\{\neg, \Rightarrow\}$. We know from the previous section that any connective may be written in terms of $\{\wedge, \vee, \neg\}$, and $\wedge$ may be eliminated because $p \wedge q$ equals $\neg((\neg p) \vee (\neg q))$. Also $\vee$ may be replaced by $\Rightarrow$ because $p \vee q$ equals $(\neg p) \Rightarrow q$.

For example, the system of Church (1956) uses the connectives $\neg$ and $\Rightarrow$, and its axioms are

$$p \Rightarrow (q \Rightarrow p),$$
$$(s \Rightarrow (p \Rightarrow q)) \Rightarrow ((s \Rightarrow p) \Rightarrow (s \Rightarrow q)),$$
$$\neg\neg p \Rightarrow p.$$

Its rules of inference are **substitution**, whereby any variable may be replaced in all its occurrences by a formula and **modus ponens** or **cut**, which allows us to derive q from the formulas p and $p \Rightarrow q$. Bearing in mind that p may be replaced by any formula, it is conceivable that the proof of a short formula q may involve a very long formula $p \Rightarrow q$. This actually happens in the proof of the formula $p \Rightarrow p$, which led Church (1956: 81) to comment that:

> The reader ... may be led to remark that this proposed theorem is
> not only obvious but more obvious than any of the axioms.

Nevertheless, modus ponens (as one can guess from its Latin name) is a classical rule of inference, and it can be shown that all valid formulas are in fact derivable from Church's axioms by his rules of inference. Moreover, *only* valid formulas are derivable, because it can be checked by truth tables that the axioms are valid and that the rules of inference produce valid formulas from valid formulas.

The first to prove these properties, called, respectively, the **completeness** and **consistency** of propositional logic, was Post (1921). He used a different axiom system, with the connectives $\vee$ and $\neg$, but the same rules of inference (appropriately rewritten in terms of $\vee$ and $\neg$).

Gentzen (1935) took a new approach to logic, motivated by the desire to present proofs more naturally. In particular, he wished to avoid modus ponens, so that long formulas were not needed to prove short ones, and to display the *shape* of a proof as a tree rather than as a straight line. The avoidance of modus ponens is now called **cut elimination**. Around 1955, several logicians seem to have noticed an even simpler approach, in which rules of inference arise as the reversals of "falsification rules." The history of this development is described in Annellis (1990).

Cut-Free Proofs

Both cut elimination and the tree structure of proofs are nicely illustrated in the case of propositional logic. To minimize notational complications we first make some simplifications that will be used without comment below:

1. We take all formulas to be written in terms of $\vee$ and $\neg$. For example,

$$p \Rightarrow (q \Rightarrow p) \quad \text{becomes} \quad (\neg p) \vee ((\neg q) \vee p).$$

2. We omit brackets where possible, using the associativity of the $\vee$ operation. In particular,

$$(\neg p) \vee ((\neg q) \vee p) \quad \text{is rewritten} \quad (\neg p) \vee (\neg q) \vee p.$$

3. We can change the order of terms connected by the $\vee$ operation, using its commutativity. In the present example,

$$(\neg p) \vee (\neg q) \vee p \quad \text{can be rewritten} \quad (\neg p) \vee p \vee (\neg q).$$

Now we are ready to find the axioms and rules of inference for a cut-free system. These are most easily discovered by working backward, by aiming instead to *falsify* a formula by splitting it into smaller parts, or by shortening it. This creates a **falsification tree**, which, if all branches end in unfalsifiable formulas, we turn upside down to create a **proof tree**. Since we are using only the connectives $\vee$ and $\neg$, every formula has the form either $P \vee Q$ or $\neg R$ for some subformulas P, Q, R.

Falsification rules. To falsify a formula $P \vee Q$ we have to falsify both P and Q, in which case we look inside P and Q to see whether they can be split or shortened. Splitting or shortening occurs only for a formula of the form $\neg R$, in which case we have either $R = P \vee Q$ or $R = \neg S$.

Splitting. To falsify a formula $\neg(P \vee Q)$ we have to falsify either $\neg P$ or $\neg Q$, which we indicate by writing P and Q on separate branches below $\neg(P \vee Q)$ in the falsification tree. Thus we aim to find a falsifiable formula at the end of *at least one branch* of the falsification tree:

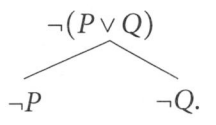

Shortening. To falsify $\neg\neg S$ we have to falsify the formula S, which we indicate by continuing the branch from $\neg\neg S$ farther down the falsification tree, to S:

$$\begin{array}{c} \neg\neg S \\ | \\ S. \end{array}$$

These rules remain valid when the formula to be split or shortened is combined by $\vee$ with another formula T. Thus the **general splitting rule** is

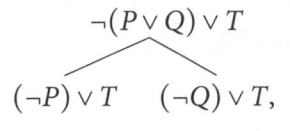

$$\neg(P \vee Q) \vee T$$

$$(\neg P) \vee T \qquad (\neg Q) \vee T,$$

and the **general shortening rule** is

$$\neg\neg S \vee T$$

$$S \vee T.$$

Now let us see what happens when we apply the rules to a formula that is unfalsifiable, or **valid**, such as

$(p \Rightarrow q) \Rightarrow ((\neg q) \Rightarrow (\neg p))$, which equals $\neg((\neg p) \vee q) \vee ((\neg\neg q) \vee (\neg p))$.

Shortening the $\neg\neg q$ to q and omitting unnecessary brackets leads to the formula $\neg((\neg p) \vee q) \vee q \vee (\neg p)$, which has the following falsification tree:

$$\neg((\neg p) \vee q) \vee q \vee (\neg p)$$

$$(\neg\neg p) \vee q \vee (\neg p) \qquad (\neg q) \vee q \vee (\neg p)$$

$$p \vee q \vee (\neg p).$$

We see that the formulas at the ends of branches *cannot be falsified*: one because it contains $p \vee (\neg p)$ and, the other because it contains $q \vee (\neg q)$. For any formula, splitting and shortening ultimately produce formulas containing only variables and negated variables, connected by $\vee$. When these variables and negated variables are *distinct* we can falsify the formula by giving each of them the value 0. But an unfalsifiable formula—that is, a valid formula—must have a falsification tree whose branches all end in formulas containing $p \vee (\neg p)$ for some variable p.

Now, looking at falsification in reverse, we get axioms and rules of inference that produce all valid formulas. The axioms are the unfalsifiable formulas occurring at the ends of branches. The rules are cut-free, and in fact they always increase the length of formulas.

Axioms. All formulas are of the form $p \vee (\neg p) \vee T$, where p is a variable and T is any formula.

Rules of inference. For any formulas P, Q, S, T, the rules are the reverses of splitting and shortening, namely,

$$\text{from } (\neg P) \vee T \text{ and } (\neg Q) \vee T, \text{ infer } \neg (P \vee Q) \vee T, \text{ and}$$
$$\text{from } S \vee T \text{ infer } \neg\neg S \vee T.$$

If one prefers simpler axioms, one can take just the formulas $p \vee (\neg p)$, where p is a variable. Then we need another rule for any formulas S and T: from S infer $S \vee T$. In any case, we need commutative and associative rules for $\vee$ to rearrange and regroup formulas.

Another convenient feature of this axiom system is that it is easily seen to be **consistent**, since it proves no formula with truth value 0. This is because the axioms clearly have truth value 1, and the rules of inference lead from formulas of truth value 1 to conclusions with truth value 1.

To sum up, we now have a consistent axiomatic approach to propositional logic for which *the proof of any valid formula T can be represented as a tree of depth no greater than the number of symbols in T*. In the next two sections we will see to what extent this idea applies to the logic of mathematics.

15.3 PREDICATE LOGIC

As von Plato (2017: 94) remarked, "Frege is the founder of contemporary logic." He refers to Frege (1879), titled *Begriffschrift* (*Concept Writing*), whose subtitle he translates as "a formula language of pure thinking, modeled upon that of arithmetic." Because of its novel two-dimensional notation,

> Frege's book made a rather bizarre impression on many, and no one else ever used it. Luckily he had Bertrand Russell among his few readers; some twenty-five years after the *Begriffschrift*, Russell rewrote Frege's formula language in a style, adopted from Giuseppe Peano, that later evolved into the standard logical and mathematical notation we have today.
>
> (von Plato 2017: 94)

What then are the nature and purpose of this language? The main distinction between propositional and predicate logic is that sentences have more than just a truth value—they have *internal structure* involving **predicates** and **individual variables**. The individual variables $x, y, z, \ldots$ range over some domain D of mathematical objects, such as numbers or points, and the predicates $P, Q, R, \ldots$ denote properties of, or relations between, objects in D.

Thus $P(x)$ may be read as "x has property P," and $R(x, y)$ as "x stands in relation R to y." Frege and many others made the equality relation = part of the language. This is convenient though not strictly necessary, since one can replace $x = y$ by a relation $E(x, y)$ and state the properties of E as axioms. Likewise, it is convenient to admit function symbols, though also not strictly necessary because $f(x_1, x_2, \ldots, x_n) = y$ is a relation $F(x_1, x_2, \ldots, x_n, y)$ with the property that there is at most one y for each n-tuple $(x_1, x_2, \ldots, x_n)$.

The important ingredients of the language that go along with individual variables are the **quantifiers** $\forall x$, "for all x," and $\exists x$, "there exists an x." Thus, for example, $\forall x \exists y R(x, y)$ is read as "for all x there exists a y such that x stands in relation R to y." The order of quantifiers is important, as one learns from a common example in analysis: the distinction between continuity and uniform continuity, first raised in section 11.6.

There we said a function f is **uniformly continuous** on domain D if, for each $\varepsilon > 0$, there is a $\delta > 0$ such that, *for all $a, x \in D$,*

$$|x - a| < \delta \quad \text{implies} \quad |f(x) - f(a)| < \varepsilon.$$

Using quantifier symbols, uniform continuity is the property

$$\forall \varepsilon \exists \delta \forall a \forall x (|x - a| < \delta \text{ implies } |f(x) - f(a)| < \varepsilon),$$

whereas continuity is

$$\forall \varepsilon \forall a \exists \delta \forall x (|x - a| < \delta \text{ implies } |f(x) - f(a)| < \varepsilon).$$

In the first case, $\forall \varepsilon \exists \delta$ says δ is a function of ε, so δ *is independent of a.* In the second case, $\forall \varepsilon \forall a \exists \delta$ says δ is a function of ε *and a.*

So much for the *language* of predicate logic; now for the logic itself. First we note that either of $\forall$ or $\exists$ can be defined in terms of the other, because $\neg \forall x \neg P(x)$ is equivalent to $\exists x P(x)$ and because $\neg \exists x \neg P(x)$ is equivalent to $\forall x P(x)$. We will use just $\forall$, together with the connectives $\vee$ and $\neg$ already known to be sufficient for the propositional part of logic. The rules of inference for $\vee$ and $\neg$ still apply, and the **axioms** are $P \vee (\neg P) \vee T$, but P, Q, S, T are now any formulas in the language of predicate logic.

Finally, there are two **rules of inference** for $\forall$ that, like those for propositional logic, come from falsification rules. They involve **constants**, which are really just "free variables," that is, variables different from any variable within the scope of a quantifier:

- To falsify $\forall x P(x) \vee S$ falsify $P(a) \vee S$ for any constant a.
- To falsify $\neg \forall x P(x) \vee S$ falsify $\neg P(a) \vee \neg \forall x P(x) \vee S$.

It seems unfortunate that the second rule does not shorten the formula, but this is unavoidable because our first choice $x = a$ to falsify $\neg \forall x P(x)$ may not be the right choice, so we have to carry along $\neg \forall x P(x)$ in case we need to try again. So the second rule can lead to an infinite branch, but only if the branch contains no formula of the form $P \vee (\neg P)$, because such a formula is unfalsifiable and causes the branch to terminate.

We will see in the next section that an infinite branch in the falsification tree of a formula F gives an **interpretation** that falsifies F, so F is not **valid**. Thus a valid formula has a falsification tree in which all branches terminate in unfalsifiable formulas, which are **axioms**, and so F is provable by **rules of inference** that are just the reverse of the falsification rules. These are the rules of inference already found for propositional logic plus the following quantifier rules:

From $P(a) \vee S$ for any constant a infer $\forall x P(x) \vee S$.
From $\neg P(a) \vee \neg \forall x P(x) \vee S$ for any constant a infer $\neg \forall x P(x) \vee S$.

15.4 GÖDEL'S COMPLETENESS THEOREM

Frege (1879) was the first to propose axioms and rules of inference for predicate logic. His axioms were of the type that became standard until the 1930s (and even later, for many logicians), and his rules included modus ponens and "from $P(a)$ infer $\forall x P(x)$." Frege apparently believed his rules were complete, but the first to prove this was Gödel (1930). For the cut-free rules given at the end of the previous section, there is an easier proof which also exposes an essential ingredient: a property of infinite trees that was proved in a more general form by König (1927). A similar idea for proving completeness was already present in Skolem (1928), though it was not noticed until later.

Weak Kőnig lemma. *If a binary tree has infinitely many vertices, then it has an infinite path.*

Proof. As in section 10.1, we take a tree to be a connected graph with no simple closed path. The tree is *binary* if there is a *root vertex v* of valency at most 2 and every other vertex has valency at most 3. Thus there are at most two vertices one edge away from *v*; for each of these there are at most two vertices two edges away from *v*; and so on. Thus paths out of *v* fork into at most two paths at each vertex, hence the term "binary." Figure 15.1 shows a typical binary tree.

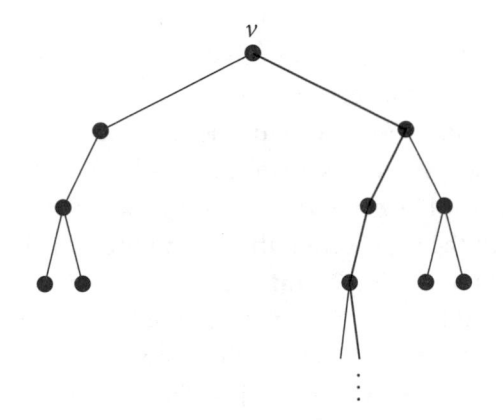

Figure 15.1 : Path in a binary tree

The argument for an infinite path is like arguments used in analysis for the Bolzano-Weierstrass and Heine-Borel theorems (sections 14.3 and 11.7, respectively). We repeatedly divide the tree in two and argue that, if there are infinitely many vertices in the whole, then there are infinitely many vertices in one half. Thus if there are infinitely many vertices in the tree, there are infinitely many vertices in one of the (at most) two subtrees connected to *v*—the one on the left or the one on the right. Take the leftmost edge into an infinite subtree and repeat the argument. If the tree is infinite, then this process can be continued indefinitely, giving an infinite path. □

Before we embark upon the proof of the completeness theorem, we recall again the falsification rules, given in diagrammatic form in section 15.2:

1. To falsify $\neg(P \vee Q) \vee S$ falsify either $\neg P \vee S$ or $\neg Q \vee S$.
2. To falsify $\neg\neg S \vee T$ falsify $S \vee T$.
3. To falsify $\forall x P(x) \vee S$ falsify $P(a) \vee S$ for any constant a.
4. To falsify $\neg\forall x P(x) \vee S$ falsify $\neg P(a) \vee \neg\forall x P(x) \vee S$.

Completeness of predicate logic. *Any valid formula F of predicate logic has a falsification tree in which all branches terminate in valid formulas of the form $P \vee (\neg P) \vee S$.*

Proof. Given a valid (hence unfalsifiable) formula F of predicate logic we apply the falsification rules in such a way that any rule that *can* be applied eventually *is* applied. Then, by the weak Kőnig lemma, either all branches terminate or there is an infinite branch. If all branches terminate then the unfalsifiable formulas at their ends are necessarily of the form $P \vee (\neg P) \vee S$.

Otherwise, there is an infinite branch that, because we missed no opportunity to apply a rule, has the following properties:

1. If $\neg(P \vee Q)$ occurs in a formula at some vertex, then it is split into $\neg P$ and $\neg Q$ at some lower vertices, one of which is on the branch.
2. If $\neg\neg S$ occurs in a formula at some vertex, then it is shortened to S at some lower vertex on the branch.
3. If $\forall x P(x)$ occurs in a formula at some vertex, so does $P(a_i)$, for some constant a_i.
4. If $\neg\forall x P(x)$ occurs in a formula at some vertex, so does $\neg P(a_i)$ for every constant a_i.

Also, because we miss no opportunity to apply rules 3 and 4, along the infinite branch every predicate term $R(x, y, \ldots)$ with quantified variables is eventually broken down to a quantifier-free term $R(a_i, a_j, \ldots)$ for constants $a_i, a_j, \ldots$. These terms may be accompanied by quantified terms, carried along by falsification rule 4, but nevertheless a term of the form $R(a_i, a_j, \ldots)$ will occur *and* $\neg R(a_i, a_j, \ldots)$ *will not.* Otherwise we would have a formula of the form $R(a_i, a_j, \ldots) \vee \neg R(a_i, a_j, \ldots) \vee S$, which cannot be falsified, and the branch would terminate.

Because of this, we can consistently assign each term $R(a_i, a_j, \ldots)$ the value "false." Then, working back up the branch, the falsification rules make every formula on the branch false, including the formula F at the top.

Therefore, since F is unfalsifiable, all branches terminate in unfalsifiable formulas, necessarily of the form $P \vee (\neg P) \vee T$. $\qquad\qquad$ □

Corollary (completeness). *The axioms $P \vee (\neg P)$ yield all the valid formulas of predicate logic by certain rules of inference.*[1]

Proof. All the formulas $P \vee (\neg P) \vee T$ occurring at the endpoints of branches in the falsification tree of an unfalsifiable formula F may be derived from the axioms $P \vee (\neg P)$ by the rule "from S infer $S \vee T$," together with commutative and associative rules for $\vee$.

Then, if all branches of the falsification tree for F terminate in formulas $P \vee (\neg P) \vee T$, we may derive all the formulas above them, including F itself, by the reverses of the falsification rules 1–4:

$$\text{From } (\neg P) \vee T \text{ and } (\neg Q) \vee T \text{ infer } \neg(P \vee Q) \vee T.$$
$$\text{From } S \vee T \text{ infer } \neg\neg S \vee T.$$
$$\text{From } P(a) \vee S, \text{ for any constant } a, \text{ infer } \forall x P(x) \vee S.$$
$$\text{From } \neg P(a) \vee \neg \forall x P(x) \vee S, \text{ for any constant } a, \text{ infer } \neg \forall x P(x) \vee S.$$

This gives a proof of any **valid** (that is, unfalsifiable) formula F with a finite falsification tree. And an F with an infinite falsification tree is falsifiable by the argument above. Thus all valid formulas of predicate logic are provable from the axioms $P \vee (\neg P)$ by the above rules of inference. $\quad$ □

When Gödel reflected in 1967 on his completeness proof (letter to Wang in Gödel 2003: 397), he ventured the opinion that he succeeded because he was willing to consider a *nonconstructive* proof, that is, one that does not try to *decide*, for a given formula F, whether F is provable or not. Rather, the decision is made by a process (in our case, building the falsification tree) whose outcome cannot always be foreseen. The provability of a formula is, so to speak, only *semi*decidable. If the formula is provable, we will find a proof in a finite time; if not, the process can run forever and we may never know that it is not going to stop. In 1930, Gödel did not know whether this situation was inevitable, but in 1936

---

1. As in the case of propositional logic, in section 15.2, there is an easy consistency proof for this system of predicate logic. The proof "forgets" the individual variables and quantifiers, treating both $P(a)$ and $\forall x P(x)$ as a propositional variable P. Then we see that the axioms have truth value 1, and the rules of inference draw conclusions of value 1 from premises of value 1, so formulas of value 0 are not provable.

Turing and Church showed that it was. We will see why in the next two sections.

15.5 REDUCING LOGIC TO COMPUTATION

After Post proved the completeness and consistency of an axiom system for propositional logic in 1921, also giving a method for deciding validity, he took aim at a much bigger problem: doing the same for *any* axiom system and *any* rules of inference. Viewing axioms simply as strings of symbols, Post began with the most general concept of "rule of inference" that he could think of: a procedure that takes a finite set of strings (the "premises") and produces, from parts of the premises, a particular string called the "conclusion."

A simple example is the modus ponens rule:

Two strings $P, P \Rightarrow Q$ produce the string Q.

Post's generalization of this was the following: finitely many *premises*

$$g_{11}P_{i_{11}}g_{12}P_{i_{12}} \cdots g_{1m_1}P_{i_{1m_1}}g_{1(m_1+1)}$$

$$g_{21}P_{i_{21}}g_{22}P_{i_{22}} \cdots g_{2m_2}P_{i_{2m_2}}g_{2(m_2+1)}$$

$$\cdots$$

$$g_{k1}P_{i_{k1}}g_{k2}P_{i_{k2}} \cdots g_{km_k}P_{i_{km_k}}g_{k(m_k+1)},$$

which together produce a *conclusion*

$$g_1 P_{i_1} g_2 P_{i_2} \cdots g_m P_{i_m} g_{m+1}.$$

Here the g_{ij} are certain specific symbols or strings of symbols, such as the $\Rightarrow$ symbol in modus ponens, and the P_{k_l} are arbitrary strings (such as the P and Q in modus ponens). Each P_{k_l} in the conclusion is in at least one of the premises. Post had tested this idea on the *Principia Mathematica* of Whitehead and Russell (1910)—at that time the only substantial attempt to derive all of mathematics in a single system—and found that its theorems could be generated by such rules. This gave him reason to believe that his rules would cover any other rules that draw conclusions from premises in a determinate way.

What Post had inadvertently described is a general concept of *algorithm*. He thought of it as a general way of generating *theorems*, but in fact any kind of output can be generated by a finite set of such rules, which Post called **production rules**. The problem he had set himself, which he later called the **general combinatorial decision problem**, was this: given a finite set of initial strings of symbols ("axioms") and a finite set of production rules ("rules of inference"), decide whether a given string X ("theorem") can be produced from the initial strings by the rules.

Since this problem potentially encompasses all of mathematics, it seems hopelessly difficult, yet Post was able to simplify the problem dramatically. He showed that the output of any set of production rules could be generated from a single initial string by a finite set of rules of the form

$$gP \quad \text{produces} \quad Pg',$$

which he called a **normal system**.[2] Thus, apparently, understanding all of mathematics reduces to understanding normal systems.

Despite the apparent simplicity of normal systems, their behavior can be complicated, as Post discovered in the case of the system with the eight rules below. This system may be described more simply by the rule, if the string begins with 0, remove the leftmost three symbols and attach 00 on the right; if the string begins with 1, remove the leftmost three symbols and attach 1101 on the right. For all initial strings that Post tried, the process either terminated with the empty string or else became periodic:

$$
\begin{aligned}
000P \quad &\text{produces} \quad P00 \\
001P \quad &\text{produces} \quad P00 \\
010P \quad &\text{produces} \quad P00 \\
011P \quad &\text{produces} \quad P00 \\
100P \quad &\text{produces} \quad P1101 \\
101P \quad &\text{produces} \quad P1101 \\
110P \quad &\text{produces} \quad P1101 \\
111P \quad &\text{produces} \quad P1101.
\end{aligned}
$$

This kind of system, in which the left end of the string advances by a constant amount at each step, as if trying to catch the right-hand end,

2. Most of Post's work was not published at the time, for reasons explained below and more fully in Urquhart (2009). Post himself wrote an interesting account of his early work (Post 1941). His reduction to normal form first appeared in Post (1943). A reduction of a typical system of symbolic logic to a normal system may be seen in Rosenbloom (1950), where Post's work is described in the preface as "the creation of a new branch of mathematics of the same fundamental importance as algebra and topology."

was called a **tag system** by Post. For the particular tag system here, it is *still unknown*[3] whether the outcome from an arbitrary initial string of 0s and 1s can be anything other than termination or periodicity.

This unexpected difficulty stopped Post in his tracks and led to a sudden reversal of his whole program: instead of trying to solve the general combinatorial decision problem, he would try to prove the problem *un*solvable. If, as he believed, the normal systems formalize all possible algorithms for generating sets of strings, then this can be done by a simple diagonal argument. We will see how in the next section.

15.6 COMPUTABLY ENUMERABLE SETS

Post came to the concept of a normal system, generating strings of symbols, by generalizing the concept of a formal system generating theorems. He also saw that there was no loss of generality in taking the strings generated by a normal system to stand for natural numbers, since the finitely many symbols used in a given normal system can be viewed as numerals in some base. Thus each normal system can be associated with a set of natural numbers that he called a *recursively enumerable set* (now often called a **computably enumerable set**).

He could also see that, if we settle on a fixed alphabet for all normal systems, then the normal systems can be enumerated; hence so too can the computably enumerable sets $W_0, W_1, W_2, \ldots$. This creates an opportunity for a diagonal argument like that of Cantor (1891), as we will see shortly.

Post had these, or similar, ideas in the 1920s, but they first appeared in Post (1944). By that time, the focus on computability had shifted from computably generated sets to computable functions, so Post defined a (nonempty) computably enumerable set W to be the range of a computable function f on the positive integers. This reflects the idea of "listing" the members of W as

first member $= f(1)$, second member $= f(2)$, third member $= f(3)$, ...

---

3. There was a flurry of excitement in 2017, when the string consisting of 110 repetitions of the string 100 initiated a process running for more than 10^{12} steps. Alas, after some more computation, termination was found to occur after 43,913,328,040,672 steps. This news was reported in Sloane (2018).

It is intuitively clear that the original idea of a generated set also fits this description. To "list" the strings generated by a normal system, one systematically applies the production rules to the initial string: first list the strings obtainable by one application of the rules, then those obtainable by two applications, and so on. Then define $f(n)$ to be the nth string on the list, and it is clear that $f(n)$ is computable by running the process just described until n strings have been listed.

Post also introduced the idea of a *recursive* set S of natural numbers (now often called a **computable** set), which is one having an algorithm for deciding membership. This leads to the following relationship between computable and computably enumerable sets:

A set $S \subseteq \mathbb{N}$ is computable if and only if S and its complement $\mathbb{N} - S$ are computably enumerable.

If S is computable, we can enumerate both S and $\mathbb{N} - S$ by applying the membership algorithm to $0, 1, 2, 3, \ldots$ in succession. Conversely, if both S and $\mathbb{N} - S$ are computably enumerable, we can decide whether $n \in S$ by listing both S and $\mathbb{N} - S$ simultaneously. Eventually n will appear in one of them, at which time we will know whether $n \in S$.

Now for the fundamental theorem about computable enumerable sets:

Computable enumerable but noncomputable set. *There is a computably enumerable set whose complement is not computably enumerable.*

Proof. It follows from the enumeration $W_0, W_1, W_2, \ldots$ of computably enumerable sets that the set

$$D = \{n \in \mathbb{N} : n \notin W_n\}$$

is not computably enumerable, because it differs from each computably enumerable set W_n with respect to the number n. However, its complement

$$\mathbb{N} - D = \{n : n \in W_n\}$$

is computably enumerable. We can generate its members by generating a list of the normal systems, say, by first listing the one-letter normal systems (if any), then the two-letter normal systems, and so on. As the normal systems are generated we set them running to produce the members of the corresponding sets $W_0, W_1, W_2, \ldots$. If $n \in W_n$ we will eventually observe this fact, at which point we list n in $\mathbb{N} - D$. □

The set D is the first example of a set for which there is no algorithm to decide membership. In other words, the problem of deciding membership in D is **unsolvable**. Later we will see that the same is true of several sets that arise naturally in logic and mathematics. But first we need to say more about Post's assumption that normal systems capture all possible algorithms for generating strings of symbols.

15.7 TURING MACHINES

Post's discovery of unsolvability in the early 1920s was a potential turning point in mathematics, but it was not immediately influential. To be certain that the result was correct, Post needed to know that his normal systems implement all possible algorithms, and this was not clear. Indeed, the statement is not provable in the mathematical sense—it is more like a law of nature, in need of continual testing. For this reason Post did not publish his discovery at the time (or its corollary on **incompleteness**, which we will see in the next chapter).

He did, however, develop a different concept of algorithm, equivalent to normal systems but closer to the intuitive idea of human computation. This was finally published in Post (1936), only to be upstaged by the independent publication of virtually the same idea by Turing (1936). Since Turing's paper was considerably richer in results than Post's—including a proof that the *Entscheidungsproblem* was unsolvable—the concept has since become known as the **Turing machine**.

As Turing explained in the introductory section of his paper:

> We may compare a man in the process of computing a real number to a machine which is only capable of a finite number of conditions $q_1, q_2, \ldots q_l$ which will be called "m-configurations". The machine is supplied with a "tape" (the analogue of paper) running through it, and divided into sections (called "squares") each capable of bearing a "symbol". At any moment there is just one square, say the r-th, bearing the symbol $\mathfrak{S}(r)$ which is "in the machine". We may call this square the "scanned square". The symbol on the scanned square may be called the "scanned symbol". The "scanned symbol" is the only one of which the machine is, so to speak, "directly aware". However, by altering its m-configuration the machine can effectively remember some of the symbols which it has "seen" (scanned).

previously. The possible behaviour of the machine at any moment is determined by the m-configuration q_n and the scanned symbol $\mathfrak{S}(r)$. This pair $q_n, \mathfrak{S}(r)$ will be called the "configuration": thus the configuration determines the possible behaviour of the machine. In some of the configurations in which the scanned square is blank (i.e. bears no symbol) the machine writes down a new symbol on the scanned square: in other configurations it erases the scanned symbol. The machine may also change the square which is being scanned, but only by shifting it one place to right or left. In addition to any of these operations the m-configuration may be changed.

(Turing 1936: 231)

This is almost exactly how we would describe a Turing machine today, except that "m-configurations" are now called *states* and the symbols are usually denoted by $S_1, S_2, \ldots, S_m$. We also like to give the machine a **reading head** that scans one square of tape at a time. Like the states, the symbols are finite in number. Turing gives reasons for this later in his paper, along with a justification for limiting computation to a one-dimensional "tape." (Bear in mind that in 1936 a "computer" was a human being who did computations.)

> Computing is normally done by writing certain symbols on paper. We may suppose this paper is divided into squares like a child's arithmetic book. In elementary arithmetic the two-dimensional character of the paper is sometimes used. But such a use is always avoidable, and I think that it will be agreed that the two-dimensional character of paper is no essential of computation. I assume then that the computation is carried out on one-dimensional paper, i.e. on a tape divided into squares. I shall also suppose that the number of symbols which may be printed is finite. If we were to allow an infinity of symbols, then there would be symbols differing to an arbitrarily small extent.... Symbols, if they are too lengthy, cannot be observed at one glance. This is in accordance with experience. We cannot tell at a glance whether 9999999999999999 and 999999999999999 are the same.
>
> The behaviour of the computer at any moment is determined by the symbols which he is observing, and his "state of mind" at that moment. We may suppose that there is a bound B to the number of symbols or squares which the computer can observe at one

moment. If he wishes to observe more, he must use successive observations. We will also suppose that the number of states of mind which need be taken into account is finite. The reasons for this are of the same character as those which restrict the number of symbols. If we admitted an infinity of states of mind, some of them will be "arbitrarily close" and will be confused.

(249–250)

I have quoted Turing at length because his argument was decisive in convincing the mathematical logic community that computability could be precisely defined, and hence that unsolvability of certain problems could be rigorously proved. Further evidence that Turing machines capture the concept of computability accumulated in succeeding decades as more and more formalizations of computation were proved equivalent to the Turing machine concept. Turing himself began this enterprise in 1936 by showing that Turing machine computability is the same as **λ-definability**, a concept proposed by Church (1936) as a definition of computability. Because of Church's contribution, the claim that Turing machines capture the concept of computability is now known as the **Church-Turing thesis**.

Machine Descriptions

Turing's idea of computing on a tape divided into squares, with each step of computation determined by the current state and scanned symbol, leads to a description of machines in terms of *quintuples*, which are of two types:

$$q_i S_j S_k R q_l, \quad \text{meaning}$$

"when in state q_i, scanning symbol S_j, replace it by S_k, move one square to the right, and go into state q_l"; and

$$q_i S_j S_k L q_l, \quad \text{meaning}$$

"when in state q_i, scanning symbol S_j, replace it by S_k, move one square to the left, and go into state q_l."

Because the numbers of states and symbols must be finite, a Turing machine is described by a finite set of quintuples. It is also assumed that the "input" to the machine is finite, so the machine starts on a tape on which all but a finite number of squares are blank. Here is an example of a machine, using the symbol □ for the blank square:

$$q_1 \square \square R q_2, \quad q_2 11 R q_2, \quad q_2 \square 1 R q_3$$

If we start this machine in state q_1 on a blank square to the left of a block of squares marked with 1s, then the snapshots in figure 15.2 show the tape and scanned square (with the current state written beneath it) at successive steps of the computation.

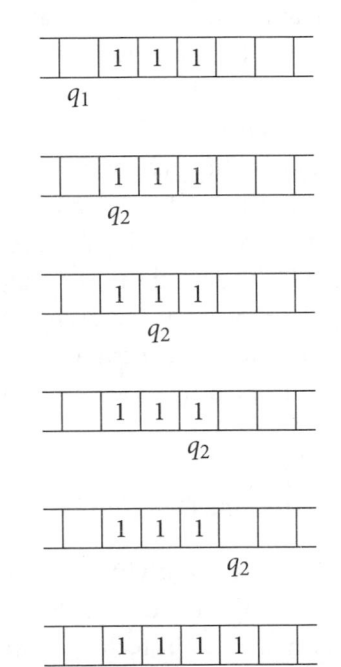

Figure 15.2 : Successive steps of computation

The last picture shows a **halting configuration**: since the quintuples do not say what to do when in state q_3 and scanning a blank square, the machine simply halts. The "output" is what remains on the tape if and when the machine halts. We see that this machine, when started anywhere to the left of an "input" of n ones, halts with an "output" of $n+1$ ones. Thus we can view it as a machine that computes the successor function.

Since a Turing machine can have any finite number of states or symbols, we are actually using infinitely many symbols $q_1, q_2, \ldots S_1, S_2, \ldots$ to describe all Turing machines. However, we can get around this by replacing q_i by the string $q''^{\ldots'}$ with i primes, and S_j by $S''^{\ldots'}$ with j primes.

Then each Turing machine description can be written as a string of symbols from the alphabet $\{q, S, R, L,'\}$. We will call this the **standard description**. Standard descriptions form a set whose members can be enumerated by listing the one-letter descriptions first (although there won't be any!), then the two-letter descriptions, and so on. In this way, each Turing machine has a position n in the list, so we can speak of the nth Turing machine, T_n.

Numbering Turing machines allows us to make diagonal arguments. Turing, with an eye toward using the classical diagonal argument for real numbers, considered machines that compute infinite decimal expansions. However, it is a little easier to do a diagonal argument for computable functions.

Definition. A function $f : \mathbb{N} \to \mathbb{N}$ is **computable** if there is a Turing machine T that, when given any block of n ones as input, halts with a block of $f(n)$ ones on its tape.

Thus, *if* we know that machine T computes function f, we can compute $f(m)$ for any natural number m by giving T a block of m ones as input and observing the number of ones on the tape when T halts. But this is a big "if," as we can see by applying the diagonal argument.

Unrecognizability of computable functions. *There is no Turing machine that, given the description of a Turing machine T, will compute whether T defines a computable function.*

Proof. To simplify the proof we assume the Church-Turing thesis. Namely, we give an intuitively computable procedure and then use the Church-Turing thesis to claim that the procedure can be implemented by a Turing machine.

Assuming there is a Turing machine S that can recognize, from the description of a Turing machine T, whether T computes a function, we apply S to the descriptions of $T_1, T_2, T_3, \ldots$ in turn. If T_n computes a function, f_n, we run T_n on input n so as to compute $f_n(n)$. This, and the ability to recognize *whether* T_n computes a function, enables us to compute the function

$$g(n) = \begin{cases} 0 & \text{if } f_n \text{ is undefined} \\ f_n(n) + 1 & \text{if } f_n \text{ is defined.} \end{cases}$$

The first case ensures that g is defined for all n, and the second case ensures that g differs from each computable function f_n.

Thus g is computable, yet different from each computable function. This contradiction shows that we were wrong to suppose there is a Turing machine that can decide, for each n, whether T_n computes a function.

□

Appealing to the Church-Turing thesis again, we can say that there is no algorithm for recognizing which Turing machines define computable functions. We also say that the problem of recognizing such Turing machines is **algorithmically unsolvable** or simply **unsolvable**. Note that unsolvability can be ascribed *only* to problems that consist of an infinite number of questions. A single question may be *unsolved*, in the sense that we do not know the answer, but if the answer is a finite string of symbols there is certainly a Turing machine that gives the correct answer—we just do not know *which* machine.

Another unsolvable problem comes to light when we try to understand why no Turing machine recognizes the T_n that define computable functions. The problem is to recognize when T_n fails to give a value for input n. If T_n halts, then we will see what is wrong with the output. So the difficulty lies in recognizing when T_n *does not halt* on a given input. The problem of deciding whether a given machine halts of a given input— the so-called **halting problem**—must be unsolvable. This result was also observed by Turing (1936).

Universal Machines

Another discovery of Turing (1936) is easily understood in light of the Church-Turing thesis: the existence of a **universal machine** U that can simulate the behavior of an arbitrary Turing machine T on an arbitrary input I. Essentially, U receives the standard description des(T) of T, plus the input I (encoded in the same alphabet used for standard descriptions), and then moves back and forth between des(T) and the encoded tape contents of T. Using some extra symbols to "mark its place" in both des(T) and the encoded tape contents, U can read the current quintuple, carry out the required change, go back to read the next quintuple, and so on.

It is intuitively clear that a human computer can carry out this process. Therefore, by the Church-Turing thesis, there is a Turing machine

U that will do the same thing. Indeed, universal machines are, well, universal today. Every desktop or laptop computer is a universal Turing machine, at least if we assume an unlimited supply of external storage ("tape").

A universal machine is also useful from a theoretical viewpoint, because it allows us to state unsolvable problems about a single machine. We can say, for example, that the halting problem is unsolvable for any universal machine U; otherwise we could decide whether machine T eventually halts after receiving input I by asking the corresponding question about the universal machine.

The Entscheidungsproblem

When Hilbert posed the *Entscheidungsproblem* in the 1920s he did so because predicate logic was already known to be strong enough to express the notion of logical consequence in mathematics. The main point of the *Entscheidungsproblem* was that its solution would decide any question of the form, Does statement S follow from axioms Σ?

But, conversely, if unsolvable problems exist they should be expressible in predicate logic—in which case the *Entscheidungsproblem* is unsolvable too. This is exactly what Turing and Church realized in 1936. In particular, Turing (1936) encoded outcomes of computations by formulas of predicate logic, so that a solution of the *Entscheidungsproblem* would solve a problem about Turing machines he had already shown to be unsolvable.

Since Turing (1936) was relating universal computation to logic for the first time (he also sketched a construction for generating the valid formulas of predicate logic), this part of his paper is quite laborious and inelegant. Today it is easier to connect Turing machines to the *Entscheidungsproblem* via an unsolvable problem in *algebra* discovered independently by Post (1947) and Markoff (1947). We describe this problem in the next section.

15.8 THE WORD PROBLEM FOR SEMIGROUPS

Post (1947) was motivated by a combinatorial problem raised by Thue (1914). It concerns strings of symbols in some finite alphabet, which we will now call **words**. Words may be transformed by a finite number of equations, and the **word problem** is to find an algorithm for deciding when words are equal according to a given finite set of equations.

For example, if the words are made from the letters a and b and we have the equation $ab = ba$, then an adjacent a and b in a word can be interchanged, so any word has a "normal form" where all the as precede all the bs. Moreover, it is clear that two words are equal if and only if they have equal numbers of as and bs.

Definitions. Given words u, v in an alphabet $\{a_1, a_2, \ldots, a_m\}$ and equations

$$u_1 = v_1, \quad u_2 = v_2, \quad \ldots, \quad u_n = v_n,$$

we call words u and v **equal** if u may be converted to v by finitely often replacing a subword of the form u_i by v_i, or vice versa. The **word problem** for a given alphabet and given equations is to find an algorithm that decides, for given words u and v, whether $u = v$.

This problem is now known as the *word problem for semigroups* because the classes $[w]$ of equal words form a **semigroup** under the "product" operation of juxtaposition:

$$[w_1][w_2] = [w_1 w_2].$$

The defining property of a semigroup is associativity of the product operation, $u(vw) = (uv)w$, which obviously holds for the product of words.

Post's (1947) idea was to use words to represent configurations of a Turing machine, and equations to make word transformations that reflect steps of computation. Hopefully, the halting problem can then be reduced to the word problem. But a slight difficulty is apparent immediately. Steps of computation are generally not reversible, whereas an equation $u_i = v_i$ can be used in either direction: to replace u_i by v_i or to replace v_i by u_i. So, initially we will use *one-way* transformations $u_i \rightarrow v_i$ (replace u_i by v_i), which better reflect computation. Later we will see that it is harmless to allow the reverse transformations, and hence equations.[4]

---

4. It is an amusing linguistic accident that the name Thue is pronounced "two-way."

From Turing Machine to Semigroup

Suppose we are given a Turing machine T with states $q_1, q_2, \ldots, q_s$ and symbols $\square = S_0, S_1, \ldots, S_t$. Then we can encode the configuration of T at any step of a computation by a word on the alphabet

$$\{q_1, q_2, \ldots, q_s, S_0, S_1, \ldots, S_t, [,]\}.$$

The string of symbols on the marked portion of tape is written using the symbols $S_0, S_1, \ldots, S_t$, with the current state q_i inserted to the left of the scanned symbol. Symbols [and] at the left and right ends of the string serve as "prompts" for the creation of extra symbols $\square$ when the reading head of T moves beyond the marked portion of tape. Figure 15.3 shows an example of a machine configuration and the **configuration word** that encodes it.

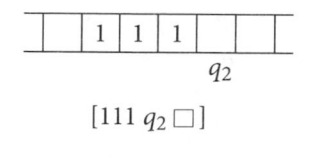

$$[111 \, q_2 \, \square]$$

Figure 15.3 : A machine configuration and its word

Each step of computation changes the configuration word, but only in the neighborhood of the state symbol q_i and the scanned symbol S_j, so the change is made by a word replacement transformation corresponding to the quintuple of T that begins with $q_i S_j$ (if any). The following table gives the word transformations corresponding to the two types of quintuple.

quintuple	replacement rules
$q_i S_j S_k R q_l$	$q_i S_j \to S_k q_l$
	$q_i] \to S_k q_l]$ if $S_j = \square$
$q_i S_j S_k L q_l$	$S_m q_i S_j \to q_l S_m S_k$ for each symbol S_m
	$[q_i S_j \to [q_l \square S_k$

Now, to exploit the unsolvability of the halting problem, we introduce a new symbol H, created by the occurrence of a halting configuration $q_i S_j$. For any word $q_i S_j$ that is *not* part of a quintuple we introduce the transformation

$$q_i S_j \to H,$$

together with the transformations

$$HS_m \to H, \quad S_m H \to H, \quad [H \to H, \quad H] \to H,$$

which enable the symbol H to "swallow" all other symbols in the configuration word, leaving just the one-letter word H. Let us call this set of word transformations (replacement rules plus symbol-swallowing) Σ_T.

Clearly, until the appearance of a halting configuration $q_i S_j$, Σ_T exactly simulates T's computation from a given initial configuration. So the symbol H appears if and only if the computation halts. And the "symbol swallowing" transformations then ensure that a word transforms to the single letter H if and only if the computation halts. This leads to the following:

Unsolvable word transformation problem. *There is no algorithm that, given a Turing machine T and a configuration word w, will decide whether the word transformation system Σ_T converts w to H.*

Proof. As we have just seen, Σ_T converts w to H if and only if the configuration of T encoded by the word w leads T to halt. Therefore a solution of the word transformation problem gives a solution of the halting problem, which is impossible. Thus the word transformation problem is also unsolvable. □

Now we convert the word transformation system Σ_T to a semigroup S_T by simply adding the *opposite* $v \to u$ to each transformation $u \to v$ in Σ_T. This amounts to changing each transformation $u \to v$ to an equation $u = v$. But any *use* of this equation is a replacement of u by v or vice versa, so we will view it as either one of the original transformations $u \to v$ or its reverse $u \leftarrow v$. This viewpoint allows us to extend the previous theorem to one about the semigroup S_T.

Detection of halting computations. *If w encodes a machine configuration of T, then $w = H$ in S_T if and only if there is a halting computation of T beginning with the configuration encoded by w.*

Proof. If there is a halting computation beginning with situation w, then

$$w \to \text{word containing } H$$

by allowing H to "swallow" symbols,

and therefore $w = H$.

Conversely, if $w = H$ then there is a series of equations

$$w = w_1 = w_2 = \cdots = w_n = \text{word containing } H, \qquad (*)$$

where w_n is the last word not containing H. Each equation $w_i = w_{i+1}$ comes from a transformation $w_i \to w_{i+1}$ or the reverse, $w_i \leftarrow w_{i+1}$, so we can rewrite each equation $w_i = w_{i+1}$ or $w_n = \text{word containing } H$ in $(*)$ with the appropriate arrow, $\to$ or $\leftarrow$, in place of the $=$ sign.

We definitely do not have $w_n \leftarrow \text{word containing } H$, because w_n does not contain H and no transformation destroys H. We cannot have $w_{n-1} \leftarrow w_n$ either, because w_n necessarily contains a halting configuration $q_i S_j$ and only the transformation $q_i S_j \to H$ applies to it.

If there is any reverse arrow at all, let

$$w_{i-1} \leftarrow w_i \to w_{i+1}$$

be its rightmost occurrence. This means that $w_{i-1} = w_{i+1}$, because each machine situation w_i has at most one successor. We can then delete $w_{i-1} \leftarrow w_i \to$ from the series, eliminating two opposite arrows. Repeating this argument, we can eliminate all reverse arrows and hence conclude that

$$w \to \text{word containing } H,$$

which means that w leads to a halting computation. $\qquad \square$

It follows immediately, by the unsolvability of the halting problem, that there is no algorithm that will decide, for a given S_T and words u and v, whether $u = v$ in S_T. We can strengthen this result to a specific semigroup by choosing T to be a universal machine.

Unsolvability of the word problem. *If U is a universal machine, then the semigroup S_U has an unsolvable word problem.*

Proof. Since U is universal, its halting problem is unsolvable. But if the word problem for S_U were solvable, we could use the solution algorithm to decide whether a given configuration word $w = H$ in S_U, thereby solving the halting problem for U, by the theorem above. Since there is no such algorithm, the word problem for S_U must be unsolvable. $\qquad \square$

A direct construction of a universal machine, and subsequent conversion of its quintuples to equations, would of course produce a very complicated semigroup. However, some ingenious simplifications have been discovered, leading to a semigroup with unsolvable word problem but remarkably simple defining equations. This semigroup was discovered by Ceĭtin (1956), then a 19-year-old student of Markoff. It is generated by the letters a, b, c, d, e, with defining equations

$$ac = ca, \quad ad = da, \quad bc = cb, \quad bd = db, \quad ce = eca, \quad de = edb, \quad cca = ccae.$$

Application to the Entscheidungsproblem

Given a finite set of equations between words

$$u_1 = v_1, \quad u_2 = v_2, \quad \ldots, \quad u_m = v_m,$$

deducing an equation $u = v$ from them is essentially a matter of logic, and it can be implemented in predicate logic quite easily if we allow constants, the equality relation, and functions in the language. If $p(x, y)$ is the function that represents the product of letters x and y, we will also need the axiom

$$p(x, p(y, z)) = p(p(x, y), z),$$

which expresses the associative law $x(yz) = (xy)z$, and axioms for equality that allow a sequence of word substitutions to be carried out, such as

$$(u = v \wedge v = w) \Rightarrow u = w.$$

Only finitely many such axioms are needed, and we can suppose them combined in a single formula φ. We assume also that the equations $u_i = v_i$ are written in terms of the function p. Then, if there were an algorithm for the *Entscheidungsproblem* we could decide whether $u = v$ follows from the equations $u_i = v_i$ by asking whether the sentence

$$[\varphi \wedge (u_1 = v_1) \wedge (u_2 = v_2) \wedge \cdots \wedge (u_m = v_m)] \Rightarrow (u = v)$$

is logically valid. But if we take a semigroup with unsolvable word problem (such as the Ceĭtin semigroup above) there is no algorithm for deciding whether such sentences are valid, and hence *there is no algorithm for the Entscheidungsproblem.*

15.9 REMARKS

This chapter has given a whirlwind tour of twentieth-century mathematical logic and computability theory, with little opportunity to follow all the avenues it opens up. Here then are some of the digressions we neglected to make on the way through.

P and NP

Section 15.1 touched on the validity and satisfiability problems in propositional logic, remarking that their computational difficulty seems to grow exponentially with the number of variables. This is clear for the truth table method of testing validity, where one must test all 2^n sequences of truth values for n variables. It is not so clear in the case of satisfiability, where one can stop as soon as a satisfying sequence of truth values is found. However, all known methods for testing satisfiability take exponential time in the worst cases.

In computational complexity theory it is usual to measure the time to answer a given question Q by the number of steps in a Turing machine computation, and to express this number as a function of the number of symbols, $|Q|$, in Q. A problem $\mathscr{P}$ consisting of questions Q is said to be in the class **P** of **polynomial-time solvable** problems if there is a Turing machine that correctly answers each question Q in $\mathscr{P}$ in time at most $f(|Q|)$, where f is a polynomial. These are reckoned to be the "feasible" problems.

The reason for choosing polynomial functions rather than, say, linear functions is that there are many models of computation other than Turing machines, and different models may give polynomial computation times of different degrees. For example, an ordinary Turing machine may take quadratic time to solve a problem solvable in linear time by a machine with two tapes. However, all physically realistic machines that have been considered can be simulated by an ordinary Turing machine with at worst a polynomial increase in computation time.[5]

Here are examples of problems in **P**:

1. Adding two n-digit numbers (in ordinary base 10, or base 2, notation), which takes linear time in n by the usual algorithm.

---

5. If we lived in a hyperbolic space this would no longer be the case. It can be seen from the pictures of tilings of the hyperbolic plane in section 9.7, and it also follows from Gauss's formula for the circumference of a circle in the hyperbolic plane in section 9.5, that the number of cells in a tiling within distance r of the origin grows *exponentially* with r. A machine that enlists all these cells in a computation, called a **cellular automaton**, can exploit their exponential growth to solve problems such as the satisfiability problem in polynomial time, though an ordinary Turing machine apparently cannot.

2. Multiplying two n-digit numbers, which takes quadratic time by the usual algorithm.

3. Testing whether a given assignment of truth values to its variables satisfies a propositional logic formula of length n, which takes polynomial time (I guess of degree 3 or 4—maybe the reader would like to think about this).

Related to these problems, in a kind of "inverse" way, are the following problems, which we do *not* know how to solve in polynomial time:

4. Given a finite set S of n natural numbers of length at most m, and a natural number k, find a subset of S whose members sum to k (this is called the **knapsack problem**).

5. Given an n-digit number k, find the prime factors of k.

6. Given a propositional logic formula φ of length n, find an assignment of truth values that satisfies φ (this is the **satisfiability problem**).

Because of the way problems 4, 5, 6 are related to problems 1, 2, 3, respectively, it is clear that 4, 5, 6 can each be solved in polynomial time with the help of a "lucky guess"—to solve 4, guess the right subset; to solve 5, guess the factors; to solve 6, guess the assignment of truth values—after which the guess can be verified by a polynomial-time computation.

The latter type of problems belong to the class **NP** of **nondeterministic polynomial time** problems, which is defined by generalizing the concept of Turing machine to allow "guesses" at certain steps of the computation. A problem $\mathscr{P}$ is in **NP** if answers to the questions Q in $\mathscr{P}$ can be found in at most $f(|Q|)$ steps by a nondeterministic Turing machine, where f is a polynomial.

What section 15.1 called "the main problem in logic and computation today" is whether **NP** = **P**. The problem was first posed in this form by Cook (1971), though there is a hint of it in a 1956 letter from Gödel to von Neumann. In it Gödel pointed out that *proofs in mathematics* are the prime example of being hard to find yet easy to verify. If it turns out that **NP** = **P**, then mathematicians may no longer be useful:

> Namely, it would obviously mean that in spite of the undecidability
> of the Entscheidungsproblem, the mental work of a mathematician
> concerning Yes-or-No questions could be completely replaced by

a machine. After all, one would simply have to choose the natural number n [the length of proof sought] so large that when the machine does not deliver a result, it makes no sense to think more about the problem.

(translation from Gödel 2003: 375)

Consistency and Existence

The proof of the completeness theorem for predicate logic in section 15.4 has some interesting corollaries on models and mathematical existence. If we repeat the construction of a falsification tree for an *arbitrary* formula F, as in section 15.4, an infinite branch still gives an interpretation the falsifies F. A finite branch may also allow us to falsify F, more trivially, by ending in a formula $P \vee Q \vee R \vee \cdots$ where $P, Q, R, \ldots$ can be falsified dependently.

Now suppose F is a formula that does not lead to a contradiction. Then $\neg F$ can be falsified, because if $\neg F$ cannot be falsified we can prove $\neg F$, which certainly contradicts F. Since $\neg F$ can be falsified, there is an interpretation that falsifies $\neg F$ and hence *satisfies F*, so we have:

Noncontradiction implies existence. *If F is a formula of predicate logic that does not lead to contradiction, then there is an interpretation of the relation symbols in F, over some domain, that satisfies F.*   □

This corollary of the completeness theorem bears out a remark about mathematical existence made by Hilbert (1902: 448):

> If contradictory attributes be assigned to a concept, I say, that mathematically the concept does not exist. So, for example, a real number whose square is −1 does not exist mathematically. But if it can be proved that the attributes assigned to the concept can never lead to a contradiction by the application of a finite number of logical processes, I say that the mathematical existence of the concept (for example, of a number or a function which satisfies certain conditions) is thereby proved.

Hilbert in 1902 probably had no concrete sense of "existence" in mind, but the proof of the completeness theorem gives a concrete interpretation, in which the variables of the noncontradictory formula F range over a certain set T of constant terms, and the relation symbols in F denote relations on T given by an infinite branch in the falsification tree for $\neg F$.

There is a generalization of this process for finding an interpretation of a noncontradictory formula that finds an interpretation of a noncontradictory *set* of formulas, such as the set of axioms for Peano arithmetic (assuming it is consistent) plus $N > 0$, $N > 1$, $N > 2$, ... mentioned in section 13.5. Thus we can find concrete (though not necessarily natural) models of "infinite numbers" and "infinitesimals."

Binary Trees

The proof of the completeness theorem is nonconstructive because there is no obvious algorithm for finding an infinite path in a binary tree. Even if the set of vertices of the tree is computable, finding an infinite path depends on the (apparently noncomputable) ability to decide which edge out of a given vertex leads to an infinite subtree. This intuition is confirmed by a result of Kreisel (1953), which constructs an infinite computable binary tree with no computable infinite path. This means that the weak Kőnig lemma used to prove the completeness theorem is *not* constructively provable.

One might hope for a proof of the completeness theorem that avoids the weak Kőnig lemma asserting the existence of an infinite path. But this idea also fails because it can be constructively shown that the completeness theorem for predicate logic *implies* the weak Kőnig lemma. Thus the completeness theorem is constructively *equivalent* to the weak Kőnig lemma and hence not constructively provable outright. This is a result of **reverse mathematics** that we discuss in section 16.5.

In fact, the weak Kőnig lemma is ubiquitous in mathematics, appearing nearly everywhere where tree arguments are common. These include the "infinite bisection" proofs of the Heine-Borel theorem and the extreme value theorem that we saw in section 11.7 and will see again in section 16.7. Reverse mathematics shows that these theorems are also constructively equivalent to the weak Kőnig lemma and hence to the completeness theorem—revealing a surprising commonality between analysis and logic.

CHAPTER 16

Incompleteness

In the previous chapter we saw how the early decades of the twentieth century brought new clarity to the concept of proof—to the point where *proof itself* could become a subject of mathematical study. In particular, a logic suitable for all mathematics, **predicate logic**, had been described and shown to be complete by Gödel (1930). In the same period the concept of computation had been formalized by Post and Turing, along with the concepts of algorithmically *solvable* and *unsolvable* problems.

Post realized, as early as 1921, that our understanding of any unsolvable problem $\mathscr{P}$ is necessarily **incomplete**. That is, if $\mathscr{P}$ consists of questions $Q_1, Q_2, Q_3, \ldots$, then no axiom system $\mathscr{A}$ can generate precisely the true propositions of the form "Q_i has answer A_i," because in doing so it would solve the unsolvable problem $\mathscr{P}$. However, Post hesitated to publish this discovery, due to his doubts about its generality, which lingered until Turing (1936) settled the concept of computation.

In the meantime, Gödel (1931) proved incompleteness of the specific system *Principia Mathematica*. He did so by a very general **arithmetization** method, ultimately applicable to any computable means of producing theorems. Arithmetization implies, in particular, that any consistent axiom system $\mathscr{A}$ for arithmetic is incomplete. Also, one of the statements that $\mathscr{A}$ cannot prove is the sentence Con($\mathscr{A}$) asserting its own consistency.

But why is the consistency of arithmetic hard to prove? Gentzen (1936) found that the complexity of arithmetic is measured by a countable ordinal called ε_0, and that ε_0-**induction** suffices to prove the consistency of first-order Peano arithmetic (PA). Since then ε_0 has been found lurking in several interesting propositions of arithmetic that are not provable in PA.

16.1 FROM UNSOLVABILITY TO UNPROVABILITY

There are several ways to view incompleteness theorems, which reflect the evolution of proof theory and computability theory from Post to Gödel to Turing (with some input from von Neumann too). This section discusses the first, unpublished, stage of this evolution in the work of Post in the 1920s. We know about this stage mainly from Post (1941), which was ultimately published in Davis (2004), and some details given in Post (1943) and Post (1944).

Post's aim, as we saw in section 15.5, was to describe the general process of computably generating theorems from "axioms" by "rules of inference." He in fact found a general description of such processes, equivalent to all other definitions proposed in the subsequent century. But, since his definition was the first, he had to wait until others had thought about the same question, with the result that Church and Turing published first and got most of the credit for the formal definitions of computability and algorithmic solvability. They also got the credit, more justifiably, for finding specific algorithmically *unsolvable* problems, such as the *Entscheidungsproblem*.

Nevertheless, Post's simple path from computability to unsolvability to incompleteness somehow escaped notice until it was published in Post (1944). As emphasized in section 15.6, the path from computability to unsolvability is essentially the computable version of the Cantor (1891) "diagonal" argument that $\mathscr{P}(\mathbb{N})$ is uncountable:

- Enumerate the computably enumerable sets $W_0, W_1, W_2, \ldots$, for example, by enumerating the normal systems that generate them.
- Observe that the set $D = \{n : n \notin W_n\}$ is *not* computably enumerable, because it differs from each W_n with respect to the number n.
- Conclude that the problem of deciding, for any n, whether $n \in D$ is unsolvable. This is because an algorithm that solves this problem, applied to $0, 1, 2, \ldots$ in turn, gives a computable enumeration of D.

It follows easily that *there is no computable process for generating all the correct theorems of the form $m \notin W_n$*. If there were, by picking out the theorems of the form $n \notin W_n$ we could computably enumerate D, which is impossible. Therefore, *any computable process for generating correct theorems of the form* $m \notin W_n$ *is incomplete*; that is, the process fails to generate some correct theorem of the form $n \notin W_n$.

In fact, for any system $\mathscr{A}$ that computably generates correct theorems we can find a *specific* true statement of the form $n_0 \notin W_{n_0}$ that $\mathscr{A}$ fails to prove. Since the theorems of $\mathscr{A}$ are computable enumerable, so is the set

$$W_{n_0} = \{n : \mathscr{A} \text{ proves the sentence } ``n \notin W_n"\}.$$

Then $n_0 \in W_{n_0}$ means, by definition of W_{n_0}, that $\mathscr{A}$ proves "$n_0 \notin W_{n_0}$." If so, it is *true* that $n_0 \notin W_{n_0}$, by our assumption of correctness, and we have a contradiction. Therefore, it is actually true that $n_0 \notin W_{n_0}$, *but $\mathscr{A}$ does not prove it*, to escape the contradiction $n_0 \in W_{n_0}$.

Since a "computable process for generating theorems" includes any axiom system with well-defined rules of inference, the statement of incompleteness we just arrived at is extremely general. It gets both generality and precision from the definition of computability, which solidified around 1936 in the **Church-Turing thesis**. Before computability was considered a well-defined notion, incompleteness could be proved only for a particular system, such as *Principia Mathematica*. This of course was achieved by Gödel (1931), and to do so he had to describe the theorem-proving process of *Principia* in the language of *Principia* itself.

16.2 THE ARITHMETIZATION OF SYNTAX

The idea of computability applies most naturally to strings of symbols, such as the formulas of a language such as *Principia Mathematica* or Peano arithmetic. As we saw in sections 15.7 and 15.8, the only operations on strings that we really need to consider are *replacement* of a subword w_1 by w_2. That means transformation of any string of the form uw_1v to uw_2v. Since a given language has only finitely many symbols, and a given computation is given by finitely many pairs $\langle w_1, w_2 \rangle$, we can view the symbols as nonzero *numerals* and the finitely many transformations as transformations of natural numbers.

This makes it at least plausible that the process of generating theorems of *Principia*, say, can be described arithmetically. There are many details to be worked out, so it was an astonishing feat when Gödel (1931) first did it, but most of the details are not interesting in themselves, and we will not examine them here. What is important is that the necessary functions can be defined in the language of first order Peano arithmetic, PA, and hence ultimately in terms of addition and multiplication. (The simplest approach to this **arithmetization of syntax** that I know is in Smullyan [1961].)

It follows that the recursively enumerable sets $W_0, W_1, W_2, \ldots$ from the previous section can be defined in PA, including the set

$$W_{n_0} = \{n : \text{PA proves } "n \notin W_n"\}. \tag{*}$$

Then, assuming that the theorems of PA are correct, we can now conclude that *the proposition $n_0 \notin W_{n_0}$ is true but not provable in* PA. In fact, it is enough to assume PA is *consistent*—that is, that it does not generate contradictory theorems—because PA can prove $n \in W_n$ when this is true, since arithmetization means PA can do all computations. So if PA proves "$n \notin W_n$" then n is really not in W_n; otherwise PA would prove "$n \in W_n$" as well.

Thus arithmetization gives the stronger theorem: *if* PA *is consistent then $n_0 \notin W_{n_0}$ is true but not provable in* PA. The unprovability of true arithmetic sentences in PA is known as **Gödel's first incompleteness theorem**.

Self-Reference

While the above argument relies on Gödel's arithmetization, it is otherwise more like Post's diagonal argument, based in turn on the original diagonal argument of Cantor (1891). Gödel used a different kind of diagonalization, based on *self-reference*. In effect, by arithmetization of formulas and proofs, he was able to construct a sentence of PA that can be interpreted as saying, "I am not provable." Since it is contradictory for this sentence to be provable, it is *not* provable—and therefore true!

Rather than take the reader through Gödel's construction of the self-referential sentence, I prefer to point out that the sentence "$n_0 \notin W_{n_0}$" asserts its own unprovability. Namely, by the definition (*) of W_{n_0}:

$$n_0 \notin W_{n_0} \Leftrightarrow \text{"}n_0 \notin W_{n_0}\text{" is not provable in PA.}$$

Now let us recapitulate the gist of the incompleteness argument.

From the assumption that PA is consistent we draw the conclusion that "$n_0 \notin W_{n_0}$" is not provable in PA, which is equivalent simply to

$n_0 \notin W_{n_0}$. The consistency of PA can be expressed in many ways, for example, by the sentence

"$0 = 1$" is not provable in PA.

By the arithmetization of syntax, the consistency sentence can be rewritten as a sentence Con(PA) of PA. So we have actually proved the implication

$$\text{Con(PA)} \Rightarrow n_0 \notin W_{n_0},$$

which is a sentence of PA. Not only that, *its proof can be carried out in* PA as well! This is an even more laborious arithmetization task, but it can be done, as Gödel mentioned as an afterthought in Gödel (1931), possibly prompted by a letter from von Neumann (1930).

So Con(PA) $\Rightarrow n_0 \notin W_{n_0}$ is provable in PA even though $n_0 \notin W_{n_0}$ is not. It follows that Con(PA) is *not* provable in PA; otherwise $n_0 \notin W_{n_0}$ would follow by modus ponens. In other words, PA *cannot prove its own consistency*. This result is known as **Gödel's second incompleteness theorem**, and it is even more of a bombshell than the first—it is perhaps the greatest discovery in logic of all time.

The Hilbert Program

This brings us back to Hilbert's question about the consistency of reasoning about infinity, which was raised in section 9.8. Given any system Σ of axioms and rules of inference, the process for generating theorems is fairly transparent and *finite*. Each sentence is a finite string of symbols; each rule of inference is a finite process for producing new sentences from old; a proof is a finite sequence of formulas, each of which follows from some predecessors by a rule of inference; and finally a *theorem* is the last line of a proof. The question of consistency asks whether a certain finite object, say, the sentence "$0 = 1$," is a theorem. Thus the question of consistency, even when it concerns sentences about infinite objects, can be reduced to a question about finite objects and finite processes.

As Hilbert realized, this makes consistency resemble a number theory question, where one asks whether a certain computation leads to a certain result in a finite number of steps. For example, the Goldbach conjecture that any even number > 2 is a sum of two primes is about the outcome of the computation that generates the even numbers, the primes, and the sums of two primes and searches for an even number that is not the sum of two primes. We *expect* that the outcome of any such process may someday be known by reasoning about finite objects and processes, though the Goldbach example is a warning that this may be difficult.

At any rate, Hilbert thought it reasonable to hope that consistency of any consistent axiom system might one day be proved by reasoning about finite objects and, indeed, in some fixed system for reasoning about finite objects, such as PA. This was the so-called **Hilbert program**, which aimed to refute objections to reasoning about infinity. If reasoning about infinity can be shown free of contradiction, what is the harm in it?

Up to a point, Hilbert's program was on the right track. As we saw in the previous section, sentences and rules of inference can be "arithmetized" so that consistency of a system Σ is *expressed* by a sentence $\mathrm{Con}(\Sigma)$ of PA. The disappointment is that $\mathrm{Con}(\Sigma)$ may not be *provable* in PA, in which case it is unlikely to be provable by any kind of reasoning about finite objects. Shockingly, Gödel's second incompleteness theorem shows that $\mathrm{Con}(\mathrm{PA})$ is not provable in PA, so any proof of $\mathrm{Con}(\mathrm{PA})$ must appeal to axioms above or beyond PA, and it is hard to imagine how any such axiom could be more plausible than PA itself. Experience has shown that almost any arithmetic statement that is provable at all is provable in PA, so PA seems not to have "missed" any obvious truths of arithmetic.

16.3 GENTZEN'S CONSISTENCY PROOF FOR PA

Granted that we have to assume some axiom $\mathscr{A}$ beyond PA to prove $\mathrm{Con}(\mathrm{PA})$, we might nevertheless learn something from a proof of $\mathrm{Con}(\mathrm{PA})$. If, say, we can prove $\mathscr{A} \Rightarrow \mathrm{Con}(\mathrm{PA})$ in some system *weaker* than PA, then it seems fair to say that $\mathscr{A}$ encapsulates the strength of PA in some sense. Gentzen (1936) did all this, and more, by taking $\mathscr{A}$ to be a statement about a countable ordinal.

Before we examine the countable ordinal in question—which is larger than any of the ordinals pictured in figure 12.7—it will be helpful to see how Gentzen himself explained the role of ordinal numbers in consistency proofs. The following passage is from a piece he wrote for a general audience soon after his consistency proof was published:

> I shall explain now how the concepts of the transfinite ordinal numbers and the rule of transfinite induction enter into the consistency

proof. The connection is quite natural and simple. In carrying out a consistency proof for elementary number theory we must consider all conceivable number-theoretical proofs and must show that in a certain sense, to be formally defined, each individual proof yields a "correct" result, in particular, no contradiction. The "correctness" of the proof depends on the correctness of certain other simpler proofs contained in it as special cases or constituent parts. This fact has motivated the arrangement of proofs in linear *order* in such a way that those proofs on whose correctness the correctness of another depends precede the latter proof in the sequence. This arrangement of proofs is brought about by correlating with each proof a certain transfinite ordinal number; the proofs preceding a given proof are precisely those proofs whose ordinal numbers precede the ordinal number of the given proof. . . . We need the transfinite ordinal numbers for the following reason: It may happen that the correctness of a proof depends on the correctness of infinitely many simpler proofs. An example: Suppose that in the proof a proposition is proved for *all* natural numbers by complete induction. In that case the correctness of the proof obviously depends on the correctness of every single one of the infinitely many individual proofs obtained by specializing to a particular natural number. Here a natural number is insufficient as an ordinal number for the proof, since each natural number is preceded by only finitely many other numbers in the natural ordering. We therefore need the transfinite ordinal numbers in order to represent the natural ordering of the proofs according to their complexity.

(translation from Gentzen 1969: 231–232)

Perhaps the only thing to add to this explanation is that the ordering of proofs must be a *well*-ordering, not merely a linear ordering, to avoid an infinite regress, where $proof_\alpha$ depends on a preceding $proof_\beta$, which depends on a preceding $proof_\gamma$, and so on ad infinitum. A well-ordering of proofs ensures that any such descending sequence $\alpha > \beta > \gamma > \cdots$ is finite, so there is no infinite regress. This does not rule out proofs that depend on infinitely *many* preceding proofs, needed for the reason given by Gentzen, since it allows $proof_\omega$ to depend on $proof_0$, $proof_1$, $proof_2$, . . ., for example. But it does mean that the ordering of proofs must be quite long.

The ordinal ε_0 of this long ordering is larger than any shown in figure 12.7; in a sense that picture represents just the first rung on the ladder to ε_0. To explain ε_0 better we need to introduce the sum, product, and exponential functions for ordinals. These are defined inductively, as for natural numbers, except that each definition has an extra clause for limit ordinals.

Recall from section 12.5 that the successor function $s(\alpha) = \alpha \cup \{\alpha\}$ applies to all ordinals, and that the union of a set of ordinals is the least upper bound of the set. Then we can write the definitions as follows:

$$\alpha + 0 = \alpha,$$
$$\alpha + s(\beta) = s(\alpha + \beta),$$
$$\alpha + \lambda = \bigcup_{\beta < \lambda} (\alpha + \beta), \quad \text{for limit } \lambda; \qquad \text{(sum)}$$

$$\alpha \cdot 0 = 0,$$
$$\alpha \cdot s(\beta) = (\alpha \cdot \beta) + \alpha,$$
$$\alpha \cdot \lambda = \bigcup_{\beta < \lambda} (\alpha \cdot \beta), \quad \text{for limit } \lambda; \qquad \text{(product)}$$

$$\alpha^0 = 1,$$
$$\alpha^{s(\beta)} = (\alpha^\beta) \cdot \alpha,$$
$$\alpha^\lambda = \bigcup_{\beta < \lambda} (\alpha^\beta), \quad \text{for limit } \lambda. \qquad \text{(exponential)}$$

As with our definitions of sum and product for natural numbers (section 13.1), these definitions have a base step defining the function for all α and for $\beta = 0$ and then an inductive step (now split into successor and limit cases) defining the function for a certain value of β in terms of previous values. The extended functions have some of the properties of the natural number functions, such as associative sum and product, but not all. For example, the limit step for sum and product implies

$$1 + \omega = \omega \neq \omega + 1, \quad 2 \cdot \omega = \omega \neq \omega \cdot 2,$$

so commutativity fails.

The definition of the exponential function allows us to define all terms of the sequence

$$\omega, \quad \omega^\omega, \quad \omega^{\omega^\omega}, \quad \omega^{\omega^{\omega^\omega}}, \quad \ldots,$$

of which we have previously seen only ω and ω^ω in figure 12.7. The whole sequence $\alpha_0, \alpha_1, \alpha_2 \ldots$ is defined inductively by

$$\alpha_0 = \omega, \quad \alpha_{n+1} = \omega^{\omega_n}.$$

Finally, $\epsilon_0 = \bigcup_{n<\omega} \omega_n$, which we sometimes write as

$$\varepsilon_0 = \omega^{\omega^{\omega^{\cdot^{\cdot^{\cdot}}}}}.$$

Although ε_0 lies far beyond the natural numbers, it is still a countable set, and we can define a relation $R_{\varepsilon_0}(m,n)$ in PA that is a well-ordering of order type ε_0. But PA cannot *prove* that $R_{\varepsilon_0}(m,n)$ is a well-ordering. In fact, the salient facts about ε_0 that Gentzen proved are these:

- The assumption that $R_{\varepsilon_0}(m,n)$ is a well-ordering implies Con(PA). (We often say that "Gentzen proved Con(PA) by ε_0-induction.")
- For all ordinals $\beta < \varepsilon_0$, and only those, there is a well-ordering $R_\beta(m,n)$ of order type β that PA can prove to be a well-ordering.

Since the presumably "hard" part of PA is the induction schema, Gentzen's proof of Con(PA) has sometimes been derided as "using ε_0-induction to prove the validity of ordinary induction." However, this misses the point that his proof that

$$\varepsilon_0\text{-induction} \Rightarrow \text{Con(PA)}$$

can be carried out in a system called **primitive recursive arithmetic**, a system much weaker than PA in which induction is asserted only for quantifier-free formulas. Thus, Gentzen's proof validates the whole induction *schema* of PA by induction on the single relation $R_{\varepsilon_0}(m,n)$. Moreover, Gentzen's second theorem, which appeared in Gentzen (1943), shows that ε_0 is the least ordinal strong enough to do this job.

In a sense, ε_0 plays a role in PA, something like the role played by inaccessibles in set theory (section 12.7). It shines "light from above" on the theory, and because of that, its existence cannot be proved within the theory. In the case of inaccessibles, their unprovability was immediate. In the case of ε_0-induction we know indirectly that it is unprovable because we know that Con(PA) is unprovable in PA, by Gödel's second incompleteness theorem. However, the proof of Gentzen's second theorem in Gentzen (1943) is relatively direct, being a "miniaturization" of a set theory proof that the union of all countable ordinals is not countable.

16.4 HIDDEN OCCURRENCES OF ε_0 IN ARITHMETIC

In this book we will not go into more detail about Gentzen's proof of Con(PA) by ε_0-induction. The most accessible proof I know is in Mendelson (1964), and a proof of a somewhat easier theorem—proving Con(PA) by transfinite induction but not finding the exact ordinal—may be found in Stillwell (2010). Instead, I would like to expose ε_0 hiding in some theorems of arithmetic that are seemingly not about ordinals or proofs.

Goodstein's Theorem

Think of a number, say, 20. First write 20 as a sum of powers of 2, which can be done by first finding the greatest power of 2 that is less than 21, namely, 2^4, then the greatest power of 2 less than $20 - 2^4$, and so on:

$$20 = 2^4 + 2^2.$$

Then write each *exponent* as a sum of powers of 2, continuing (if necessary) until the topmost exponents are either 0 or 1:

$$2^{2^{2^1}} + 2^{2^1}.$$

We will call this the *base* 2 *normal form* of 20. Now replace each 2 by a 3, and subtract 1 from the resulting number, in this case, obtaining

$$3^{3^{3^1}} + 3^{3^1} - 1 = 7625597485013.$$

Now write this number in *base* 3 *normal form*, by first subtracting the largest possible power of 3, then subtracting the largest possible power of 3 from what remains, and so on. This time some powers may be repeated, at most twice, in which case we write the power times 2. Do the same with the exponents, until the topmost exponents are either 0, 1, or 2. This amounts to rewriting $3^3 - 1$ as a sum of powers of 3, and we get

$$3^{3^{3^1}} + 3^2 \cdot 2 + 3^1 \cdot 2 + 3^0 \cdot 2.$$

Next, replace each 3 by a 4, subtract 1, and convert the resulting number to *base* 4 *normal form*, which will be the much larger number

$$4^{4^{4^1}} + 4^2 \cdot 2 + 4^1 \cdot 2 + 4^0 \cdot 2 - 1 = 4^{4^{4^1}} + 4^2 \cdot 2 + 4^1 \cdot 2 + 4^0$$

$$= 340282366920938463463374607431768211497.$$

I hope you can guess the next step: replace each 4 by a 5, subtract 1, convert the resulting number to *base* 5 *normal form*, and continue. We will call this process the *Goodstein process* on the number 20, and we can clearly apply it to any natural number n.

Now we ask, Do the numbers created by the Goodstein process on n increase indefinitely? Except when $n = 0, 1, 2, 3$ the numbers quickly grow too large to write down, yet the amazing truth is that *the numbers produced by the Goodstein process eventually decrease, ending with the number* 0. Goodstein (1944) found this theorem by a kind of mimicry of ordinals by natural numbers. To prove it he used the well-ordering of ordinals less than ε_0.

We can see ordinals less than ε_0 hidden in the Goodstein process if we replace each symbol 2 in the base 2 normal form by ω, as follows:

$$2^{2^{2^1}} + 2^{2^1} \quad \rightarrow \quad \omega^{\omega^{\omega^1}} + \omega^{\omega^1} = \omega^{\omega^\omega} + \omega^\omega.$$

We will call $\omega^{\omega^\omega} + \omega^\omega$ the *ordinal of step* 1 of the Goodstein process on $n = 20$. We similarly replace each 3 by ω in the base 3 normal form at step 2 of the process to obtain the *ordinal of step* 2, then each 4 by ω in the base 4 normal form at step 3 to obtain the *ordinal of step* 3, and so on. Here are the replacements giving the ordinals for the first three steps of the Goodstein process on the number 20:

$$2^{2^{2^1}} + 2^{2^1} \quad \rightarrow \quad \omega^{\omega^\omega} + \omega^\omega$$

$$3^{3^{3^1}} + 3^2 \cdot 2 + 3^1 \cdot 2 + 3^0 \cdot 2 \quad \rightarrow \quad \omega^{\omega^\omega} + \omega^2 \cdot 2 + \omega \cdot 2 + 2$$

$$4^{4^{4^1}} + 4^2 \cdot 2 + 4^1 \cdot 2 + 4^0 \quad \rightarrow \quad \omega^{\omega^\omega} + \omega^2 \cdot 2 + \omega \cdot 2 + 1.$$

Notice now that the subtraction of 1 at each step, which fails to make any noticeable impression on the numbers, makes a significant difference to the ordinals: *it decreases them*. A strict proof of this fact involves some theory of the sum, product, and exponential functions for ordinals, and the so-called *Cantor normal form* for ordinals less than ε_0, but I think the decrease is clear enough. The point is that *since the ordinals less than ε_0 are well-ordered, any decreasing sequence of them ends in a finite number of steps.* It follows that *the Goodstein process also ends, necessarily at the number* 0.

Goodstein's theorem seems tailor-made for ε_0-induction, but could it be proved without it? The answer was not known for several decades after Goodstein's proof, but finally Kirby and Paris (1982) showed that the answer is no: Goodstein's theorem is not provable in PA.

Fusible Numbers

Fusible numbers get their name from a recreational mathematics problem about *fuses*. We suppose there is an unlimited supply of fuses, each of which burns for one minute after being lit, or for half a minute if it is lit simultaneously at both ends. We are given no time-keeping device other than the fuses themselves, and the question is, Which time intervals of t minutes can we create, using only observations of burning fuses?

For example, is it possible to create an event occurring at time $t = 3/4$? The answer is yes, if we employ two fuses as follows. At time $t = 0$ light $fuse_1$ at both ends and $fuse_2$ at one end. Then $fuse_1$ will burn out at time $t = 1/2$, at which time half of $fuse_2$ will remain. Therefore, if we light the other end of $fuse_2$ at this moment, it will burn for another 1/4 minute, and we will see it burn out at time $t = 3/4$.

A number r is called *fusible* if we can create a burnout at time r by using a finite number of fuses, each of which can be lit at one or both ends at time 0 or at a time when one of the fuses burns out. It is fairly clear that the fusible numbers are rational (in fact, binary fractions) and greater than or equal to 1/2, but it seems hard to survey them in their totality. The definition of fusible numbers is simple, but their totality is astonishingly complex. The uncovering of their structure by Erickson et al. (2020) produced the following two theorems, which show that the fusible numbers exceed the grasp of PA:

Order type. *The fusible numbers are well-ordered by the $<$ relation, with order type ε_0.*

This theorem is not provable in PA, because of the Gentzen (1943) theorem that the provable well-orderings of PA have order type less than ε_0. In fact, Erickson et al. (2000), show that the following weaker statement is not provable in PA:

Minimal fusible numbers. *For each natural number n there is a least fusible number greater than n.*

Given the rather simple concept of fusible number, which can be defined in PA, the latter theorem is perhaps the best candidate currently known for a natural statement about numbers not provable in PA. However, it remains to be seen whether it will be of interest to mainstream mathematicians.

16.5 CONSTRUCTIVITY

Gödel's first incompleteness theorem fails to interest many mathematicians because the *known* unprovable statements of PA so far do not include any sentences that ordinary mathematicians have tried to prove. It is actually somewhat surprising that the only known examples, such as Con(PA) and Goodstein's theorem, were devised by logicians, and their connection to logic is either obvious or only slightly hidden. In areas of mathematics outside number theory, such as geometry or set theory, unprovable sentences came to light *because* mathematicians failed to prove them from given axioms. This experience brought to light the unprovability of the parallel axiom in neutral geometry and the unprovability of the axiom of choice in the Zermelo-Fraenkel axioms (ZF).

Another such area has emerged in recent decades: analysis, or *second-order* arithmetic. The arithmetization of analysis, described in chapter 11, reduces a large part of analysis to the theory of natural numbers and sets of natural numbers. The question then arises, What axioms should be the starting point for such a theory? A good choice turns out to be **constructive analysis**, in which infinite objects (such as sets of natural numbers) can be asserted to exist only if they are computable. Since computability is a precise mathematical concept, and indeed, one that can be defined in PA, constructive analysis is both a close relative of PA and, it turns out, one in which incompleteness is both natural and remarkably well structured. Not only can we show that theorems such as extreme value and Bolzano-Weierstrass are unprovable in constructive analysis, we can also identify **set existence axioms** suitable for proving them.

Brouwer's Intuitionism

It appears that Brouwer, around 1907, foresaw that the diagonal argument would disrupt the axiomatic approach to mathematics, and he believed that mathematics should be radically reconstructed to give a greater role to intuition. To him, this meant a rejection of pure existence proofs, rejection of nonconstructive arguments, and even rejection of classical logical principles, such as the law of excluded middle.

At the same time, he realized that he needed to become an established mathematician in order to be a credible advocate for these radical ideas, and fortunately he had more than enough mathematical talent to do so. In the years 1910–1912 he blazed a new path in topology by proving what we now call the **Brouwer fixed point theorem** and the related theorems **invariance of domain** and **invariance of dimension**. The latter resolved the crisis provoked by Cantor's bijection between the line and the plane by showing that no bijection between $\mathbb{R}^m$ and $\mathbb{R}^n$ ($m \neq n$) can be continuous.

These proofs, while accepted and admired by other mathematicians, used nonconstructive methods, and in lectures in Berlin in 1927 Brouwer rejected his own fixed point theorem, as well as several standard theorems of analysis proved by similar methods: the intermediate and extreme value theorems, and the Bolzano-Weierstrass theorem. Of course, most mathematicians do not adhere to Brouwer's intuitionism, partly from conviction that nonconstructive methods are valid and partly because they do not want to lose standard theorems. Nevertheless, it is of interest to know which theorems require nonconstructive proofs and exactly which nonconstructive axioms are needed to prove them. This can conceivably be done by isolating some constructive axioms for analysis and then searching for the "right" nonconstructive axioms to prove theorems such as extreme value or Bolzano-Weierstrass. This is the program of what we call **reverse mathematics**.

Reverse Mathematics: The Base System

Reverse mathematics grew from the MIT thesis of Friedman (1967) on systems of second-order arithmetic, and it began to take shape as a search for the "right" axioms to prove basic theorems in the survey of Friedman (1975: 235), in which he declared:

> When a theorem is proved from the right axioms, the axioms can
> be proved from the theorem.

In recent years, reverse mathematics has found the "right" axioms to prove many basic theorems of analysis, topology, combinatorics, and algebra. The axioms in question are "right" in the same sense that the parallel axiom is the "right" axiom to prove that the angle sum of a triangle is π or that the axiom of choice is the "right" axiom to prove the well-ordering theorem in ZF. And, just as the parallel axiom has many equivalents in neutral geometry and the axiom of choice has many equivalents in ZF, many theorems of second-order arithmetic can be proved equivalent in constructive analysis. In particular, Brouwer's three theorems turn out to be equivalent—to each other and to the nonconstructive principle called the **weak König lemma**.

What makes reverse mathematics possible is also what makes Gödel's incompleteness theorems possible: the existence of noncomputable objects. In particular, computably enumerable but noncomputable sets make constructive analysis a weak system, and computable infinite trees with no computable infinite paths make the weak König lemma stronger than the axioms of constructive analysis. We will say more about this shortly, but first, What are these axioms?

We start with a system of constructive analysis called RCA_0. This system has two types of variables, lowercase for natural numbers and uppercase for sets of natural numbers. The axioms for natural numbers include the first four Peano axioms of section 13.1. The remaining axioms for numbers are the induction axiom schema, but restricted to *computably enumerable* properties Φ (RCA acquired the subscript 0 when the induction axiom was restricted in this way):

> *If Φ holds for 0, and if Φ holds for $s(n)$ whenever it holds for n,*
> *then Φ holds for all numbers.*

This is in keeping with the intention to deal only with computable objects, and it aligns with the convenient fact that computably enumerable properties have a simple description in the language of PA. They are precisely those properties $\Phi(n)$ expressible in the form

$$\Phi(n) \Leftrightarrow \exists n_1 \exists n_2 \cdots \exists n_k R(n_1, n_2, \ldots, n_k, m),$$

where $R(n_1, n_2, \ldots, n_k, m)$ is *quantifier-free*. It follows that R is a *computable* property because it boils down a combination of equations in the numbers $n_1, n_2, \ldots, n_k, m$ using only the computable functions successor, sum, and product.

The letters RCA stand for **recursive comprehension axiom**, which refers to the set existence axiom of RCA_0, asserting the existence of computable sets. ("Recursive" is the old term for "computable," in common use when RCA_0 was set up and still used occasionally today.) It is a schema, for each recursive property Ψ, saying that there is a set whose members are the natural numbers with property Ψ:

$$\exists X \forall n [n \in X \Leftrightarrow \Psi(n)]. \qquad \text{(RCAx)}$$

The property Ψ can contain free set variables $Y, Z, \ldots$, so as to allow computation *relative to* given sets $Y, Z, \ldots$. For example, given a set Y, RCAx allows us to assert the existence of the set of even members of Y, since it is computable whether or not a number is even.

With its adherence to computable objects, RCA_0 is close to Brouwer's idea of constructive mathematics. It differs mainly in its logic, which is classical predicate logic; in particular the law of excluded middle is present. This does not impair its ability to identify the nonconstructiveness of theorems such as the Brouwer fixed point theorem, as we will see.

RCA_0 is a weak system, but it is able to prove two notable theorems: the intermediate value theorem and its consequence, the fundamental theorem of algebra. Also, and this is what makes RCA_0 a good base system, it can prove many *equivalences* between statements it cannot prove outright. Moreover, these equivalences put a large number of statements into a small number of equivalence classes. We will look at two notable equivalence classes in the next two sections.

16.6 ARITHMETIC COMPREHENSION

To prove that a certain sentence σ is not provable in RCA_0, we do as we did with the parallel axiom in neutral geometry: find a model of RCA_0 in which σ does not hold. A suitable model of RCA_0 consists of the natural numbers, together with all computable subsets of $\mathbb{N}$. We call this the *minimal model*, because any model must include a collection of objects like the natural numbers and a collection of their subsets that includes all computable sets.

A simple sentence not provable in RCA_0 is the **monotonic convergence theorem**: *every bounded increasing sequence of rational numbers has a limit*. The theorem does not hold in the minimal model of RCA_0 because we can construct a computable bounded monotonic sequence $a_0, a_1, a_2, \ldots$ of rational numbers whose limit is not computable.

Namely, let W be a computably enumerable but noncomputable set (as found in section 15.6), let f be a computable function whose values are the members of W, and let

$$a_n = \sum_{i=0}^{n} 2^{-f(i)}.$$

Notice that $a_0, a_1, a_2, \ldots$ is an increasing sequence whose limit has the binary expansion with 1 in the mth place if and only if $m \in W$. Since W is not computable, neither is this binary expansion. But the sequence is the set of pairs $\langle n, a_n \rangle$, which *is* computable: to decide whether the pair $\langle s, t \rangle$ is in the sequence, compute a_s and see whether $t = a_s$. Thus the sequence (or some coded version of it) is in the minimal model of RCA_0, but its limit is not. Consequently, the monotonic convergence theorem is false in the minimal model and hence not provable in RCA_0.

The reason that monotonic convergence is false in the minimal model of RCA_0 is clearly because the minimal model contains only computable sets, which in turn is because the recursive comprehension axiom guarantees only that computable sets exist. If we had an **arithmetic comprehension axiom** then the minimal model would have to include all arithmetically definable sets. These would include the limit of any computable sequence, since it requires only a few quantifiers to define a limit in the language of arithmetic (assuming appropriate coding of real numbers by sets):

a is the limit of $a_0, a_1, a_2, \ldots \Leftrightarrow \forall p \exists n \forall m [m > n \Rightarrow |a_m - a| < 1/p]$.

If we take the position that theorems not provable in RCA_0 are not constructive, then the monotonic convergence theorem is not constructive, and hence neither is the arithmetic comprehension axiom used to prove it.

A system based on the arithmetic comprehension axiom is called ACA_0, and its axioms are the full set of Peano axioms plus the arithmetic comprehension axiom (schema):

$$\exists X \forall n [n \in X \Leftrightarrow \Phi(n)], \hspace{2cm} \text{(ACAx)}$$

where $\Phi(n)$ is any property in the language of arithmetic, possibly with free set variables other than X. The monotonic convergence theorem is provable in ACA_0, by the argument just sketched, and in fact the implication $ACAx \Rightarrow$ monotonic convergence theorem is provable in RCA_0. More surprising, the monotonic convergence theorem is *equivalent* to $ACAx$ in RCA_0, as are the following well-known theorems of basic analysis:

Sequential Bolzano-Weierstrass theorem. Any bounded infinite sequence of real numbers contains a convergent subsequence.

Cauchy convergence criterion. A sequence $a_0, a_1, a_2, \ldots$ of real numbers is convergent if and only if, for any $\varepsilon > 0$, there is an n such that
$$m > n \Rightarrow |a_m - a_n| < \varepsilon.$$

Another interesting equivalent of $ACAx$ is the following theorem of infinite graph theory, due to König (1927):

Kőnig infinity lemma. If $\mathcal{T}$ is an infinite tree[1] in which each vertex has finite valency then $\mathcal{T}$ contains an infinite simple path.

A proof of the Kőnig infinity lemma is clear but nonconstructive. Choose any vertex u_0 of $\mathcal{T}$ with valency at least 2. Since $\mathcal{T}$ is infinite, at least one edge e_0 out of u_0 leads to an infinite subtree of $\mathcal{T}$. If u_1 is the vertex at the other end of e_0 then at least one edge $e_1 \neq e_0$ out of u_1 leads to an infinite subtree of $\mathcal{T}$, so we can repeat the argument. Continuing in this way, we will not revisit any vertex, because $\mathcal{T}$ contains no simple closed path, so we obtain an infinite simple path in $\mathcal{T}$.

This proof is nonconstructive because there is no algorithm for recognizing which edges lead to infinite subtrees, but it can be carried out in ACA_0 because the infinite path is arithmetically definable. The proofs of all three theorems above in ACA_0 are basically the classical proofs, given the necessary coding of real numbers and continuous functions by sets of natural numbers. The proofs that each theorem implies $ACAx$ are more involved and may be found in Simpson (2009) or Stillwell (2018).

---

1. Notice that an infinite tree $\mathcal{T}$ in which each vertex has finite valency is necessarily countable. Since $\mathcal{T}$ is connected, each vertex is only finitely many edges away from some fixed vertex u_0, and since vertices have finite valency there are only finitely many vertices of $\mathcal{T}$ within n edges of u_0. Because of this, the tree can be coded by a set of natural numbers and hence discussed in ACA_0.

16.7 THE WEAK KŐNIG LEMMA

One of the achievements of reverse mathematics, in my opinion, has been to highlight the role of *trees* in mathematics, and especially in analysis. In particular, a number of key theorems are equivalent in RCA_0 to the following special case of the Kőnig infinity lemma (previously mentioned in slightly different language in section 15.4).

Weak Kőnig lemma. If $\mathcal{T}$ is an infinite tree in which each vertex has valency at most 3 then $\mathcal{T}$ contains an infinite simple path.

The restriction to valency 3 is sufficient to cover a number of proofs that involve repeated *bisection* of a closed interval, such as the proof of the Heine-Borel theorem in section 11.7. In these proofs the tree $\mathcal{T}$ is a subtree of the *full binary tree* shown in figure 16.1, whose vertices are the initial interval $[a, b]$ and its subintervals resulting from bisection, with an edge from each vertex to its left and right halves.

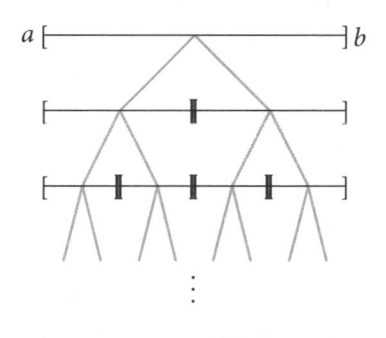

Figure 16.1 : A full binary tree

In this case the infinite path guaranteed by the lemma is a nested sequence of intervals containing a single point, whose existence is the key to the theorem in question.

Thus the weak Kőnig lemma implies the Heine-Borel theorem and its well-known consequences that we saw in section 11.7:

Uniform continuity. Any function continuous on a closed interval is uniformly continuous there.

Riemann integrability. Any function continuous on a closed interval is Riemann-integrable there.

Another theorem provable from the weak Kőnig lemma, as we already saw in section 15.4, is:

Completeness of predicate logic. Any valid formula of predicate logic is provable. (Equivalently, any invalid formula can be falsified by the rules given in section 15.4.)

Another remarkable group of consequences of the weak Kőnig lemma are the following theorems of Brouwer:

Brouwer fixed point theorem. Any continuous map of the n-dimensional ball, $\mathbb{B}^n = \{\langle x_1, x_2, \ldots, x_n \rangle \in \mathbb{R}^n : x_1^1 + x_2^2 + \cdots + x_n^2 \leq 1\}$, into itself has a fixed point.

Invariance of domain. Any continuous injection of $\mathbb{R}^n$ into itself has an open image.

Invariance of dimension. There is no continuous bijection $\mathbb{R}^m \to \mathbb{R}^n$ when $m \neq n$.

All of these theorems are in fact equivalent to the weak Kőnig lemma in RCA_0. Proofs for the first four may be seen in Simpson (2009), and the equivalence of the Brouwer theorems was proved recently by Kihara (2020).

The weak Kőnig lemma is not provable in RCA_0, because Kreisel (1953) showed that there is a computable binary tree with no computable infinite path. Thus Kreisel's tree is in the minimal model of RCA_0, but its infinite paths are not. This means that the theory RCA_0 + weak Kőnig lemma, which is called WKL_0, is stronger than RCA_0, and the theorems listed above are nonconstructive to exactly the same extent as the weak Kőnig lemma. This confirms that the Brouwer fixed point theorem is nonconstructive, as Brouwer thought. But it also confirms that the Brouwer invariance theorems are equally nonconstructive, which Brouwer may not have known.

It is also true that the weak Kőnig lemma is weaker than the full Kőnig lemma, so WKL_0 is weaker than ACA_0. This can be proved by finding a model of WKL_0 whose sets do not include all of the sets in the minimal model of ACA_0. The construction of this model of WKL_0 requires some fairly sophisticated computability theory, which can be found in Simpson (2009: 318–319).

16.8 THE BIG FIVE

The three systems of second-order arithmetic studied so far, RCA_0, WKL_0, and ACA_0, are incomplete, like any system that includes enough of PA to express computation. They are bound to be incomplete, by Gödel's incompleteness theorem, but (unlike PA, apparently) their unprovable sentences include many that are well known and natural. We find these sentences most easily among the *second-order* sentences, which can express properties of real numbers and continuous functions, as well as properties of other infinite structures, such as graphs.

For example, WKL_0 proves that a continuous function on a closed interval is Riemann-integrable, which RCA_0 cannot; ACA_0 proves the full Kőnig infinity lemma, which WKL_0 cannot.

When it comes to *first-order* sentences, however, the three systems do not have much to tell us. The first-order theorems of ACA_0 are exactly the same as the theorems of PA—a remarkable theorem proved by Friedman (1976)—and as far as we know, most of the interesting theorems of PA are already provable in RCA_0. We know some first-order theorems not provable in PA, such as Goodstein's theorem, so these are also not provable in ACA_0. But again it seems that the most natural unprovable sentences of ACA_0 are second order.

Two of these are theorems of Cantor about *closed* and *perfect* sets. We will define these in a way (equivalent to the usual definition) that makes it clear they can be encoded by sets of natural numbers and hence expressed in second-order arithmetic. For simplicity we consider just subsets of $\mathbb{R}$, but the definitions and theorems generalize to $\mathbb{R}^n$.

Definitions. A subset of $\mathbb{R}$ is **open** if it is a countable union of open intervals (r_i, s_i), where r_i and s_i are rational. A subset of $\mathbb{R}$ is **closed** if it is the complement of an open set. A closed set $\mathscr{F}$ is called **perfect** if every open interval containing a point of $\mathscr{F}$ contains infinitely many points of $\mathscr{F}$.

Then the two theorems in question are as follows:

Perfect set theorem. Every uncountable closed subset of $\mathbb{R}$ contains a perfect set.

Cantor-Bendixson theorem. Every closed subset of $\mathbb{R}$ is the union of a countable set and a perfect set.

The perfect set theorem is equivalent, in RCA_0, to the set existence axiom of a system called ATR_0. The Cantor-Bendixson theorem is equivalent, in RCA_0, to the set existence axiom of a theory called $\Pi_1^1\text{-}CA_0$. Together with RCA_0, WKL_0, and ACA_0, these two new systems make up what are called the **big five**. Each of the big five is strictly stronger than the one before, and together they cover most of the basic theorems in analysis. For more information, especially about ATR_0 and $\Pi_1^1\text{-}CA_0$, see Simpson (2009).

The set existence axiom of ATR_0 is too technical to explain here, but ATR stands for "arithmetic transfinite recursion." As this term suggests, ATR_0 is a system that can express properties of countable ordinals and carry out transfinite induction. In particular, ATR_0 admits the use of ε_0-induction, so it can prove Goodstein's theorem and Con(PA). This shows that ATR_0 exceeds the reach of ACA_0 even in first-order theorems.

The set existence axiom of $\Pi_1^1\text{-}CA_0$ is "Π_1^1-comprehension," which means the property defining the set can take the form called Π_1^1, in which the arithmetic expression is preceded by a universal set quantifier:

$$\exists X[n \in X \Leftrightarrow \forall Y \varphi(Y, n)], \qquad (\Pi_1^1\text{-}CAx)$$

where $\varphi(Y, n)$ can have free set variables other than X. This axiom takes us into territory logicians call **impredicative**, where a set X of natural numbers may be defined in terms of a totality ("all sets Y") that includes itself. Many logicians consider that ATR_0 takes us to the limit of predicativity, so the Cantor-Bendixson theorem is an example of a theorem not provable by predicative means, since it is not provable in ATR_0. It is interesting to note that the first published paper on reverse mathematics, Friedman (1969), is actually about a system similar to $\Pi_1^1\text{-}CA_0$. Thus reverse mathematics began at the top of the big five, where analysis shades into set theory, and only later separated out the simpler systems that capture the basic theorems of analysis with more nuance.

The big five systems illustrate incompleteness in a nicely structured way, since each of the last four systems contains natural theorems not provable in the one before. There is also an interesting theorem beyond the big five: the **graph minor theorem** of Robertson and Seymour (2004), a long-conjectured theorem about infinite sequences of finite graphs whose proof finally appeared at the end of a sequence of papers spanning twenty years! It was shown to be unprovable in $\Pi_1^1\text{-}CA_0$ by Friedman et al. (1987). For more about the graph minor theorem, see Diestel (2010).

16.9 REMARKS

In the twentieth century mathematical proofs grew, in both length and depth, to an extent that stretched human capacity to the limit. Some eminent mathematicians, finding mistakes in their own work, decided it was time to make proofs **computer checkable** and to use **computer assistance** in writing computer-checkable proofs. See Voevodsky (2014) for an account by a Fields-Medal-winning mathematician of a series of mistakes and how they changed his mathematical life. Voevodsky wrote:

> But to do the work at the level of rigor and precision I felt was necessary would take an enormous amount of effort and would produce a text that would be very difficult to read. And who would ensure that I did not forget something and did not make a mistake, if even the mistakes in much more simple arguments take years to uncover?
>
> I think it was at this moment that I largely stopped doing what is called "curiosity driven research" and started to think seriously about the future. . . . It soon became clear that the only real long-term solution to the problems that I encountered is to start using computers in the verification of mathematical reasoning.

While it seems clear that some kind of computer verification will be more common in the future, it is not yet clear what form it will take. There are several rival systems, and none is yet widely used by mathematicians.

For this reason, in this book I have not attempted to cover either the most advanced proofs of the twentieth century or the forms of computer assistance that have been developed so far. (However, I recommend the splendid article Avigad [2018].)

Instead, I have focused on results about *proof itself*, which are comparatively accessible yet not well known. As I said in the preface, the theory of proof has not yet caught the attention of many mathematicians. I hope this book will convince more of them, particularly the younger generation, that the concept of proof is mathematically interesting—and perhaps fundamental to understanding the mathematics of the future.

Bibliography

■ ■ ■ ■ ■

Abel, N. H. (1826). Mémoire sur une propiété générale d'une classe très étendue de fonctions transcendantes. *Mémoire des Savants Étrangers, 1841, 7,* 176–264. *Oeuvres Complètes* 2: 145–211.

Annellis, I. H. (1990). From semantic tableaux to Smullyan trees: A history of the development of the falsification tree method. *Modern Logic 1,* 36–69.

Argand, J. R. (1806). *Essai sur une manière de représenter les quantités imaginaires dans les constructions géométriques.* Paris.

Artin, E. (1957). *Geometric Algebra.* Interscience Publishers Inc., New York–London.

Aubrey, J. (1898). *Brief Lives.* Edited by Andrew Clark. Oxford: Clarendon Press. Set down by John Aubrey between the years 1669 and 1696.

Avigad, J. (2018). The mechanization of mathematics. *Notices Amer. Math. Soc.* 65(6), 681–690.

Baer, R. (1928). Zur Axiomatik der Kardinalarithmetik. *Mathematische Zeitschrift 29,* 381–396.

Banach, S., and A. Tarski (1924). Sur la décomposition des ensembles de points en parties respectivement congruents. *Fundamenta Mathematicae 6,* 244–277.

Beltrami, E. (1865). Risoluzione del problema: Riportare i punti di una superficie sopra un piano in modo che le linee geodetiche vengano rappresentate da linee rette. *Ann. Mat. pura appl., ser. 1 7,* 185–204. In his *Opere Matematiche* 1: 262–280.

Beltrami, E. (1868a). Saggio di interpretazione della geometria non-euclidea. *Giorn. Mat. 6,* 284–312. In his *Opere Matematiche* 1: 262–280. English translation in Stillwell (1996).

Beltrami, E. (1868b). Teoria fondamentale degli spazii di curvatura costante. *Ann. Mat. pura appl., ser. 2 2,* 232–255. In his *Opere Matematiche* 1: 406–429. English translation in Stillwell (1996).

Berkeley, G. (1734). *The Analyst.* London: J. Tonson.

Bernoulli, D. (1753). Réflexions et éclaircissemens sur les nouvelles vibrations des cordes exposées dans les mémoires de l'académie de 1747 & 1748. *Hist. Acad. Sci. Berlin 9,* 147–172.

Bernoulli, J. (1694). Curvatura laminae elasticae. *Acta. Erud. 13,* 262–276.

Bernoulli, J. (1696). Positionum de seriebus infinitis pars tertia. In his *Werke,* 4, 85–106.

Bernoulli, J. (1704). Positionum de seriebus infinitis . . . pars quinta. In his *Werke,* 4, 127–147.

Bézout, E. (1779). *Théorie générale des équations algébriques.* Paris: PhD. Pierres. English translation in *General Theory of Algebraic Equations,* by Eric Feron, Princeton University Press, Princeton, NJ, 2006.

Biggs, N. L., E. K. Lloyd, and R. J. Wilson (1976). *Graph Theory: 1736–1936*. Oxford: Oxford University Press.

Blass, A. (1984). Existence of bases implies the axiom of choice. In *Axiomatic Set Theory (Boulder, Colorado, 1983)*, Volume 31 of *Contemporary Mathematics*, 31–33. American Mathematical Society, Providence, RI.

Bolyai, F. (1832a). *Tentamen juventutem studiosam in elementa matheseos purae, elementaris ac sublimioris, methodo intuitiva, evidentiaque huic propria, introducendi.* Marosvásárhely.

Bolyai, J. (1832b). Scientiam spatii absolute veram exhibens: A veritate aut falsitate Axiomatis XI Euclidei (a priori haud unquam decidanda) independentem. Appendix to Bolyai (1832a). English translation in Bonola (1912).

Bolzano, B. (1817). *Rein analytischer Beweis des Lehrsatzes dass zwischen je zwey Werthen, die ein entgegengesetztes Resultat gewähren, wenigstens eine reelle Wurzel der Gleichung liege.* Ostwald's Klassiker, vol. 153. Engelmann, Leipzig, 1905. English translation in Russ (2004), 251–277.

Bolzano, B. (1851). *Paradoxien der Unendlichen.* Leipzig: Bei C. H. Reclam Sen. English translation in Russ (2004), 591–678.

Bombelli, R. (1572). *L'algebra. Prima edizione integrale. Introduzione di U. Forti. Prefazione di E. Bortolotti.* Reprint by Biblioteca scientifica Feltrinelli. Milano: Giangiacomo Feltrinelli Editore (1966).

Bonola, R. (1912). *Noneuclidean Geometry.* Chicago: Open Court. Reprinted by Dover, New York, 1955.

Boole, G. (1847). *Mathematical Analysis of Logic.* Reprinted by Basil Blackwell, London, 1948.

Borel, É. (1895). Sur quelques points de la théorie des fonctions. *Ann. Sci. École Norm. Sup. 12*, 9–55.

Borel, É. (1898). *Leçons sur la théorie des fonctions.* Paris: Gauthier-Villars.

Bosse, A. (1653). *Moyen Universel de Pratiquer la Perspective sur les Tableaux, ou Surfaces Irrégulières.* Paris: Chez ledit Bosse.

Bourgne, R., and J.-P. Azra (1962). *Ecrits et mémoires mathématiques d'Évariste Galois: Édition critique intégrale de ses manuscrits et publications.* Gauthier-Villars & Cie, Imprimeur-Éditeur-Libraire, Paris. Préface de J. Dieudonné.

Bressoud, D. M. (2019). *Calculus Reordered.* Princeton University Press, Princeton, NJ.

Bring, E. S. (1786). *Meletemata quaedam mathematica circa transformationem aequationum algebraicarum.* Lund University, Promotionschrift.

Brouwer, L. E. J. (1912). Beweis der Invarianz des n-dimensionalen Gebiets. *Mathematische Annalen 71*, 305–315.

Byrne, O. (1847). *The First Six Books of the Elements of Euclid.* London: William Pickering.

Cantor, G. (1874). Über eine Eigenschaft des Inbegriffes aller reellen algebraischen Zahlen. *J. reine und angew. Math. 77*, 258–262. In his *Gesammelte Abhandlungen*, 145–148. English translation by W. Ewald in Ewald (1996), vol. 2, 840–843.

Cantor, G. (1878). Ein Beitrag zur Mannigfaltigskeitlehre. *J. reine und angew. Math. 84*, 242–258. In his *Gesammelte Abhandlungen*, 119–133.

Cantor, G. (1883). *Grundlagen einer allgemeinen Mannigfaltigkeitslehre.* Leizig: Teubner. In his *Gesammelte Abhandlungen*, 165–204. English translation by W. Ewald in Ewald (1996), vol. 2, 878–919.

Cantor, G. (1891). Über eine elementare Frage der Mannigfaltigkeitslehre. *Jahresber. deutsch. Math. Verein. 1*, 75–78. English translation by W. Ewald in Ewald (1996), vol. 2, 920–922.

Cardano, G. (1545). *Ars magna.* Translated in *The great art or the rules of algebra* by T. Richard Witmer, with a foreword by Oystein Ore. MIT Press, Cambridge, MA–London, 1968.

Cauchy, A.-L. (1821). *Cours d'Analyse de l'École Royale Polytechnique.* Paris. Annotated English translation by Robert E. Bradley and C. Edward Sandifer, *Cauchy's Cours d'analyse: An Annotated Translation,* Springer, 2009.

Cauchy, A.-L. (1823). *Résumé des lecons données à de l'École Royale Polytechnique sur le Calcul Infinitésimal.* Paris: l'Imprimérie Royale. In his *Oeuvres,* série II, tome IV.

Cayley, A. (1859). A sixth memoir on quantics. *Phil. Trans. Roy. Soc. 149,* 61–90. In his *Collected Mathematical Papers* 2: 561–592.

Cayley, A. (1878). The theory of groups. *Amer. J. Math. 1,* 50–52. In his *Collected Mathematical Papers* 10: 401–403.

Ceĭtin, G. S. (1956). Associative calculus with insoluble equivalence problem. *Dokl. Akad. Nauk SSSR (n.s.) 107,* 370–371.

Church, A. (1936). An unsolvable problem in elementary number theory. *American Journal of Mathematics 58,* 345–363.

Church, A. (1956). *Introduction to Mathematical Logic. vol. 1.* Princeton University Press, Princeton, NJ.

Clagett, M. (1968). *Nicole Oresme and the Medieval Geometry of Qualities and Motions.* University of Wisconsin Press, Madison, WI.

Clay, M., and D. Margalit (2017). *Office Hours with a Geometric Group Theorist.* Princeton University Press, Princeton, NJ.

Clebsch, A. (1864). Über einen Satz von Steiner und einige Punkte der Theorie der Curven dritter Ordnung. *J. reine und angew. Math. 63,* 94–121.

Cohen, P. (1963). The independence of the continuum hypothesis I, II. *Proc. Nat. Acad. Sci. 50, 51,* 1143–1148, 105–110.

Cohen, P. J. (1966). *Set Theory and the Continuum Hypothesis.* W. A. Benjamin Inc., New York–Amsterdam.

Cook, S. A. (1971). The complexity of theorem-proving procedures. *Proceedings of the 3rd Annual ACM Symposium on the Theory of Computing,* 151–158. Association of Computing Machinery, New York.

Corry, L. (2004). *David Hilbert and the Axiomatization of Physics (1898–1918).* Vol. 10 of *Archimedes: New Studies in the History and Philosophy of Science and Technology.* Kluwer Academic Publishers, Dordrecht.

Cramer, G. (1750). *Introduction à l'analyse des lignes courbes algébriques.* Geneva.

d'Alembert, J. l. R. (1746). Recherches sur le calcul intégral. *Hist. Acad. Sci. Berlin 2,* 182–224.

Darboux, G. (1875). Mémoire sur les fonctions discontinues. *Ann. Sci. de l'École Norm. Super., Sér. 2 4,* 57–112.

Davis, M. (Ed.) (2004). *The Undecidable.* Mineola, NY: Dover Publications Inc. Corrected reprint of the 1965 original, Raven Press, Hewlett, NY.

de la Vallée Poussin, C. J. (1896). Recherches analytiques sur la théorie des nombres premiers. *Ann. Soc. Sci. Bruxelles 20,* 183–256.

Dedekind, R. (1871). Supplement X. In Dirichlet's *Vorlesungen über Zahlentheorie,* 2nd ed., Vieweg, 1871.

Dedekind, R. (1872). *Stetigkeit und irrationale Zahlen.* Braunschweig: Vieweg und Sohn. English translation in *Essays on the Theory of Numbers,* Dover, New York, 1963.

Dedekind, R. (1877a). Schreiben an Herrn Borchardt über die Theorie der elliptischen Modulfunctionen. *J. reine und angew. Math. 83,* 265–292.

Dedekind, R. (1877b). *Theory of Algebraic Integers.* Cambridge: Cambridge University Press. Translated from the 1877 French original and with an introduction by John Stillwell.

Dedekind, R. (1888). *Was sind und was sollen die Zahlen?* Braunschweig: Vieweg und Sohn. English translation in *Essays on the Theory of Numbers,* Dover, New York, 1963.

Dedekind, R. (1890). *Letter to Keferstein.* In van Heijenoort, 1967, 98–103.

Dedekind, R. (1894). Supplement 11. In Dirichlet's *Vorlesungen über Zahlentheorie,* 4th ed., Vieweg, 1894.

Dedekind, R., and H. Weber (1882). Theorie der algebraischen Functionen einer Veränderlichen. *Journal für die reine und angewandte Mathematik 92,* 181–291. English translation in Dedekind and Weber (2012).

Dedekind, R., and H. Weber (2012). *Theory of Algebraic Functions of One Variable.* Vol. 39 of *History of Mathematics.* American Mathematical Society, Providence, RI; London Mathematical Society, London. Translated from the 1882 German original and with an introduction, bibliography, and index by John Stillwell.

Dehn, M. (1900). Über raumgleiche Polyeder. *Gött. Nachr. 1900,* 345–354.

Dehn, M. (1910). Über die Topologie des dreidimensionalen Raumes. *Math. Ann. 69,* 137–168.

Dehn, M. (1912). Über unendliche diskontinuierliche Gruppen. *Mathematische Annalen 71,* 116–144.

Desargues, G. (1639). *Brouillon projet d'une atteinte aux évènements des rencontres du cône avec un plan.* In Taton (1951), 99–180.

Descartes, R. (1637). *The geometry of René Descartes (with a facsimile of the first edition, 1637).* New York, NY: Dover Publications Inc., 1954. Translated by David Eugene Smith and Marcia L. Latham.

Diestel, R. (2010). *Graph Theory* (4th ed.). Vol. 173 of *Graduate Texts in Mathematics.* Springer, Heidelberg.

Dirichlet, P. G. L. (1829). Sur la convergence des séries trigonométriques qui servent à représenter une fonction arbitraire entre des limites données. *J. reine und angew. Math 4,* 157–169. In his *Werke* 1: 117–132.

Dirichlet, P. G. L. (1863). *Vorlesungen über Zahlentheorie.* Braunschweig: F. Vieweg und Sohn. English translation *Lectures on Number Theory,* with supplements by R. Dedekind. Translated from the German and with an introduction by John Stillwell, American Mathematical Society, Providence, RI, 1999.

Dombrowski, P. (1979). *150 Years after Gauss' "Disquisitiones generales circa superficies curvas."* Paris: Société Mathématique de France. With the original text of Gauss.

du Bois-Reymond, P. (1875). Über asymptotische Werte, infinitäre Approximationen und infinitäre Auflösung von Gleichungen. *Math. Ann. 8,* 363–414.

Ebbinghaus, H.-D., H. Hermes, F. Hirzebruch, M. Koecher, K. Mainzer, J. Neukirch, A. Prestel, and R. Remmert (1990). *Numbers.* Vol. 123 of *Graduate Texts in Mathematics.* Springer-Verlag, New York. With an introduction by K. Lamotke. Translated from the second German edition by H. L. S. Orde, translation edited and with a preface by J. H. Ewing.

Edwards, H. M. (1974). *Riemann's Zeta Function.* Pure and Applied Mathematics, vol. 58. Academic Press, New York–London.

Eisenstein, G. (1844). Allgemeine Auflösung der Gleichungen von der ersten vier Graden. *Journal für die reine und angewandte Mathematik 27,* 81–83.

Erickson, J., G. Nivasch, and J. Xu (2020). Fusible numbers and Peano arithmetic. https://arxiv.org/abs/2003.14342.

Euler, L. (1734). De summis serierum reciprocarum. *Comm. Acad. Sci. Petrop. 7.* In his *Opera Omnia*, ser. 1, 14: 73–86.

Euler, L. (1736). Solution problematis ad geometriam situs perttinentis. *Comm. Acad. Sci. Petrop. 8.* English translation in Biggs et al. (1976).

Euler, L. (1748). *Introductio in analysin infinitorum, I.* Volume 8 of his *Opera Omnia*, series 1. English translation in *Introduction to the Analysis of the Infinite. Book 1*, Springer-Verlag, 1988.

Euler, L. (1752). Elementa doctrinae solidorum. *Novi Comm. Acad. Sci. Petrop. 4*, 109–140. In his *Opera Omnia*, ser. 1, 26: 71–93.

Euler, L. (1760). Recherches sur la courbure des surfaces. *Mém. Acad. Sci. Berlin 16*, 119–143. In his *Opera Omnia*, ser. 1, 28: 1–22.

Euler, L. (1770). *Elements of Algebra.* Translated from the German by John Hewlett. Reprint of the 1840 edition, with an introduction by C. Truesdell, Springer-Verlag, New York, 1984.

Ewald, W. (1996). *From Kant to Hilbert: A Source Book in the Foundations of Mathematics.* Vols. 1, 2. Clarendon Press, Oxford University Press, New York.

Fagnano, G. C. T. (1718). Metodo per misurare la lemniscata. *Giorn. lett. d'Italia 29.* In his *Opere Matematiche*, 2: 293–313.

Feferman, S., and A. Levy (1963). Independence results in set theory by Cohen's method ii. *Notices Amer. Math. Soc. 10*, 593.

Fermat, P. (1657). Letter to Frenicle, February 1657. In his *Œuvres* 2: 333–334.

Ferreirós, J. (2007). *Labyrinth of Thought* (2nd ed.). Birkhäuser Verlag, Basel.

Fibonacci (1202). *Liber abaci.* In *Scritti di Leonardo Pisano*, edited by Baldassarre Boncompagni, Rome 1857–1862. English translation in *Fibonacci's Liber abaci*, by L.E. Sigler, Springer, New York, 2002.

Fibonacci (1225). *Liber Quadratorum.* English translation in *The Book of Squares*, by L.E. Sigler, Academic Press, 1987.

Fourier, J. (1822). *La théorie analytique de la chaleur.* Paris: Didot. English translation in *The Analytical Theory of Heat*, Dover, New York, 1955.

Fraenkel, A. (1914). Über die Teiler der Null und die Zerlegung von Ringen. *J. Reine und Angew. Math. 145*, 139–176.

Fraenkel, A. (1922). Zu den Grundlagen der Cantor-Zermeloschen Mengenlehre. *Mathematische Annalen 86*, 230–237.

Franchella, M. (1997). On the origins of Dénes König's infinity lemma. *Arch. Hist. Exact Sci. 51*(1), 3–27.

Fréchet, M. R. (1906). Sur quelques points du calcul fonctionnel. *Rendiconti del Circolo Matematico di Palermo 22*, 1–72.

Frege, G. (1879). *Begriffschrift.* English translation in van Heijenoort (1967), 5–82.

Fricke, R., and F. Klein (2017). *Lectures on the Theory of Automorphic Functions.* Vols. 1 and 2. Classical Topics in Mathematics. Higher Education Press, Beijing. Translated from the German originals by Arthur M. DuPre.

Friedman, H. (1967). *Subsystems of Set Theory and Analysis.* PhD thesis, MIT Department of Mathematics, Cambridge MA.

Friedman, H. (1969). Bar induction and Π_1^1 – CA. *J. Symbolic Logic 34*, 353–362.

Friedman, H. (1975). Some systems of second order arithmetic and their use. In *Proceedings of the International Congress of Mathematicians (Vancouver, B. C., 1974)*, vol. 1, 235–242.

Friedman, H. (1976). Systems of second order arithmetic with restricted induction I, II. *Journal of Symbolic Logic 41*, 557–559.

Friedman, H., N. Robertson, and P. Seymour (1987). The metamathematics of the graph minor theorem. In *Logic and Combinatorics*, pp. 229–261. American Mathematical Society, Providence, RI.

Gabbay, D. M., and J. Woods (Eds.) (2009). *Handbook of the History of Logic*, vol. 5. Elsevier, Amsterdam.

Galois, E. (1831). Mémoire sur les conditions de résolubilité des équations par radicaux. In Bourgne and Azra (1962), 43–71.

Gauss, C. F. (1799). Demonstratio nova theorematis omnem functionem algebraicum rationalem integram unius variabilis in factores reales primi vel secundi gradus resolvi posse. Helmstedt dissertation, in his *Werke* 3: 1–30.

Gauss, C. F. (1801). *Disquisitiones arithmeticae*. Translated and with a preface by Arthur A. Clarke. Revised by William C. Waterhouse, Cornelius Greither, and A. W. Grootendorst and with a preface by Waterhouse. Springer-Verlag, New York, 1986.

Gauss, C. F. (1816). Demonstratio nova altera theorematis omnem functionem algebraicum rationalem integram unius variabilis in factores reales primi vel secundi gradus resolvi posse. *Comm. Recentiores (Gottingae) 3*, 107–142. In his *Werke* 3: 31–56.

Gauss, C. F. (1819). Die Kugel. *Werke* 8: 351–356.

Gauss, C. F. (1825). Die Seitenkrümmung. *Werke* 8: 386–395.

Gauss, C. F. (1827). *Disquisitiones generales circa superficies curvas*. Göttingen: König. Ges. Wiss. Göttingen. English translation in Dombrowski (1979).

Gauss, C. F. (1831). Letter to Schumacher, July 12, 1831. In his *Werke* 8: 215–218.

Gauss, C. F. (1832). Theoria residuorum biquadraticorum. *Comm. Soc. Reg. Sci. Gött. Rec. 4*. In his *Werke* 2: 67–148.

Gentzen, G. (1935). Untersuchungen über das logische Schliessen. *Mathematische Zeitschrift 19*, 176–210. English translation by M. E. Szabo in Gentzen (1969), 68–131.

Gentzen, G. (1936). Die Widerspruchsfreiheit der reinen Zahlentheorie. *Mathematische Annalen 112*, 493–565. English translation by M. E. Szabo in Gentzen (1969), 132–213.

Gentzen, G. (1943). Beweisbarkeit und Unbeweisbarkeit von Anfangsfällen der transfiniten Induktion in der reinen Zahlentheorie. *Mathematische Annalen 119*, 140–161. English translation by M. E. Szabo in Gentzen (1969), 287–308.

Gentzen, G. (1969). *The Collected Papers of Gerhard Gentzen*. Edited by M. E. Szabo. Amsterdam: North-Holland Publishing Co.

Girard, A. (1629). *Invention nouvelle en l'algèbre*. Chez Guillaume Iansson Blaeuw, Amsterdam.

Gödel, K. (1930). Die Vollständigkeit der Axiome des logischen Funktionenkalküls. *Monatshefte für Mathematik und Physik 37*, 349–360.

Gödel, K. (1931). Über formal unentscheidbare Sätze der Principia Mathematica und verwandter Systeme. I. *Monatshefte für Mathematik und Physik 38*, 173–198.

Gödel, K. (1938). The consistency of the axiom of choice and the generalized continuum hypothesis. *Proceedings of the National Academy of Sciences 25*, 220–224.

Gödel, K. (1956). Letter to von Neumann, 20 March, 1956. In Gödel (2003), p. 375.

Gödel, K. (2003). *Collected Works.* Vol. 5. *Correspondence H.–Z.* Edited by Solomon Feferman, John W. Dawson, Jr., Warren Goldfarb, Charles Parsons, and Wilfried Sieg. The Clarendon Press, Oxford University Press, Oxford.

Gödel, K. (2014). *Collected Works.* Vol. 5. *Correspondence H–Z.* Edited by Solomon Feferman, John W. Dawson, Jr., Warren Goldfarb, Charles Parsons, and Wilfried Sieg. The Clarendon Press, Oxford University Press, Oxford. Paperback edition of the 2003 original.

Goodstein, R. L. (1944). On the restricted ordinal theorem. *The Journal of Symbolic Logic 9*, 33–41.

Grabiner, J. V. (1981). *The Origins of Cauchy's Rigorous Calculus.* MIT Press, Cambridge, MA–London.

Grassmann, H. (1844). *Die lineale Ausdehnungslehre.* Otto Wigand, Leipzig. English translation in Grassmann (1995), 1–312.

Grassmann, H. (1847). *Geometrische Analyse geknüpft an die von Leibniz gefundene Geometrische Charakteristik.* Weidmann'sche Buchhandlung, Leipzig. English translation in Grassmann (1995), 313–414.

Grassmann, H. (1861). *Lehrbuch der Arithmetic.* Enslin, Berlin.

Grassmann, H. (1862). *Die Ausdehnungslehre.* Enslin, Berlin. English translation of 1896 edition in Grassmann (2000).

Grassmann, H. (1995). *A New Branch of Mathematics.* Open Court Publishing Co., Chicago, IL. Translated from the German and with a note by Lloyd C. Kannenberg. With a foreword by Albert C. Lewis.

Grassmann, H. (2000). *Extension Theory.* American Mathematical Society, Providence, RI; London Mathematical Society, London. Translated from the 1896 German original and with a foreword, editorial notes, and supplementary notes by Lloyd C. Kannenberg.

Gray, J. (2015). *The Real and the Complex: A History of Analysis in the 19th Century.* Springer Undergraduate Mathematics Series. Springer, Cham.

Gregory, J. (1668) *Geometriae pars Universalis.* Padua: Paolo Frambotto.

Gromov, M. (1981). Hyperbolic manifolds (according to Thurston and Jørgensen). In *Bourbaki Seminar, vol. 1979/80.* Vol. 842 of *Lecture Notes in Math.*, 40–53. Springer, Berlin–New York.

Grothendieck, A. (1957). Sur quelques points d'algèbre homologique. *Tohoku Math. J. (2) 9*, 119–221.

Guggenheimer, H. (1977). The Jordan curve theorem and an unpublished manuscript by Max Dehn. *Arch. Hist. Exact Sci. 17*(2), 193–200.

Hadamard, J. (1896). Sur la distribution des zéros de la fonction $\zeta(s)$ et ses conséquences arithmétiques. *Bull. Soc. Math. France 24*, 199–220.

Hamel, G. (1905). Eine Basis aller Zahlen und die unstetigen Lösungen der Funktionalgleichung $f(x+y) = f(x) + f(y)$. *Mathematische Annalen 60*, 459–462.

Hamilton, W. R. (1837). Theory of Conjugate Functions or Algebraic Couples. *Trans. Roy. Irish Acad. 17*, 393–422.

Harriot, T. (1631). *Artis Analyticae Praxis.* Robert Barker, London. English translation in Seltman and Goulding (2007).

Hartshorne, R. (2000). *Geometry: Euclid and Beyond.* Undergraduate Texts in Mathematics. Springer-Verlag, New York.

Hausdorff, F. (1914). *Grundzüge der Mengenlehre.* Leipzig: Von Veit.

Heath, T. L. (1897). *The Works of Archimedes*. Cambridge University Press, Cambridge. Reprinted by Dover, New York, 1953.

Heath, T. L. (1910). *Diophantus of Alexandria: A Study in the History of Greek Algebra*. 2nd ed. Cambridge University Press. Reprint, Dover Publications Inc., New York, 1964.

Heath, T. L. (1925). *The Thirteen Books of Euclid's Elements*. Cambridge University Press, Cambridge. Reprinted by Dover, New York, 1956.

Heine, E. (1872). Die Elemente der Functionenlehre. *J. reine und angew. Math. 74*, 172–188.

Hermite, C. (1873). Sur la fonction exponentielle. *C. R. 77*, 18–24, 74–79, 226–233, 285–293. In his *Oeuvres* 3: 150–181.

Hessenberg, G. (1905). Beweis des *Desargues*schen Satzes aus dem *Pascal*schen. *Mathematische Annalen 61*, 161–172.

Hilbert, D. (1899). *Grundlagen der Geometrie*. Leipzig: Teubner. English translation in *Foundations of Geometry*, Open Court, Chicago, 1971.

Hilbert, D. (1900). Mathematische Probleme. Vortrag, gehalten auf dem internationalen Mathematiker-Congress zu Paris 1900. *Gött. Nachr.* 1900, 253–297.

Hilbert, D. (1901). Über Flächen von constanter Gaussscher Krümmung. *Trans. Amer. Math. Soc. 2*, 87–89. In his *Gesammelte Abhandlungen* 2: 437–438.

Hilbert, D. (1902). Mathematical problems. *Bulletin of the American Mathematical Society 8*, 437–479. Translated by Frances Winston Newson.

Hobbes, T. (1656). Six lessons to the professors of mathematics. *The English Works of Thomas Hobbes*, 7: 181–356. Scientia Aalen, Aalen, West Germany, 1962.

Hobbes, T. (1672). Considerations upon the answer of Doctor Wallis. *The English Works of Thomas Hobbes*, 7: 443–448. Scientia Aalen, Aalen, West Germany, 1962.

Huygens, C. (1693a). Appendix to Huygens (1693b). In his *Oeuvres Complètes* 10: 481–482.

Huygens, C. (1693b). Letter to H. Basnage de Beauval, February 1693. In his *Oeuvres Complètes* 10: 407–417.

Jacobi, C. G. J. (1829). *Fundamenta nova theoriae functionum ellipticarum*. Königsberg: Bornträger. In his *Werke* 1: 49–239.

Jacobi, C. G. J. (1834). De usu theoriae integralium ellipticorum et integralium abelianorum in analysi diophantea. *J. reine und angew. Math. 13*, 353–355. In his *Werke* 2: 53–55.

Jordan, C. (1887). *Cours de Analyse de l'École Polytechnique*. Gauthier-Villars, Paris.

Katz, V. J., and K. H. Parshall (2014). *Taming the Unknown*. Princeton University Press, Princeton, NJ.

Kihara, T. (2020). The Brouwer invariance theorems in reverse mathematics. https://arxiv.org/abs/2002.10715.

Kirby, L., and J. Paris (1982). Accessible independence results for Peano arithmetic. *The Bulletin of the London Mathematical Society 14*(4), 285–293.

Klein, F. (1871). Über die sogenannte Nicht-Euklidische Geometrie. *Math. Ann. 4*, 573–625. In his *Gesammelte Mathematische Abhandlungen* 1: 254–305. English translation in Stillwell (1996).

Klein, F. (1928). *Vorlesungen über Nicht-Euklidische Geometrie*. Berlin: Springer.

Klein, F., and R. Fricke (1890). *Lectures on the Theory of Elliptic Modular Functions*. Vol. 1. Volume 1 of *Classical Topics in Mathematics*. Higher Education Press, Beijing, 2017. Translated from the 1890 German original by Arthur M. DuPre.

Klein, F., and R. Fricke (2017). *Lectures on the Theory of Elliptic Modular Functions.* Vols. 1 and 2. *Classical Topics in Mathematics.* Higher Education Press, Beijing, 2017. Translated from the German originals by Arthur M. DuPre.

Kolmogorov, A. N. (1933). *Grundbegriffe der Wahrscheinlichkeitsrechnung.* Berlin: Springer. English translation in *Foundations of the Theory of Probability,* Chelsea, New York, 1956.

König, D. (1927). *Uber eine Schlussweise aus dem Endlichen ins Unendliche. Acta Litterarum ac Scientiarum 3,* 121–130.

König, D. (1936). *Theorie der endlichen und unendlichen Graphen.* Leipzig: Akademische Verlagsgesellschaft. English translation by Richard McCoart, *Theory of Finite* and *Infinite Graphs,* Birkhäuser, Boston 1990.

Kreisel, G. (1953). A variant to Hilbert's theory of the foundations of arithmetic. *British J. Philos. Sci. 4,* 107–129; errata and corrigenda, 357 (1954).

Krömer, R. (2007). *Tool and object.* Vol. 32 of *Science Networks. Historical Studies.* Birkhäuser Verlag, Basel.

Kronecker, L. (1887). Ein Fundamentalsatz der allgemeinen Arithmetik. *Journal für die reine und angewandte Mathematik 100,* 490–510.

Krull, W. (1929). Idealtheorie in Ringen ohne Endlichkeitsbedingungen. *Mathematischen Annalen 101,* 729–744.

Lagrange, J.-L. (1770). Nouvelle méthode pour résoudre les équations littérales par le moyen des séries. *Histoire de l'Académie Royale des Sciences et Belles-Lettres de Berlin,* 251–326.

Lambert, J. H. (1758). Observationes variae in mathesin puram. *Acta Helveticae physicomathematico-medica 3,* 128–168.

Lambert, J. H. (1766). Die Theorie der Parallellinien. *Magazin für reine und angewandte Mathematik (1786),* 137–164, 325–358.

Laplace, P.-S. (1795). Lecons de mathématiques donnée à lécole normale en 1795. *Journale de École Polytechnique, VIIe et VIIIe Cahiers, 1812.* In *Oeuvres Complètes de Laplace,* Gauthier-Villars, Paris, 1912.

Lebesgue, H. (1902). Intégrale, longuer, aire. *Annali di matematica pura ed applicata 7,* 231–359.

Leibniz, G. W. (1684). Nova methodus pro maximis et minimis. *Acta Erud. 3,* 467–473. In his *Mathematische Schriften* 5, 220–226. English translation in Struik (1969).

Leibniz, G. W. (1702). Specimen novum analyseos pro scientia infiniti circa summas et quadraturas. *Acta Erud. 21,* 210–219. In his *Mathematische Schriften* 5: 350–361.

Levi ben Gershon (1321). *Maaser Hoshev.* German translation by Gerson Lange in *Sefer Maasei Choscheb,* Frankfurt 1909.

l'Hôpital, G. F. A. d. (1696). *Analyse des infiniment petits.* English translation in *The Method of Fluxions both Direct and Inverse,* William Ynnis, London, 1730.

Liouville, J. (1844). Sur des classes très étendues de quantités dont la valeur n'est ni algébrique, ni même réductible à des irrationnelles algébriques. *C. R. Acad. Sci. Paris 18,* 883–885.

Lobachevsky, N. I. (1829). *On the foundations of geometry.* Kazansky Vestnik (Russian).

Lohne, J. A. (1979). Essays on Thomas Harriot. *Arch. Hist. Exact Sci. 20*(3-4), 189–312.

Markoff, A. (1947). On the impossibility of certain algorithms in the theory of associative systems. *C. R. (Doklady) Acad. Sci. URSS (n.s.) 55,* 583–586.

Marquis, J.-P. (2009). *From a Geometrical Point of View.* Vol. 14 of *Logic, Epistemology, and the Unity of Science.* Springer, Dordrecht.

Mendelson, E. (1964). *Introduction to Mathematical Logic*. D. Van Nostrand Co., Inc., Princeton, NJ.

Mercator, N. (1668). *Logarithmotechnia*. London: William Godbid and Moses Pitt.

Minding, F. (1839). Wie sich entscheiden lässt, ob zwei gegebene krumme Flächen auf einander abwickelbar sind oder nicht; nebst Bemerkungen über die Flächen von unveränderlichem Krümmungsmasse. *J. reine und angew. Math. 19*, 370–387.

Minkowski, H. (1908). Raum und Zeit. *Jahresbericht der Deutschen Mathematiker-Vereinigung 17*, 75–88.

Möbius, A. F. (1863). Theorie der Elementaren Verwandtschaft. In his *Werke* 2: 433–471.

Moufang, R. (1933). Alternativkörper und der Satz der vollständigen Vierseit. *Abh. Math. Sem. Hamburg 9*, 207–222.

Muir, T. (1960). *The Theory of Determinants in the Historical Order of Development*. Dover Publications, Inc., New York.

Mumford, D., C. Series, and D. Wright (2002). *Indra's Pearls*. Cambridge University Press, New York.

Nash, J. (1956). The imbedding problem for Riemannian manifolds. *Ann. of Math. (2) 63*, 20–63.

Needham, T. (1997). *Visual Complex Analysis*. Clarendon Press, Oxford.

Nelson, E. (1987). *Radically Elementary Probability Theory*. Vol. 117 of *Annals of Mathematics Studies*. Princeton University Press, Princeton, NJ.

Neugebauer, O., and A. Sachs (1945). *Mathematical Cuneiform Texts*. Yale University Press, New Haven, CT.

Newton, I. (1665). The geometrical construction of equations. *Mathematical Papers 1*, 492–516.

Newton, I. (1667). Enumeratio curvarum trium dimensionum. *Mathematical Papers 12*, 10–89.

Newton, I. (1670). De resolutione quaestionum circa numeros. *Mathematical Papers*, 4: 110–115.

Newton, I. (1671). De methodis serierum et fluxionum. *Mathematical Papers*, 3, 32–353.

Newton, I. (1676a). Letter to Oldenburg, June 13, 1676. In Turnbull (1960), 20–47.

Newton, I. (1676b). Letter to Oldenburg, October 24, 1676. In Turnbull (1960), 110–149.

Newton, I. (1687). *Philosophiae naturalis principia mathematica*. William Dawson & Sons, Ltd., London. Facsimile of first edition of 1687.

Newton, I. (1728). *Universal Arithmetick*. J. Senex, London.

Newton, I. (1736). *The Method of Fluxions and Infinite Series*. Henry Woodfall, London. Translation by John Colson of Newton (1671).

Oresme, N. (1350). *Tractatus de configurationibus qualitatum et motuum*. English translation in Clagett (1968).

Pascal, B. (1654). Traité du triangle arithmétique, avec quelques autres petits traités sur la même manière. English translation in *Great Books of the Western World*, Encyclopedia Britannica, London, 1952, 447–473.

Peano, G. (1888). *Calcolo Geometrico secondo l'Ausdehnungslehre di H. Grassmann, preceduto dalle operazioni della logica deduttiva*. Bocca, Turin. English translation in Peano (2000).

Peano, G. (1890). Sur une courbe, qui remplit toute une aire plane. *Math. Ann. 36*, 157–160.

Peano, G. (2000). *Geometric Calculus*. Birkhäuser Boston Inc., Boston, MA. Translated from the Italian by Lloyd C. Kannenberg.

Playfair, J. (1795). *Elements of Geometry*. Bell and Bradlute, London.

Plofker, K. (2009). *Mathematics in India*. Princeton University Press, Princeton, NJ.

Poincaré, H. (1881). Sur les applications de la géométrie non-euclidienne à la théorie des formes quadratiques. *Association française pour l'avancement des sciences 10*, 132–138. English translation in Stillwell (1996), 139–145.

Poincaré, H. (1882). Théorie des groupes fuchsiens. *Acta Math. 1*, 1–62. English translation in Poincaré (1985), 55–127.

Poincaré, H. (1895). Analysis situs. *J. Éc. Polytech., ser. 2, 1*, 1–121. English translation in Poincaré (2010).

Poincaré, H. (1902). Du rôle de l'intuition et de la logique en mathématiques. *Compte rendu du Dieuxième Congrès International des Mathématiciens, Paris*.

Poincaré, H. (1904). Cinquième complément à l'analysis situs. *Palermo Rend. 18*, 45–110. English translation in Poincaré (2010).

Poincaré, H. (1985). *Papers on Fuchsian Functions*. New York: Springer-Verlag. Translated from the French and with an introduction by John Stillwell.

Poincaré, H. (2010). *Papers on Topology*, Volume 37 of *History of Mathematics*. American Mathematical Society, Providence, RI; London Mathematical Society, London. Translated and with an introduction by John Stillwell.

Post, E. L. (1921). Introduction to a general theory of elementary propositions. *Amer. J. Math. 43*, 163–185.

Post, E. L. (1936). Finite combinatory processes-formulation 1. *Journal of Symbolic Logic 1*, 103–105.

Post, E. L. (1941). Absolutely unsolvable problems and relatively undecidable propositions—an account of an anticipation (1941). In Davis (2004), 338–433.

Post, E. L. (1943). Formal reductions of the general combinatorial decision problem. *Amer. J. Math. 65*, 197–215.

Post, E. L. (1944). Recursively enumerable sets of positive integers and their decision problems. *Bull. Amer. Math. Soc. 50*, 284–316.

Post, E. L. (1947). Recursive unsolvability of a problem of Thue. *J. Symbolic Logic 12*, 1–11.

Reidemeister, K. (1927). Elementare Begründung der Knotentheorie. *Abh. Math. Sem. Univ. Hamburg 5*, 24–32.

Riemann, G. F. B. (1851). Grundlagen für eine allgemeine Theorie der Functionen einer veränderlichen complexen Grösse. In his *Werke*, 2: 3–48.

Riemann, G. F. B. (1854a). Über die Darstellbarkeit einer Function durch eine trigonometrische Reihe. In his *Werke* 2: 227–264.

Riemann, G. F. B. (1854b). Über die Hypothesen, welche der Geometrie zu Grunde liegen. In his *Werke* 2: 272–287.

Riemann, G. F. B. (1857). Theorie der Abel'schen Functionen. *J. reine und angew. Math. 54*, 115–155. In his *Werke* 2: 82–142.

Riemann, G. F. B. (1859). Über die Anzahl der Primzahlen unter einer gegebenen Grösse. In his *Werke* 2: 145–153. English translation in Edwards (1974), 299–305.

Robertson, N., and P. D. Seymour (2004). Graph minors. XX. Wagner's conjecture. *Journal of Combinatorial Theory Series B 92*(2), 325–357.

Robinson, A. (1966). *Non-standard Analysis*. North-Holland Publishing Co., Amsterdam.

Rosenbloom, P. C. (1950). *The Elements of Mathematical Logic*. Dover Publications Inc., New York.

Rosenfeld, B. A. (1988). *A History of non-Euclidean Geometry*. Vol. 12 of *Studies in the History of Mathematics and Physical Sciences*. Springer-Verlag, New York. Translated from the Russian by Abe Shenitzer.

Russ, S. (2004). *The Mathematical Works of Bernard Bolzano*. Oxford University Press, Oxford.

Russell, B. (1903). *The Principles of Mathematics*. Cambridge University Press, Cambridge.

Saccheri, G. (1733). *Euclid Vindicated from Every Blemish*. Classic Texts in the Sciences. Birkhäuser/Springer, Cham, 2014. Dual Latin-English text, edited and annotated by Vincenzo De Risi, translated from the Italian by G. B. Halsted and L. Allegri.

Schwarz, H. A. (1872). Über diejenigen Fälle, in welchen die Gaussische hypergeometrische Reihe eine algebraische Function ihres vierten Elementes darstellt. *J. reine und angew. Math.* 75, 292–335. In his *Mathematische Abhandlungen* 2: 211–259.

Seltman, M., and R. Goulding (2007). *Thomas Harriot's "Artis Analyticae Praxis."* Springer, New York.

Shelah, S. (1984). Can you take Solovay's inaccessible away? *Israel J. Math.* 48(1), 1–47.

Simpson, S. G. (2009). *Subsystems of Second Order Arithmetic* (Second ed.). Perspectives in Logic. Cambridge University Press, Cambridge; Association for Symbolic Logic, Poughkeepsie, NY.

Skolem, T. (1928). Über die mathematische Logik. *Norsk matematisk tidsskrifft 10*, 125–142. English translation in van Heijenoort (1967), 508–524.

Sloane, N. J. A. (2018). The on-line encyclopedia of integer sequences. *Notices Amer. Math. Soc.* 65(9), 1062–1074.

Smullyan, R. M. (1961). *Theory of Formal Systems* (rev. ed.). Princeton University Press, Princeton, NJ.

Snapper, E., and R. J. Troyer (1971). *Metric Affine Geometry*. Academic Press, New York–London.

Solovay, R. M. (1970). A model of set-theory in which every set of reals is Lebesgue measurable. *Ann. of Math. (2) 92*, 1–56.

Stevin, S. (1585a). *De Thiende*. Christoffel Plantijn, Leiden. English translation by Robert Norton in *Disme: The Art of Tenths, or Decimall Arithmetike Teaching*, London, 1608.

Stevin, S. (1585b). *L'Arithmetique*. Christoffel Plantijn, Leiden.

Stillwell, J. (1996). *Sources of Hyperbolic Geometry*. American Mathematical Society, Providence, RI.

Stillwell, J. (2010). *Roads to Infinity*. A K Peters Ltd, Natick, MA.

Stillwell, J. (2013). *The Real Numbers*. Undergraduate Texts in Mathematics. Springer, Cham.

Stillwell, J. (2016) *Elements of Mathematics*. Princeton University Press, Princeton, NJ.

Stillwell, J. (2018). *Reverse Mathematics*. Princeton University Press, Princeton, NJ.

Stirling, J. (1717). *Lineae tertii ordinis Neutonianae*. Edward Whistler, Oxford.

Struik, D. (1969). *A Source Book of Mathematics 1200–1800*. Harvard University Press, Cambridge, MA.

Tao, T. (2013). *Compactness and Contradiction*. American Mathematical Society, Providence, RI.

Tarski, A. (1948). *A Decision Method for Elementary Algebra and Geometry*. RAND Corporation, Santa Monica, CA.

Taton, R. (1951). *L'œuvre mathématique de G. Desargues.* Presses universitaires de France, Paris.

Thim, J. (2003). *Continuous Nowhere Differentiable Functions.* Masters Thesis, Luleå University of Technology, Department of Mathematics, http://ltu.diva-portal.org /smash/get/diva2:1022983/FULLTEXT01.pdf.

Thomae, J. (1879). Ein Beispiel einer unendlich oft unstetigen Function. *Zeit. f. Math. und Physik 24*, 64.

Thue, A. (1914). *Probleme über Veränderungen von Zeichenreihen nach gegebenen Regeln.* J. Dybvad, Kristiania.

Tietze, H. (1908). Über die topologische Invarianten mehrdimensionaler Mannigfaltigkeiten. *Monatsh. Math. Phys. 19*, 1–118.

Tomkowicz, G., and S. Wagon (2016). *The Banach-Tarski Paradox* (2nd ed.). Vol. 163 of *Encyclopedia of Mathematics and Its Applications.* Cambridge University Press, New York. With a foreword by Jan Mycielski.

Turing, A. (1936). On computable numbers, with an application to the Entscheidungsproblem. *Proceedings of the London Mathematical Society 42*, 230–265.

Turnbull, H. W. (1960). *The Correspondence of Isaac Newton, Vol. 2: 1676–1687.* Cambridge University Press, New York.

Urquhart, A. (2009). Emil Post. In Gabbay and Woods (2009), 617–666.

Van Brummelen, G. (2009). *The Mathematics of the Heavens and the Earth.* Princeton University Press, Princeton, NJ.

Van Brummelen, G. (2013). *Heavenly Mathematics.* Princeton University Press, Princeton, NJ.

van Heijenoort, J. (1967). *From Frege to Gödel. A Source Book in Mathematical Logic, 1879–1931.* Harvard University Press, Cambridge, MA.

Viète, F. (1591). De aequationum recognitione et emendatione. In his *Opera*, 82–162. English translation in Viète (1983).

Viète, F. (1593). Variorum de rebus mathematicis responsorum libri octo. In his *Opera*, 347–435.

Viète, F. (1983). *The Analytic Art.* Kent, OH: The Kent State University Press. Nine studies in algebra, geometry and trigonometry from the *Opus Restitutae Mathematicae Analyseos, seu Algebra Nova*, translated by T. Richard Witmer.

Vilenkin, N. Y. (1995). *In Search of Infinity.* Boston, MA: Birkhäuser Boston Inc. Translated from the Russian original by Abe Shenitzer with the editorial assistance of Hardy Grant and Stefan Mykytiuk.

Vitali, G. (1905). *Sul problema della misura dei gruppi di punti di una retta.* Bologna.

Voevodsky, V. (2014). The origins and motivations of univalent foundations. https: //www.ias.edu/ideas/2014/voevodsky-origins.

von Koch, H. (1904). Sur une courbe continue sans tangente, obtenue par une construction géométrique élémentaire. *Archiv för Matemat., Astron. och Fys. 1.*

von Neumann, J. (1923). Zur Einführung der transfiniten Zahlen. *Acta lit. acad. sci. Reg. U. Hungar. Fran. Jos. Sec. Sci. 1*, 199–208. English translation in van Heijenoort (1967), 347–354.

von Neumann, J. (1930). Letter to Gödel, November 20, 1930, in Gödel (2014), 337.

von Plato, J. (2017). *The Great Formal Machinery Works.* Princeton University Press, Princeton, NJ.

Wallis, J. (1655). Arithmetica infinitorum. *Opera* 1: 355–478. English translation in Wallis (2004).

Wallis, J. (2004). *The arithmetic of infinitesimals*. Springer-Verlag, New York. Translated from the Latin and with an introduction by Jaqueline A. Stedall.

Wapner, L. M. (2005). *The Pea & the Sun*. A K Peters, Ltd., Wellesley, MA.

Weber, H. (1893). Die allgemeinen Grundlagen der Galois'schen Gleichungstheorie. *Mathematische Annalen 43*, 521–549.

Weber, H. (1896). *Lehrbuch der Algebra, Zweiter Band*. Vieweg, Braunschweig.

Weil, A. (1950). The future of mathematics. *Amer. Math. Monthly 57*, 295–306.

Weil, A. (1984). *Number Theory. An Approach through History, from Hammurapi to Legendre*. Birkhäuser Boston Inc., Boston, MA.

Whitehead, A. N., and B. Russell (1910). *Principia Mathematica*, vol. 1. Cambridge University Press, Cambridge.

Wiles, A. (1995). Modular elliptic curves and Fermat's last theorem. *Ann. of Math. (2) 141*(3), 443–551.

Zermelo, E. (1904). Beweis dass jede Menge wohlgeordnet werden kann. *Mathematische Annalen 59*, 514–516. English translation in van Heijenoort (1967), 139–141.

Zermelo, E. (1908). Untersuchungen über die Grundlagen der Mengenlehre I. *Mathematische Annalen 65*, 261–281. English translation in van Heijenoort (1967), 200–215.

Index

■ ■ ■ ■ ■

Riemann, Bernhard (*continued*)
  mapping theorem, 143
  zeta function, 189
Riemann-Hurwitz formula, 262
Riemannian manifold, 221
  smoothly embeds in some $\mathbb{R}^n$, 227
ring, 61, 77
  axioms, 80
  finite, 148
  ideal of, 178
  of congruence classes, 149
  polynomial, 175
Rosenbloom, Paul C., 362
$\mathbb{RP}^2$ *see* real projective plane
rules of inference, 347
  for propositional logic, 351
  from falsification rules, 352
Russell, Bertrand
  read Frege and Peano, 355

$\mathbb{S}^2$ *see* sphere
S_2, 75
S_5, 75
S_n *see* symmetric group
Saccheri, Girolamo, 202
  tried to prove parallel axiom, 203
SAS, 20, 44
  in Byrne's *Elements*, 38
satisfiability, 350
  in propositional logic, 350
scalar multiple, 83
Schwarz, Hermann Amandus, 224
second order, 319
Seki, Takakazu, 82
self-reference, 384
self-similarity, 275
semigroup, 372
  word problem, 371
sequence
  of continuous functions, 276
    with discontinuous limit, 276
  of numbers, 264
    convergence of, 264
    limit of, 264
  uniformly convergent, 278
series
  binomial, 117
  geometric, 32
  harmonic, 113
  infinite, 113
  of continuous functions, 277
    with discontinuous sum,
      277

power, 110
  uniformly convergent, 278
set
  actually infinite, 291
  closed, 288, 401
  computable, 364
  computably enumerable, 363
  constructible, 344
  countable, 294, 295
  embodiment of, 293
  existence axioms, 393
  hereditarily finite, 309
  inaccessible, 310
  infinite, 212
    Dedekind definition, 294
  Mandelbrot, 288
  nonmeasurable, 335
  of real numbers, 286
  open, 288, 401
    axioms, 289
  perfect, 401
  potentially infinite, 291
  power, 299
  rank of, 308
  representing ordinal number,
    305
  uncountable, 291, 297
set theory, 261, 291
  and AC, 337
  as arithmetic plus infinity, 327
  axioms, 316, 324
  Zermelo-Fraenkel, 324
Shelah, Saharon, 344
simple
  graph, 230
  path, 231
sine, 115
  addition formula, 116, 122
  as function of arc length, 120
  infinite product, 120, 121
  limit property, 122
  power series, 115, 120
    by Newton, 137
  rate of change, 122, 123
singularities, 102, 250
  described by knots, 250
skew field, 91
Skolem, Thoralf, 357
Solovay, Robert, 344
solution
  by Cardano formula, 67
  by radicals, 74
  of cubic equation, 65